M. See

Greenhouse Gas Emissions

Springer-Verlag Berlin Heidelberg GmbH

Michael See

Greenhouse Gas Emissions

Global Business Aspects

With 43 Figures and 26 Tables

Springer

Dr. Michael See
Euro-Asia Pacific Corporation Ltd.
St. Alphage House
2 Fore Street
EC2Y 5DA London
United Kingdom

ISBN 978-3-642-63227-3

Library of Congress Cataloging-in-Publication Data
See, Michael.
Greenhouse gas emissions : global business aspects / Michael See.
p. cm.
Includes bibliographical references and index.
ISBN 978-3-642-63227-3 ISBN 978-3-642-56908-1 (eBook)
DOI 10.1007/978-3-642-56908-1
1. Emissions trading–Developing countries. 2. Climatic changes–Developing countries.
3. Greenhouse effect, Atmospheric–Developing countries. I. Title.

http://www.springer.de

Originally published by Springer-Verlag Berlin Heidelberg New York in 2001
Softcover reprint of the hardcover 1st edition 2001

Cover design: de'blik, Berlin
Typesetting: Best-set Typesetter Ltd., Hong Kong
Printed on acid-free paper SPIN 10760115 30/3130/as 5 4 3 2 1 0

Dedication

This book is dedicated to men and organisations actively engaged in the battle to mitigate the effects of global warming.

Economic progress must be balanced with sustainable supply of the earth's resources - governments must implement and reinforce legislative measures to protect the environment.

Mankind, however, is motivated by profits. But neither money nor technology can reverse climate change - a dying planet is not a legacy our children should inherit.

The environment must take precedence, and developed countries must take the lead.

Data Accuracy, and Notes for Readers

The author wishes to state that information provided and the basis and assumptions upon which the information is provided in this publication has been adequately researched, and accepts no legal responsibility or any liability, in any form or means whatsoever, in regard to the standing of any government, countries, firms, companies, authorities, organisations, authorities or any individual mentioned herein as the entities, or the value or implied value of the entities or their assets; nor can the author make any representation or warranty, express or implied, with regard to the accuracy, correctness or completeness of the information that is contained in the publication herein with regard to the entities; nor can the author accept any legal responsibility or liability of any form whatsoever for any errors or omissions that may have been made to any entity in the publication.

The author makes clear that this publication does not constitute an investment prospectus in any part or whole, nor meant to be an investment document of any form. This publication is neither an offer or invitation or encouragement to sell; nor a solicitation to buy or sell any securities; nor does any information contained herein constitute investment advice of any form or recommendation with respect to any corporation or entity or person described herein.

The information in this publication is presented solely for informational purposes only and should not be relied upon as authoritative or as a substitute for such research and investigation as may be deemed appropriate for a given purpose.

The author, including Euro-Asia Pacific Corporation Limited and its member firms, shall assume no liability whatsoever for any loss or damage arising from any use or application of this publication or the information contained herein.

Foreword by the President of the Maldives

Our islands, the Maldives, are located south west of India. It has a population of around 270,000 whose population growth rate is 1.9% per annum and which enjoys an annual GDP growth of over 4%. Our major exports are fish, fishery products, apparel and clothing accessories. Our major imports are machinery, building materials, fuel and textiles. The literacy rate, at almost 98%, is one of the highest in the world. Our people originate from a diverse ethnic mixture of Indian to African and Arab backgrounds.

We practise a strict code of environmental conservation – the Law on the Preservation and Protection of the Environment safeguards coral reefs, sand shores and endangered species such as dolphins and turtles. Yet, our country is now facing a very bleak future – as one peers across the horizon at the sea-walls that are being built around many of our islands.

The average height of our islands is less than six feet above the sea level – and falling as I write this foreword. Regional Impacts of Climate Change: An Assessment of Vulnerability, a special report of the Intergovernmental Panel on Climate Change, published in 1997, and the Third Assessment Report completed this month in Accra, warn that the cultural diversity of Maldivians is seriously threatened by the rising waters. It could very well end in the extinction of a unique civilisation.

Our beautiful and fragile collection of more than a thousand coral islands could indeed disappear into the mighty ocean as the sea level rises. We have spared no effort in alerting the international community to the grave dangers facing us from global warming and sea-level rise. We are determined to stand up and fight, for no Maldivian wants to become an environmental refugee. The world community could still save the Maldives and other low-lying islands by adopting stringent measures to curb global warming and stop the rise in sea and ocean levels.

This book is a comprehensive guide to those who either are experts or novices in the issues of global warming – it provides the essential instruments which industrialised countries could utilise to remedy a world weakened by climate change.

The environmental hourglass is rapidly running out. With the very survival of our beautiful islands – truly the last paradise – at stake, I urge every government and every individual to act now to save the world environment and our beloved country.

His Excellency Mr Maumoon Abdul Gayoom
President of the Republic of Maldives

30 March 2001

Foreword by the Honorary Consul of Tuvalu

Tuvalu, one of the smallest island in this world, is located about 100 kilometres north of Fiji in the South Pacific Ocean. The capital centre is Fongafale on Funafuti Atoll, and the population of 10,000 are Polynesian in ancestry.

The island economy is dependent on deep sea fishing, subsistence farming on the nine coral atolls, foreign aid, earnings of expatriate Tuvaluans, and a moderate tourist industry. Copra, textiles, metal and petroleum products, and handicraft constitute the main industries onshore. Life expectancy is amongst the lowest in the region and infant mortality is very high. Energy resources are non-existent, resulting in rising imports of diesel oil and equipment. In GNP terms, Tuvalu ranks 190th – amongst the poorest 5% of least developing countries.

A victim of global warming

Prolonged drought periods due to the El Niño effect, higher surface temperatures, increased frequencies of hurricanes, wetter-than-average conditions and rising sea levels are the most significant impacts of the greenhouse effect on Tuvalu. Experts from the Intergovernmental Panel on Climate Change rate Tuvalu as one of the ten most vulnerable small islands-at-risk from sea-level rise, in terms of average elevation of the islands.

Programmes for contingency evacuation and re-settlement of small-island-states should now form part of medium and long-term climate change mitigation strategies. The island state of Tuvalu – on behalf of all members of the Association of Small Island States AOSIS – appeals to the international community to take immediate measures to preserve global security.

The Kyoto Protocol

Our country and other islands at high risk from sea-level rise must receive fair amounts of proceeds from the Joint Implementation, Clean Development Mechanism and Emissions Trading initiatives of the Protocol, and increased humanitarian aid from donor countries. Although the US Congress has announced its negativity to the Protocol, we hope that the treaty can be re-negotiated by the international community on terms agreeable to all parties.

The book, *Greenhouse Gas Emissions*, is a most useful resource of facts, data and strategies to reduce the impact of global warming. We hope that readers will use the information in this book and work together to avert a most catastrophic event.

His Excellency Dr Iftikhar Ayaz
Honorary Consul of Tuvalu

30 March 2001

Author's Notes

The Clean Development Mechanism (CDM) initiative of the Kyoto Protocol was officially activated in January 2000, per clause 10 of Article 12 of the Protocol. At July 2000, potential CDM projects are currently being undertaken between large European enterprises and China, South Africa, Malaysia and Brazil – the pace of which is matched by a number of bilateral Joint Implementation projects between the USA and Canada; Switzerland and the Philippines; Japan and Australia; and Japan and Russia.

A ten percent discount rate has been used for calculating the average cost of electricity in developed and developing countries. The author considers this combined expected return to debt and equity financiers as commercially realistic to compensate for political and currency risks in emerging markets. A lower "minimum rate of return" may be specified by UNFCCC for different types of technologies under different market conditions.

The economies of least developed countries and islands-at-risk from sea-level rise will be severely disadvantaged by the Kyoto Protocol – as foreign investment flows are diverted to those countries which have more fossil-fuel resources and higher emission levels. To address this imbalance, a LDC and IRISK (Islands-at-RISK) Climate Change Fund was proposed to UNEP and UNFCCC in late 1999 by Euro-Asia Pacific Corp to ensure that these countries are compensated by a mandatory 1% contribution of the value of each and every emissions contract traded to end of the first commitment period, 2012. In mid-2000, UNFCCC put forward the concept of an "Adaptation Fund" at the 12th Meeting of Subsidiary Bodies.

The author's analysis of financial benefits to Annex I corporate investors, global emissions traders and host CDM countries suggest that very high cost savings can be realised by Annex I entities through their investments in CDM countries in terms of greenhouse gas reductions and that extremely profitable commissions can be earned by emissions traders – compared to disproportionately low revenues from CER (certified emission reductions) proceeds to the CDM host government, domestic joint-venture partners and stakeholders. This huge inequity stems from lower costs of carbon reduction per invested dollar in developing countries and minority shares of CER proceeds to the host government and stakeholders. As a result of these findings, Euro-Asia Pacific Corp also proposes a mandatory 1% contribution of the value of every emissions contract (based on CERs generated in a CDM host country) traded to be awarded to the CER country of origin, local stakeholders of the project, and poverty inflicted regions of the community.

To receive support from the international community, and perhaps more importantly, from ordinary citizens of the world who do not gain from political decisions, the Kyoto Protocol must ensure equitable treatment of all parties in its quest to combat global warming.

Michael See
Euro-Asia Pacific Corporation Ltd
St. Alphage House
Fore Street
London EC2Y 5DA
e-mail: michael@greenpac.freeserve.co.uk
euro_asiapac@email.com

2 April 2001

About Euro-Asia Pac Corp Ltd

Euro-Asia Pacific Corp is a consulting firm founded by Michael See, specialising in energy and the environment. It has a wide network in the Asia Pacific Region. The firm has an open dialogue with the United Nations Environmental Programme (UNEP) and its Division on Technology, Industry & Economics on issues relating to flexible mechanisms of the Kyoto Protocol. In 1999, Euro-Asia submitted *Proposal to Conference of Parties and UN Intergovernmental Panel on Climate Change on Practical Implementation of the Clean Development Mechanism.* In the same year, Euro-Asia teamed up with Shantanu Banerjee's Glonet Technology, based in Peterborough, to evaluate energy and environmental projects in the Asian market.

Acknowledgements

The work on the manuscript gathered pace in mid-1999 as a result of encouraging responses to climate change project proposals from the governments of Brazil, India, Indonesia, Kyrgyz Republic, Kazakhstan, Mozambique and the Secretariat of the Caribbean Community and Common Market. The proposals were put forward by Euro-Asia Pac Corp, a UK firm which prioritised the mitigation of global warming as its millennium objective.

By mid-2000, Belarus, Estonia, Kenya, Latvia, Mauritania, Sri Lanka, Philippines and Thailand joined the number of non-OECD countries that were keen attracting foreign direct investments to reduce greenhouse gas emissions. The strategic focus of the book is partly based on feedback to these project proposals, perceptions of "North-South" capital transfers, and critical appraisal of the impact of emissions trading on developing countries.

The author is grateful to Dr Klaus Töpfer, Executive Director; and Ms Jacqueline Aloisi de Larderel, Director of the Division on Technology, Industry and Economics; of the United Nations Environmental Programme for their helpful comments and suggestions, and for raising the issue of financial additionality requirement of projects. UNEP has included key initiatives – which address equitable treatment of project proceeds – described in this book, for deliberation by the Conference of the Parties.

Emissions data, energy and electricity generation statistics were primarily quoted from publications of the International Energy Agency and Organisation for Economic Cooperation and Development – for which the author is thankful.

Project data on Asia, particularly South-Asia, was obtained through Mr Shantanu Banerjee, Director of Glonet Technology & Trade, a UK firm based in Peterborough, and its associate, International Trade Links, based in India.

Much appreciation is extended to major global relief organisations – Aga Khan Foundation, CAFOD, CARE, International Tree Foundation, Islamic Relief, Water Aid, MERLIN, MSF and International Red Cross & Red Crescent Movements – for information on humanitarian projects in Least Developed Countries, islands-at-risks and underdeveloped countries.

Mr Adam Parkin of Integra International Ltd participated in meetings on the financing of climate change projects; Dr Qin He-Qing of CHAM Consulting assisted with her research on China as a potential market for investments; and Ms Rachel Jackson of the Association of Chartered Certified Accountants made available a paper on the proposed treatment of carbon assets in corporate accounts. Discussions with Mr Angus Macpherson, Managing Director of OM Environmental Exchange London provided helpful insights to the operation of trading platforms.

Last but not least, the author would like to thank Dr Christian Witschel, the Geosciences Editor; Ms Marion Schneider, Personal Assistant to Dr Witschel; Ms Theodora Krammer, Copy Editor; Ms Janet Sterritt-Brunner, Geosciences Desk Editor; Ms Almas Schimmel, Production Editor, and to the Board of Springer-Verlag GmbH & Co. KG for their encouragement and support for this timely publication.

Michael See
Euro-Asia Pacific Corporation Ltd
St. Alphage House
Fore Street
London EC2Y 5DA
e-mail: michael@greenpac.freeserve.co.uk
euro_asiapac@email.com

2 April 2001

Acronyms and Abbreviations

AA	Assigned amounts – emissions allowances or permits granted to Annex I nations during their commitment period of 2008-2012 under the Kyoto Protocol. The Clean Development Mechanism, Joint Implementation or Emissions Trading initiatives of the Protocol enables units to be added to or subtracted from AA.
AOSIS	Alliance of Small Island States.
AIJ	Activities Implemented Jointly, a pilot scheme of the UNFCCC to reduce greenhouse gas emissions – to be completed by the year 2000.
CARE	An international relief agency, Secretariat in Brussels.
CARICOM	Caribbean Community and Common Market.
CER	Certified Emission Reduction – greenhouse gas reductions achieved under the Clean Development Mechanism.
CO_2	Carbon dioxide, the primary greenhouse gas in terms of volume composition.
CO_2 equiv.	Carbon dioxide equivalent units of emissions according to the global warming potential of specific greenhouse gas.
CH_4	Methane, a greenhouse gas.
CIS	Commonwealth of Independent States.
COP	Conference of the Parties of the UNFCCC.
CDM	Clean Development Mechanism, an initiative of the Kyoto Protocol to reduce greenhouse gas emissions – primarily via "North-South" investment projects.
CHP	Combined Heat and Power (also known as Cogen or cogeneration).
DETR	Department of Environment, Transport and the Regions, UK.
EfW	Energy from Waste (also known as WTE, Waste to Energy).
EMAS	Eco-Management and Audit Scheme – European Community initiative.
EBRD	European Bank for Reconstruction and Development.
ERU	Emission Reduction Unit – specifically refers to emissions reductions achieved under Joint Implementation.
ESP	Electrostatic Precipitator.
EU	European Union.
FGD	Flue-Gas Desulphurisation.
GEF	Global Environmental Facility – a fund created by the World Bank.
GHG	Greenhouse Gases.
HFC	Hydrofluorocarbons, a generic series of greenhouse gases.
SF_6	Sulphur hexafluoride, a greenhouse gas.
IEA	International Energy Agency.
IGCC	Integrated Gasification Combined Cycle.

IPCC	Intergovernmental Panel on Climate Change.
IPE	International Petroleum Exchange, based in London.
IPPC	Integrated Pollution Prevention and Control.
IRISK	Islands-at-Risk from sea-level rise, five classed as top priority.
IRR	Internal rate of return.
ISO 14001	International Standards Organisation 14001 – An internationally-recognised standard for Environmental Management Systems.
JI	Joint Implementation, an initiative of the Kyoto Protocol to reduce greenhouse gas emissions post-AIJ – primarily via "North-North" investment projects.
LDC	Least Developed Countries.
LE	Large Enterprises.
LNB	Low NO_x Burner.
LNG	Liquefied Natural Gas.
LUCF	Land Use Change/Forestry changes: activities regulated under Kyoto Protocol.
MSF	Médecins Sans Frontières.
NGO	Non-Governmental Organisation, e.g. Greenpeace, Friends of the Earth.
NPI	National Provident Institution, UK financial institution.
NYMEX	New York Mercantile Exchange.
NO_X	Nitrogen oxides – nitrous oxide, nitrogen dioxide and nitric oxide.
N_2O	Nitrous oxide, a greenhouse gas.
OECD	Organisation for Economic Cooperation and Development.
OM	Options Market Group, Swedish group which owns the OM Exchange (OMEX) in Stockholm, and OM Environmental Exchange in London.
OXFAM	An international relief agency, headquarters in the United Kingdom.
PFC	Perfluorocarbons, a generic series of greenhouse gases.
QELRC	Quantified Emission Limitation and Reduction Commitment – commitments that were adopted under Annex B of the Kyoto Protocol.
SCR	Selective Catalytic Reduction – a form of NO_x reduction technique.
SICA	Central American Integration System, grouping of Central American countries.
SIMEX	Singapore International Monetary Exchange.
SME	Small and Medium-Size Enterprises.
SO_X	Sulphur oxides – sulphur dioxide and sulphur trioxide.
UN	United Nations, headquarters in New York.
UNHCR	United Nations High Commissioner for Refugees, headquarters in Geneva.
UNICEF	United Nations Children's Fund, headquarters in New York.
UNCTAD	United Nations Conference for Trade & Development, headquarters in Geneva.
UNEP	United Nations Environmental Programme, headquarters in Nairobi.
UNFCCC	United Nations Framework Convention on Climate Change, Secretariat in Bonn.
WRI	World Resources Institute, headquarters in New York.

Note: Other acronyms and abbreviations are used in specific tables in the book – these short forms are explained in the section preceding the tables.

Update on the Kyoto Protocol – COP VI, 13–25 November 2000, The Hague, and President Bush's Statement 28 March 2001

Preamble

The sixth Conference of the Parties was opened by Mr Wim Kok, Prime Minister of the Netherlands and President Jacques Chirac of France on 20 November in the Hague. The Prime Minister stressed the need for effective implementation of the Kyoto Protocol and equitable treatment of all stakeholders - specifically Least Developed Countries - in the financial mechanisms of the Protocol. The Dutch Environment Minister, Mr Jan Pronk (COP President); Mr Michael Zammit Cutajar, UNFCCC Executive Secretary; and Dr Klaus Töpfer, Executive Director of the United Nations Environmental Programme, chaired the sessions. Dr Töpfer drew sharp attention to the plight of underdeveloped nations, which will experience severe economic hardship as a result of global warming.

The update is compiled and edited by the author based on authoritative media sources which include the Financial Times, The Times, the International Institute for Sustainable Development and non-governmental organisations' press releases throughout and after the Hague conference.

An unexpected political agenda

The major outcome of the talks was general disagreement between the European Union and the USA with regard to the US proposal of using carbon sinks to reduce its greenhouse gas emissions obligation - domestically and via projects implemented abroad through the Joint Implementation and Clean Development Mechanism initiatives.

The US had originally predicted that forestry projects could absorb about half its carbon emissions from fossil-fuel combustion of about 600 million metric tonnes on an annual basis, i.e. about 300 million metric tonnes could be "stored" yearly within planted forests. A 125 million metric tonne absorption ceiling per annum proposal underpinned the US position in the negotiations. Several heated discussions took place on the issue of emissions, financial and investment additionality of projects conducted under the aegis of the Clean Development Mechanism. Whilst high transaction costs were amplified by the corporate sector, including the World Bank's Prototype Carbon Fund, the external costs of global climatic damage were hardly addressed by these entities.

The UK Deputy Prime Minister, Mr John Prescott put forward a last minute proposal to the USA negotiating team on 24/25 November - apparently allowing the USA to use carbon sinks to achieve its emissions target abroad and some percentage of its target via forestry projects at home. The European Union, functioning as a "community of nations" Party of the Conference, rejected the UK proposal.

In the afternoon of 24 November, Dr Klaus Töpfer, the Executive Director of the United Nations Environmental Programme transmitted a facsimile to the author's office. The fax outlined the procedures for the setting up of an adaptation fund for countries vulnerable to effects of climate change. Since then, the author has commenced communications with CARE International, Henderson Investors AMP, N.M. Rothschild & Sons, Friends Ivory Sime, ABN Amro, Dexia-FondElec, Schroders Investment Management, the National Climate Change Secretariat Kenya, and Tyndall Centre for Climate Change Research at the University of East Anglia. An adaptation fund task force, headed by CARE, is expected to submit an application for observer status in UNFCCC negotiating sessions before spring 2001.

Analysis of the carbon sinks proposal

The issue of accepting forestry and agricultural projects – abbreviated LULUCF, land use and land use/forest changes in the negotiations – as carbon sinks for the reduction of greenhouse gases during the first commitment period, 2008–2012 – was widely debated in COP VI. The European Union and the developing countries grouping G77/China opposed carbon sinks projects, while the Umbrella Group – which consisted of JUSSCANNZ countries without Switzerland, JUSSCANNZ composition being Japan, US, Switzerland, Canada, Australia, Norway and New Zealand; Russia and Ukraine – are firmly adamant that carbon sinks project acceptability is one of their preconditions of ratifying the Kyoto Protocol. To further stoke disagreement in the negotiating process, there is no general consensus in the carbon sinks issue between the Umbrella Group and EU on the one hand, and the G77/China grouping on the other with regard to Clean Development Mechanism projects.

On carbon sequestration projects, the scientific community and pressure groups agree that there is no guarantee that carbon is absorbed permanently in the tissues of trees, crops and plants, as such schemes are prone to damage by fires, defoliation and natural hazards. The contention against non-permanent reduction or storage of greenhouse gases via carbon sinks is frequently invoked as "compromising the environmental integrity of the Kyoto Protocol".

Developments on the flexible mechanisms of the Kyoto Protocol

As will be further elaborated in the book, the flexible mechanisms relate to market-based project initiatives known as Joint Implementation (JI, article 6), the Clean Development Mechanism (CDM, article 12) and Emissions Trading (article 17), which enable developed countries to comply with their emissions reduction targets.

Three central issues emerged during the negotiations with reference to the mechanisms – supplementarity, CDM project eligibility and emissions trading rules. The supplementarity condition refers to the amount of emissions reductions an obligated party is allowed to acquire abroad, i.e. supplementary to domestic action, provided it has met a legal percentage of its obligations "at home". Domestic emissions ceilings were opposed by the Umbrella Group, which aims to achieve most of its emissions reductions abroad. If the USA could implement JI and CDM projects in other countries and purchase emissions rights without restrictions, then net global emissions of greenhouse gases would increase.

On CDM project eligibility, the Umbrella Group are pressing for the inclusion of nuclear power projects in addition to carbon sequestration schemes – both opposed by the European Union. Renewable energy, energy efficiency and fuel switching projects are strongly favoured by the EU. Developing countries, as represented by the G77/China grouping, have expressed support for emissions additionality, environmental additionality and investment additionality as preconditions of CDM projects. To ensure equitable treatment, G77/China is campaigning for Protocol statutory or Protocol non-statutory bilaterally negotiated percentages of shares of proceeds from all flexible mechanisms projects which utilise their sovereign territory.

Financing operations, Adaptation Fund, COP VII

The governments of the developing world raised sharp concerns on the adaptation needs of small island states, low lying basins and least developed countries in responding to adverse effects of climate change. Mr Jan Pronk, President of COP VI, proposed the formation of an Adaptation Fund, a Climate Change Resources Committee and a Framework on Climate Change Convention Fund to finance the adaptation initiative and administrative expenses associated with UNFCCC's work on global warming.

The G77/China grouping welcomed the proposal for an Adaptation Fund, but stated that the resources of the Fund should be created regardless of the Kyoto Protocol or its ratification – a view strongly shared by the author.

The next talks are scheduled in Bonn, 16–27 July 2001 – resumption of the suspended COP VI Session and 14th Meeting of the Subsidiary Body for Scientific and Technical Advice. COP VII is planned for 29 October – 9 November 2001 in Marrakesh, Morocco.

President Bush's statement, How to save the Kyoto Protocol

On 28 March 2001, President Bush announced to the news media that he would not be willing to ratify the Kyoto Protocol. Several industrialised countries, particularly from the European Union, and Japan, responded with sharp criticism of the Bush statement. The EU warned that it might consider trade sanctions, as the US would gain from lower costs of industrial production if energy saving measures were not implemented to meet the US obligatory emissions reduction target.

The author proposes that a Pro-Kyoto Protocol Group, led by the EU. Japan and Russia; with Bulgaria, Czech Republic, Estonia, Hungary, Latvia, Norway, Poland, Romania, Slovakia and Switzerland, whose carbon dioxide emissions totalled 7,938.619 million metric tonnes or 57.82% of Annex I Parties' emissions in 1990, ratify the Kyoto Protocol. Together with G77/China's support, the Protocol will enter into force ninety days after ratification is confirmed.

The author hopes that the contents of this book can now unite those negotiators who may be directly involved in bringing an end to a most advanced civilisation.

Contents

1
Introduction

The world is getting hotter as it experiences the extremes of global climate change. In 1999, catastrophic storms hit Honduras, China and East India, bringing severe devastation to lives and national economies. El Niño swept across the Pacific in early 2000, inflicting the worst floods on Mozambique and neighbouring countries. Industrialised nations are not immune to global warming – cases of encephalitis, a disease transmitted by mosquitoes, were reported in the State of New York. In Antarctica, an iceberg seven times the size of Manhattan island broke loose and floated towards Cape Horn.

The melting of Arctic glaciers also continues – huge volumes of fresh-water will disrupt the warm conveyor-belt from Central America to Europe. The net effect of convergent glacial drifts from the polar regions to the equator is expected to intensify cloud formation in the tropics – hence exacerbating global warming. As the destructive forces of nature intensify, so does the rhetoric from environmental organisations – as evidenced by the disruption of the last World Trade Organisation conference in Seattle.

It is now up to civilisation to challenge climate change. It can achieve this by command and control as well as flexible mechanisms at home and abroad, before the process of global warming becomes totally irreversible.

1.1
Overview

The book principally addresses climate change and describes the remedial strategies for developing countries based on the Clean Development Mechanism of the Kyoto Protocol. It provides a very comprehensive account of the array of proposals and economic instruments devised by the international community – including the Joint Implementation and Emissions Trading initiatives of the Protocol – to abate global warming. Chapter 3, Sections 2.1 and 2.2 describe the objectives and key working platforms of the Protocol. The effects of other major atmospheric, land and water pollutants from industries and domestic sources are also covered. It is realistic to state that if current efforts by the international community fail then the book could form a practical guide for domestic action by individual countries.

1.2 How to Use This Book

Apart from the original and ongoing revisions of the text of the Kyoto Protocol, which can be referred to in www.unfccc.de, readers of the book will find the following guidelines useful in the application of information and data in project development:

Chapter	Usefulness to reader
4	Relevant articles of the Kyoto Protocol
5	Emissions characteristics for different countries
6, 7	Clean Development Mechanism (CDM) opportunities and project criteria
8	Cost-benefit analysis of remedial technologies
9	Emissions trading techniques and risk management

2
Global Warming

This chapter deals with the causative factors and effects of global warming, the greenhouse gases targeted by the Kyoto Protocol and their relative contributions to temperature rise.

Global warming, an increase in global surface temperatures, is an intensification of the "natural" greenhouse effect from human activities. Absorption of infra-red radiation and reflection to earth by a layer of gases in the atmosphere provides this "natural heating" effect – an optimal temperature environment for man's existence – but industrial and economic pressures have resulted in higher releases of gases to the atmosphere and increased reflection of solar radiation. The latter process, termed the "enhanced greenhouse effect", is the result of higher levels of commercial and business activity.

2.1
Impact on the Global Environment

The most significant environmental damages caused by global warming are:

1. Ecological damage to agricultural land – hence reducing crop yields.
2. Hydrological disruption – reducing water supplies, desertification of semi-arid land.
3. Increased human mortality rates from water-borne diseases; in the absence of clean or sterilised water.
4. Rise in sea-levels, and sinking of lowlands and small island states.
5. Spread of diseases in tropical regions, and from tropical to temperate climate regions, as mosquitoes become more prevalent.

The frequency of more violent hurricanes and tropical storms are expected to increase, nothwithstanding other destructive weather and ecological disasters which may become self evident in the millennium. At end 1999, meterological disasters continued to plague the south part of the USA, Caribbean and Central America, while massive flooding hit the Philippines and north-eastern India.

A total of about half a million inhabitants of the five most vulnerable islands-at-risk from sea-level rise – Kiribati, Maldives, Marshall Islands, Micronesia and Tuvalu – will become the world's first "environmental refugees". The industrialised nations, the largest emitters of carbon dioxide, should assume the responsibility of meeting the full costs of evacuation and resettlement of the islanders. In a separate analysis in Appendix 10, about a billion dollars will be needed in the first year of evacuation to meet the costs of adaptation; food, shelter and basic living essentials of displaced islanders including the elderly and children.

2.2 Greenhouse Gases: Primary Sources and Global Warming Potential of Each Gas

The six greenhouse gases targeted by the Kyoto Protocol are

1. Carbon dioxide
2. Methane
3. Nitrous oxide
4. Hydrofluorocarbons
5. Perfluorocarbons
6. Sulphur hexafluoride

These gases are responsible for about 76% of the global "enhanced greenhouse effect". The Kyoto Protocol requires a number of developed countries – as listed in Annex I of the UNFCCC – to reduce these gas emissions to 5.2% below 1990 levels. Countries may adopt 1995 as the datum for reduction of trace greenhouse gases. The control of other greenhouse gases – chlorofluorocarbons (CFCs), which are responsible for destruction of the ozone layer – are mandated under the Montreal Protocol.

The global warming potential of each greenhouse gas in the atmosphere is expressed in terms of carbon dioxide equivalents in the following section, for a residence period of 100 years. Carbon dioxide is the reference greenhouse gas assuming the value of 1 unit.

Largest emitters of carbon dioxide

Major industries:	Power stations burning coal, oil and gas (in descending order of emissions per unit mass of fuel). Cement and iron/steel manufacturing plants.
Other industries/ sources:	Building industry – brick manufacture, lime production and use. Chemical industry – ammonia and carbides production. Fuel industry – solid and liquid fuel transformation; fuel extraction, production and distribution. Gas venting and flaring from reservoirs. Metal industry – Ferroalloys, aluminium and magnesium. Waste management – landfill sites. Other industries, e.g. automotive and paper plants.
Private/ commercial sources:	Transport vehicles – vehicles that use gasoline and diesel fuels. In China, coal-stoves are used by a large proportion of the population.
Land use change:	Burning or clearing of forests and forest cover for commercial use.
Global Warming Potential:	1

Largest emitters of methane

Major industries /sources:	Oil/gas/coal extraction processes. Leakages from production (e.g. refineries, coking plants) and distribution systems of these fuels. Wet paddy planting, fertiliser application and land use change. Gas venting and flaring from reservoirs. Combustion (incomplete) of fossil fuels in power stations. Waste management – landfill sites. Agricultural – enteric fermentation and animal wastes. Bunker fuel storage discharges.
Other industries/ sources:	Chemical industry – ammonia, carbides and petrochemical production. Metal industry – iron, steel, ferroalloys and aluminium.
Private/ commercial sources:	Transport vehicles, depending on engine technology. Private farming – release from soils, burning of agricultural wastes. Burning of biomass, fuel wood and cleared forests. Waste emissions from landfills.
Global Warming Potential:	16–26

Largest emitters of nitrous oxide

Industry:	Nitrogenous fertilisers: application and excess usage. Chemical industry – nitric acid, adipic acid, urea and caprolactam. Refinery operations and petrochemicals. Burning of biomass, fuel wood and cleared forests. Combustion of fossil fuels – power stations, transport vehicles, industries. Atmospheric fluidised bed combustion.
Private/ commercial sources:	Private farming, fertiliser use and animal wastes. Transport vehicles – emissions from catalytic converters.
Global Warming Potential:	310

Other main sources of greenhouse gases emissions

Hydrofluorocarbons:	Halocarbons production and use: HCFC-22; HFC types – 125, 134, 134a, 143, 143a, 152a, 227ea, 23, 236fa, 245ca, 32, 41 and 43. Refrigerants and vehicle air conditioning; electrical switching gear.
Global Warming Potential:	140–11,700

Perfluorocarbons:	Aluminium and magnesium production; halocarbons production and use: PFC types – perfluorobutane, perfluoromethane, perfluoropropane, perfluoropentane, perfluorocyclobutane, perfluoroethane and perfluoro-hexane.
Global Warming Potential:	6,500–23,900
Sulphur hexafluoride:	Electrical switching gear, aluminium and magnesium industries.
Global Warming Potential:	23,900

3
The Kyoto Protocol's Flexible Mechanisms

This chapter traces the history of the "parent" of the Kyoto Protocol – the United Nations Framework Convention of Climate Change; discusses the economic advantages of the Clean Development Mechanism; and lists in detail the original text of the three flexible mechanisms of the Protocol.

3.1
United Nations Framework Convention on Climate Change (UNFCCC)

The United Nations Environmental Programme (UNEP) and the World Meteorological Organisation (WMO) set up the United Nations Framework Convention on Climate Change (UNFCCC), a global treaty to address the effects of climate change, which lead to the Earth Summit in Rio de Janeiro in June 1992. The treaty, which came into force in 1994, was followed by the second Earth Summit in 1997. Governments which have ratified UNFCCC are designated "Conference of the Parties" (COP) and participate annually in meetings to discuss and adopt policies on climate change.

The most recent COPs were COP IV at Buenos Aires in 1998 and COP V at Bonn in 1999. COP VI will be held in Hague in November 2000. One of the major problems FCCC faces is the failure of governments to put environment as first priority – for example, the relative weightings of thermal power projects in a CDM national portfolio may dwarf nenewables.

3.2
Three Main "Pillars": Market-Based Flexible Mechanisms

The Kyoto Protocol is the most ambitious UNFCCC initiative to mitigate global warming-proposed and agreed in December 1997 in Kyoto, Japan. Although regarded as an "eleventh hour" breakthrough for the international community, it is a challenge to global warming that has been launched on an unprecedented scale.

Paragraph 1 of Article 3 represents the "time and delivery" targets of the governments of industrialised countries. The market-based mechanisms through which reductions could be achieved are Clean Development Mechanism, Joint Implementation and Emissions Trading.

3.3 Clean Development Mechanism (Article 12)

The Clean Development Mechanism enables countries listed in Annex I of the Convention to invest in projects in countries not included in Annex I of the Convention – any reductions in greenhouse gases emissions in the host country can be credited to the investor country (company) to meet its own emissions targets or sold to any country (company) which is listed in Annex B of the Kyoto Protocol.

The Kyoto Protocol aims to reduce the volumes of six greenhouse gases by the year 2012 in thirty six countries, as described in Annex I of the Convention, to about five percent below 1990 levels in these countries. A scheme for cross-border emissions trading between countries listed in Annex B of the Kyoto Protocol was proposed to facilitate the reduction of the gases.

Under Article 12 of the Protocol, which specifically relates to the Clean Development Mechanism, there is no obligation for the COP or governments to provide any financing for the Mechanism. Any Annex I country financing, hence, could be voluntary; but may be limited in the case of OECD countries which have quotas on grants for overseas projects.

Private sector financing represents a potentially huge market for both investors and lenders. In the case of large projects, an investor company may not have sufficient capital to meet total project costs, hence supplementary financing from the private sector and development banks could make up the shortfall.

Commercial benefits of CDM

1. New market penetration or market expansion for investor company. Diversification of domestic risk in overseas markets. For Annex I countries, follow-up exports of goods, services and ancillary products to the host country.
2. Greenhouse gases emissions reductions in the host country can be used to meet investor company's own target emissions, or traded for hard currency. The cost of emissions reductions in a developing country (non-Annex I country) may be substantially smaller than in an Annex I developed country under the Clean Development Mechanism – by perhaps a straightforward application of an advanced pollution reduction technology.
3. Developing countries benefit from lower fiscal expenditure, inward investments and technology transfer. Future impacts of global warming on GDP averted.
4. Governments which invest in CDM projects in the form of equity stakes could utilise emissions reductions to meet their own emissions targets, perhaps at lower costs, if their own companies are not successful in bidding for CDM projects or are unable to afford carbon or energy taxes, or permits.

A win-win strategy

The Clean Development Mechanism is an initiative enable Annex I countries to meet (part of) their emission targets under the Kyoto Protocol, resulting in:

- Promoting foreign investments from developed into developing countries, specifically in the area of improving energy efficiency and use of renewables.

- Effective and efficient transfer of technology, capital and resources, without compromising aid programmes to non-energy sectors and sustainability.
- Sharing of proceeds with the host CDM country.

3.4 Joint Implementation (Article 6)

An initiative to enable developed countries to participate in joint-venture projects to off-set their GHG reduction obligations through mutual cooperation. JI has been extended to cover projects between developed and developing countries.

A pilot scheme, Activities Implemented Jointly (AIJ) started in 1995 and will finish in the year 2000. It successfully encouraged several joint-venture projects between JI participants. Credits were disallowed in this scheme, but provided valuable experience to many firms.

The main objectives of Joint Implementation are:

- Bilateral cooperation between participants;
- Offsetting the emissions obligations of developed countries, when a framework for monitoring projects and certifying credits is approved by UNFCCC.

3.5 Emissions Trading (Article 17) – Current Experiences and Global Reaction

The USA's sulphur dioxide emissions trading system under the Acid Rain Programme has been remarkably successful. Phase 1 has resulted in emissions reduction of 7 million tonnes, 3.4 million tonnes in excess of the target of 3.6 million tonnes.

The sulphur reductions were achieved by an effective combination of internal controls and external trading of sulphur dioxide allowances. More efficient companies have lower marginal costs of sulphur dioxide reduction, and sell off excess units of reduction to less efficient companies. The price per allowance ranged from $ 300 to $ 1000, while penalties of $ 2000 per tonne were imposed for non-compliance.

The USA is seeking to attain (most) of its emissions targets by emissions trading. The Protocol needs ratification by countries with at least 55% of global GHG emissions in 1990 – USA's share was 36%; EU and incumbent EU countries 47% and Japan 8.5%. If the USA chooses not to ratify the Protocol, then the EU, Japan and some other countries may cast the necessary vote to render the agreement effective. The Russian Federation, which emitted 17% of the global total, may hold a strategic card in negotiating the fate of the Protocol.

During the Kyoto Protocol negotiations, several of developing and LDCs voiced opposition to emissions trading. The USA, the leading proponent of this "flexible mechanism", was seen to favour external rather than domestic solutions. India and China – which were amongst the strongest critics of emissions trading – may have to contend with USA's preferences as both countries are the largest beneficiaries of the Clean Development Mechanism.

3.6 Original Text from the Protocol (Articles 3.1, 6, 12 and 17)

The four key "drivers" of the Protocol are Article 3, paragraph 1 (which defines obligatory emissions reduction targets of "North" countries) and the three "flexible mechanisms" of Joint Implementation, Clean Development Mechanism and Emissions Trading. The full text of the Protocol can be referred to via the website www. unfccc.de.

Article 3, paragraph 1

The Parties included in Annex I shall, individually or jointly, ensure that their aggregate anthropogenic carbon dioxide equivalent emissions of the greenhouse gases listed in Annex A do not exceed their assigned amounts, calculated persuant to their quantified emission limitation and reduction commitments inscribed in Annex B and in accordance with the provisions of this Article, with a view to reducing their overall emissions of such gases by at least 5 percent below 1990 levels in the commitment period 2008 to 2012.

Joint Implementation (Article 6)

1. For the purpose of meeting its commitments under Article 3, any Party in Annex I may transfer to, or acquire from, any such Party emission reduction units resulting from projects aimed at reducing anthropogenic emissions by sources or enhancing anthropogenic removals by sinks of greenhouse gases in any sector of the economy, provided that:
   a) Any such project has the approval of the Parties involved.
   b) Any such project provides a reduction in emissions by sources, or an enhancement of removals by sinks, that is additional to any that would otherwise occur.
   c) It does not acquire any emission reductions if it is not in compliance with its obligations under Articles 5 and 7.
   d) The acquisition of emission reduction units shall be supplemental to domestic actions for the purposes of meeting commitments under Article 3.
2. The Conference of the Parties serving as the meeting of the Parties to this Protocol may, at its first session or as soon as practical thereafter, further elaborate guidelines for the implementation of this Article, including for verification and reporting.
3. A Party included in Annex I may may authorise legal entities to participate, under its responsibility, in actions leading to the generation, transfer or acquisition under this Article of emission reduction units.
4. If a question of implementation by a Party included in Annex I of the requirements referred to in this Article is identified in accordance with the relevant provisions of Article 8, transfers and acquisitions of emission reduction units may continue to be made after the question has been identified, provided that any such units may not be used by a Party to meet its commitments under Article 3 until any issue of compliance is resolved.

Clean Development Mechanism (Article 12)

1. A clean development mechanism is hereby defined.
2. The purpose of the clean development mechanism shall be to assist Parties not included in Annex I in achieving sustainable development and in contributing to the ultimate objective of the Convention, and to assist Parties in Annex I in achieving compliance with their quantified emission limitation and reduction commitments under Article 3.
3. Under the clean development mechanism:
   a) Parties not included in Annex I shall benefit from project activities resulting in certified emission reductions.
   b) Parties included in Annex I may use the certified emission reductions accruing from such project activities to contribute to compliance with part of their quantified emission limitation and reduction commitments under Article 3, as determined by the Conference of the Parties serving as the meeting of the Parties to this Protocol.
4. The clean development mechanism shall be subject to the authority and guidance of the Conference of the Parties serving as the meeting of the Parties to this Protocol and be supervised by an executive board of the clean development mechanism.
5. Emission reductions resulting from each activity shall be certifed by operational entities to be designated by the Conference of the Parties serving as the meeting of the Parties to this Protocol, on the basis of:
   a) Voluntary participation by each Party involved.
   b) Real, measurable, and long-term benefits related to the mitigation of climate change.
   c) Reductions in emissions that are additional to any that would occur in the absence of the certified project activity.
6. The clean development mechanism shall assist in arranging funding of certified project activities as necessary.
7. The Conference of the Parties serving as the meeting of the Parties to this Protocol shall, at its first session, elaborate modalities and procedures with the objective of ensuring transparency, efficiency and accountability through independent auditing and verification of project activities.
8. The Conference of the Parties serving as the meeting of the Parties to this Protocol shall ensure that a share of the proceeds from certified project activities is used to cover administrative expenses as well as to assist developing country Parties that are particularly vulnerable to the effects of climate change to meet the costs of adaptation.
9. Participation under the clean development mechanism, including in activities mentioned in paragraph 3(a) above and in the acquisition of certifed emission reductions, may involve private and/or public entities, and is to be subject to whatever guidance may be provided by the executive board of the clean development mechanism.
10. Certified emission reductions obtained during the period from the year 2000 up to the beginning of the first commitment period (2008–2012) can be used to assist in achieving compliance in the first commitment period.

Emissions Trading (Article 17)

The Conference of the Parties shall define the relevant principles, modalities, rules and guidelines, in particular for verification, reporting and accountability for emissions trading. The Parties included in Annex B may participate in emissions trading for the purposes of fulfilling their commitments under Article 3. Any such trading shall be supplemental to domestic actions for the purpose of meeting quantified emission limitation and reduction commitments under that Article.

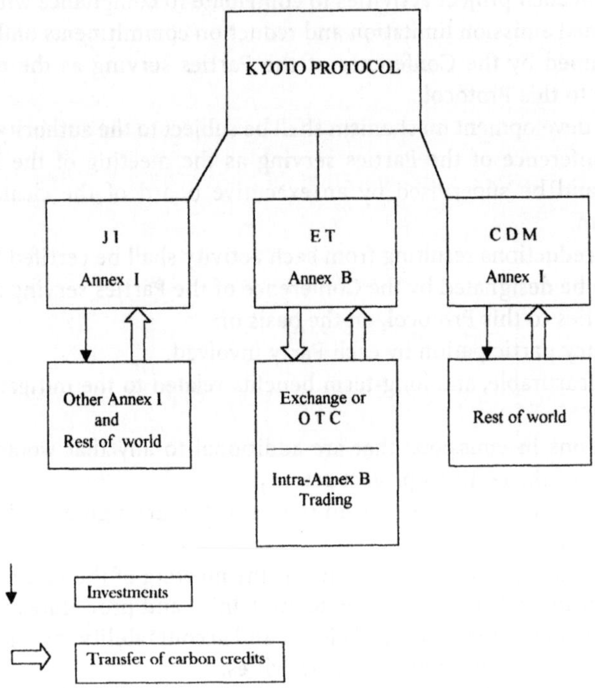

Fig. 3.1. The Kyoto Protocol – Three market-driven "pillars" (flexible mechanisms)

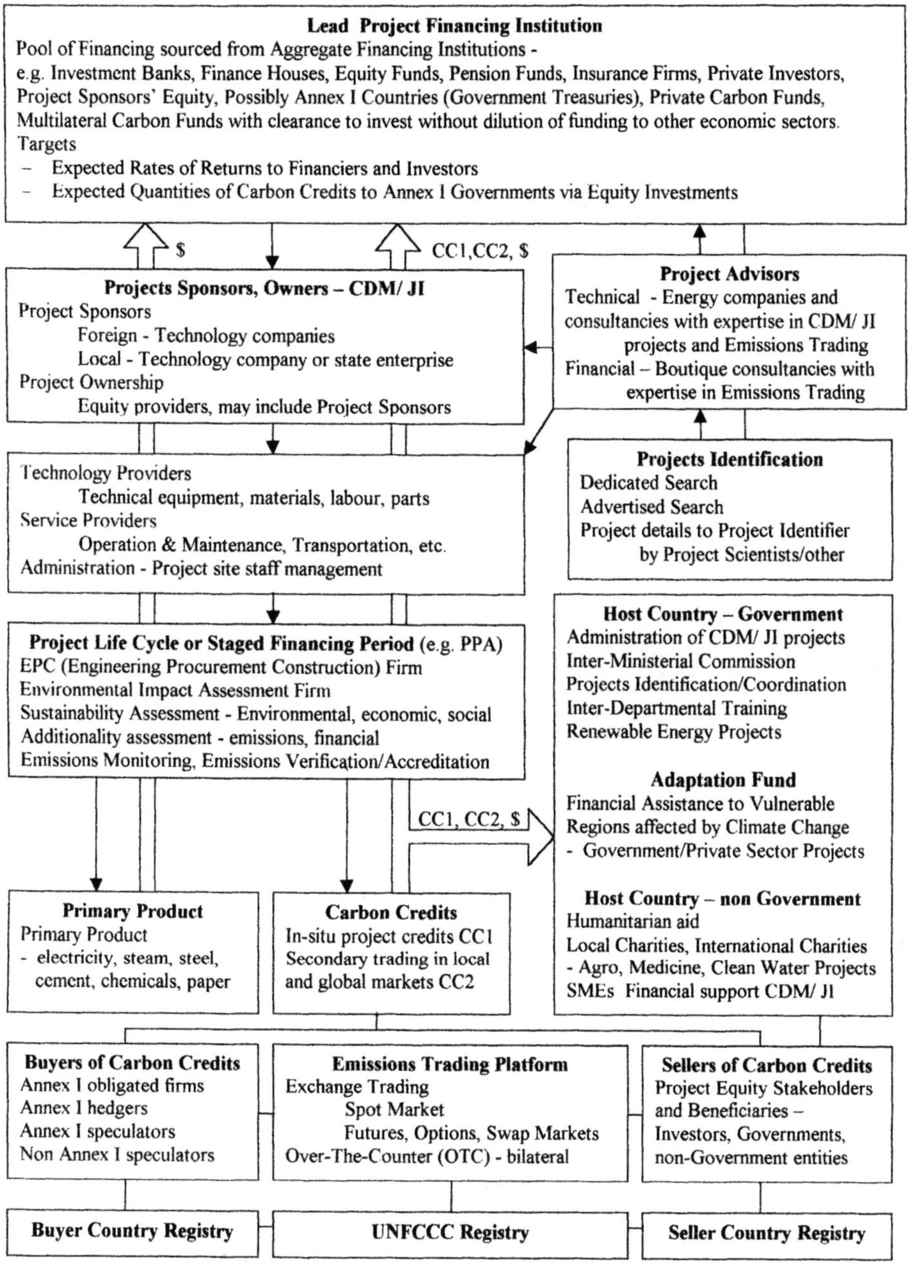

Fig. 3.2. The Kyoto Protocol – general scheme of implementing emissions reduction projects

The general schematic diagram shows the functional linkages and roles of the different entities in projects undertaken under the Kyoto Protocol. The figure shows how an "adaptation fund", proposed in meetings of the subsidiary bodies in 2000, may be incorporated in the flow of proceeds from emissions trading, JI and CDM projects.

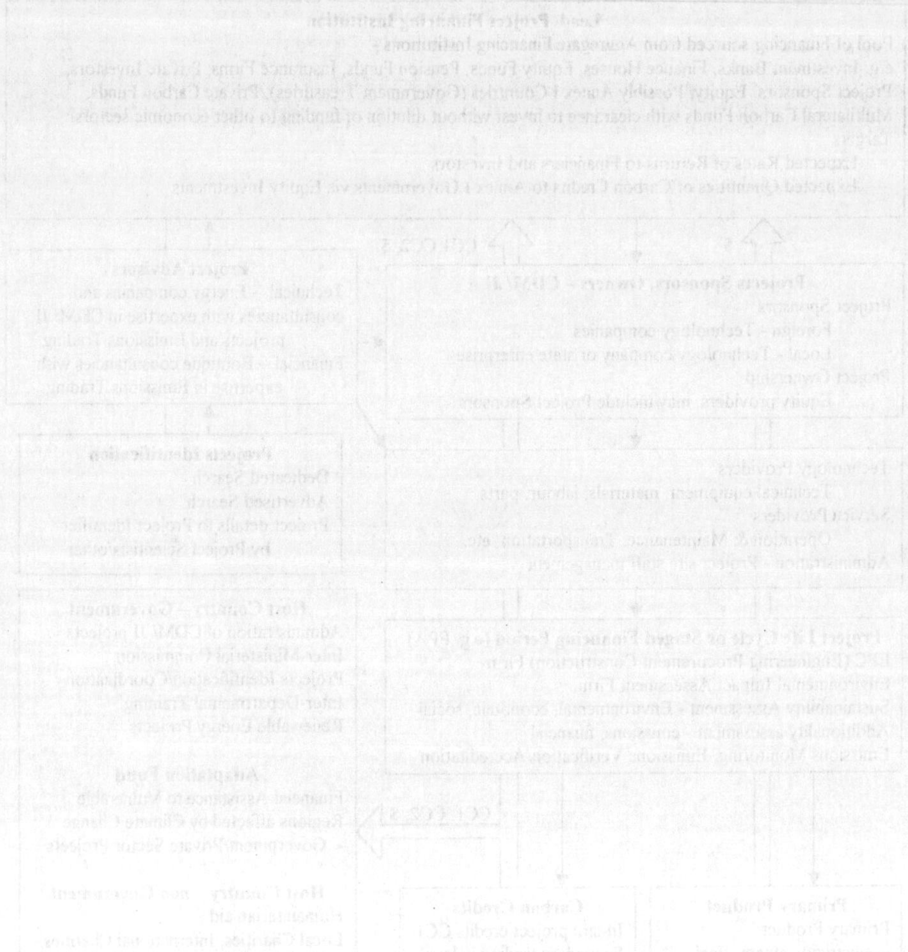

4
Institutional Issues

This chapter examines the voting system upon which the Kyoto Protocol becomes legally binding; explains why the Protocol may not stop global warming; proposes an alternative strategy if it is not ratified; and addresses the economic needs of vulnerable countries.

4.1
Kyoto Protocol: Signing and Ratification

For developed countries, reduction of six greenhouse gas emissions to 5.2% below 1990 levels is required for signatory and ratifying countries. Collective agreements are accepted to meet regional targets – the EU's "Bubble Agreement" is the most exemplary.

The EU initiative aims to reduce greenhouse gases by 2012 to 8% below 1990 levels – a target higher than the average of 5.2% for developed countries. Of the EU's obligation, the UK assumed a legal target of 12.5% below its 1990 levels. The country has also declared a national "climate challenge target" of reducing *carbon dioxide* emissions to 20% below 1990 levels by the year 2010. To achieve this, it has to achieve a reduction of about 30 million tonnes of carbon in excess of reductions that current measures and policies aim to deliver by the year 2010.

However, to render the Protocol effective, countries responsible for at least 55% of the global total of greenhouse gas emissions in 1990 must ratify the treaty. After ratification, financial penalities are expected to be imposed on those countries which fail to meet their emission reduction obligations. At 25 September 2000, 84 countries have signed and 30 have ratified the Kyoto Protocol since its acceptance as a framework to reduce greenhouse gas emissions in December 1997.

The USA, which emitted 36% of global greenhouse gases in 1990, is the pivotal player in the Kyoto Protocol. If it does not ratify the treaty (because US companies fear the loss of competitivenesss), other major economies are unlikely to cooperate in casting the majority vote as US exports would benefit from lower environmental investment costs. Ratification, if it proceeds as planned, will require ratifying countries to meet emissions reductions targets over the 2008–2012 period based on an annual quota system.

4.2
Why Flexible Mechanisms May Not Prevent Global Warming

CDM, JI and Emissions Trading may not reduce global warming, why?

Only developed countries listed in Annex I of the Convention have GHG emissions targets. The reduction target is 5.2% below 1990 levels by 2012. Data for carbon dioxide emissions in 1990 and forecasts for 2010 are currently the most reliable to work with.

The reduction figures for Annex I countries under the "business as usual" and energy-saving scenarios are estimated at 2.7 bn. and 1 bn. tonnes of carbon dioxide, respectively. Even if Annex I countries are able to meet all their emissions targets internally or with CDM/JI/emissions trading by 2012, total emissions, relative to 1990, would still increase as a result of emissions from the developing countries from 1990–2012. "South" countries are expected to increase their carbon dioxide emissions by 7.6 bn. and 7 bn. tonnes under the business-as-usual and energy saving scenarios, respectively.

Would global warming decrease if Annex I countries used the CDM to meet all their reduction obligation?

No. Under the "business as usual scenario", 2.7 bn. t. of carbon dioxide could be credited to Annex I countries (from the 7.6 bn. t. output of developing countries). But global warming will still increase, as relative to 1990, the increase in emissions in the developing countries still exceed the amount to be reduced by Annex I countries. After all CDM transactions, the global total would be about 28 bn. t. of carbon dioxide in 2010, vs 21 bn. t. in 1990.

Notes
Annex I countries may *not* be able use CDM to solve all their emissions obligations because

- not all projects may meet the criteria of bona-fide CDM projects (e.g. additionality conditions)
- sovereign risk exists for all projects.

4.3
"Ceiling" and "Check" Issues

"Ceilings" – Emissions Trading, JI and CDM

Developing countries oppose unrestricted access to emissions reductions outside developed countries – but should welcome investments with high added value to the economy. India and China are adamant that developed countries should achieve most of their emissions reductions "at home" and not increase pollution in Third World countries, because:

- the Kyoto Protocol specifies no emissions limits for developing countries. Emission limits, which may exist for some developing countries are generally less stringent than those in developed countries.
- cost for reducing emissions in developing countries may be half to a one-third of costs in developed countries suggesting "cheap control solutions" in developing countries.

However, financial and environmental additionality criteria of CDM projects – explained in Chapter 7 – require advanced technology and financial inputs into the host economy. If Annex I countries do not want "ceilings" on emissions reductions which can be achieved abroad at less cost, then they have choose the right kinds of projects to make these savings.

"Checks" – verification, certification and trading of credits

The major issues which need attention when an investor reduces greenhouse gas emissions relate to authentication of reductions and the selling of emissions credits.

- Verification and certification

Investor preference: Annex I countries prefer voluntary reporting on emissions monitoring and reduction, as this lowers the overall cost of compliance.

Regulatory concerns: NGOs advocate stringent verification procedures. The decision is likely to be taken at the sovereign level, but procedures may be streamlined by UNFCCC.

- Trading of emission credits

Trading outlets: Chicago Board of Trade, Sydney Futures Exchange, NYMEX, SIMEX, IPE and OM Group/OM Environmental Exchange are the main contenders.

Investor concerns: Initial prices of emissions credits are likely to be highly volatile in CDM host countries. Standards and systems for trading and settlement need to be implemented.

4.4
Remedial Measures in the Event of Non-Ratification of the Kyoto Protocol

Alternative strategies for governments and regional economic alliances

Regional initiatives by industrialised and higher income generating developing countries should not be dismissed as they are likely to be implemented to maintain regional security. Governments of regional initiatives could then convene to organise an international treatise to combat global warming – with each region's experience of command-and-control as well as market-based instruments.

If the Kyoto Protocol is not ratified by USA, then the European Union may either follow suit or assume the role of global leader – with the possible co-operation of Japan – in abating greenhouse gas emissions. The EU has already issued directives for GHG reduction to its member states, and would stand to gain via technological improvements in energy efficient equipment and supporting systems. Eastern Europe, the Baltic countries and the CIS would benefit as a whole as new EU members receive aid under the Agenda 2000 programme and financing via the EBRD's environmental programmes.

The USA, if dictated by its commercial enterprises and trade unions, could implement a lower cost solution to the Protocol by leading Asia Pacific Economic Cooperation countries in terms of an APEC initiative. In Central and South America, common market institutions CARICOM, SICA and MERCOSUR could join as CASIMER with the same objective.

The threat to global security by increasing climatic catastrophies and continuing reductions of investments in LDCs and non-energy abundant economies may result in destabilisation of regional governments – this could materialise in the form of mass migrations of inhabitants from small islands-at-risk from sea level rise and low-lying deltas of developing countries to neighbouring countries with higher GNP per capita.

Renegotiating the Kyoto Protocol?

A lower global emissions target, say, 2.6% of GHG emissions compared to 1990 levels – half the original target of the Kyoto Protocol – may increase ratification prospects by the US Congress. The compromised target would substantially reduce the number of forecast job losses in the United States to about 1–2 million according to independent research sources.

It is unlikely that neither India nor China would adopt voluntary emissions targets in any renegotiation of the Kyoto Protocol. At best, a revised Protocol with a lower emissions target is better than a free-for-all pollute-as-you-wish global economy, but version II must set one important requirement, that of minimum ceilings on external mechanisms to reduce emissions.

4.5 LDC/IRISK Climate Change Fund

The grouping of 48 LDCs (Least Developed Countries) and islands-at-risk (the majority are AOSIS member states) do not have the financial resources to invest in energy technologies. Instead, they rely on external aid programmes and multilateral debt to finance more urgent humanitarian and medical needs. As innocent victims of global warming – countries of the Sahel or sub-Sahara region and South Pacific are already experiencing increasing levels of poverty – voluntary contributions to non-energy dedicated country funds such as Euro-Asia Pac Corporation's LDC/IRISK Climate Change Fund (see 7.5.2 and Appendix 10) would greatly assist these economies.

Under international law, countries affected by climate change may seek recourse to those countries which are responsible for loss of national assets such as depletion of economic and human resources. The Cook Islands and Niue have intimated their rights for compensation in their signing and ratification, respectively, of the Kyoto Protocol. Whilst this is a matter to be resolved by international judicial bodies, contributions to non-energy country funds may attract some discounts on awarded damages to those countries found liable for such damages. The case for (anthropogenic or man-made) climate change damages against Annex I countries - by historical precedent - would be much stronger than for non-Annex I countries.

5
Global Emissions of Carbon Dioxide (Table 5.1, Table 5.2)

This chapter, in the form of two tables, Table 5.1 and Table 5.2, provides essential country data for project and financial decision-makers to implement Clean Development Mechanism or Joint Implementation projects in non-industrialised countries. Table 5.1, an extensive table, provides statistical and industry data on carbon dioxide emissions, electricity generation, energy consumption, and fossil fuel availability in 152 developing and underdeveloped countries. Table 5.2 provides data on country and anthropogenic sources of methane and nitrous oxide emissions.

The data tables are helpful guides to project developers and financiers to perform initial feasbility studies on country and industry potential of Clean Development Mechanism and Joint Implementation projects. In addition, the tables identify those countries with lack of fossil fuels or with relatively small emissions of carbon dioxide – mostly Least Developed Countries and small islands. These countries, which will be ignored by carbon investors, should be prioritised in the Kyoto Protocol's Adaptation Fund and other non-Protocol global aid programmes and initiatives.

Table 5.1. Global emissions of carbon dioxide: Non-Annex I Countries. Comparative Survey of 152 Developing Countries (non-Annex I countries, including non-Annex I FCCC countries in Kyoto Protocol Annex B list)

Country	CO_2 emissions from source. Estimates are made for countries which do not have specific data. Total incl. cement manufacturing and gas flaring. 1995 (est)					Electricity and heat generation (hydro, nuclear and other energy types incl. in total)/ Comments 1996 (est), latest fig.					CO_2 from cement manuf. 1995	CO_2 from iron/ steel 1995 (est)	Primary energy consumption (hydro and nuclear incl. in in total). Part of fossil fuels used for thermal electric gen. 1997, latest fig.				Fossil fuel availability, in order of proved reserves 1997, latest fig.
	Fuel	Coal	Oil/ pet.	Gas	Total		Coal	Pet. pdts.	Gas	Total			Coal	Oil	Gas	Total	Fuels: proved recoverable basis/ Comments
	Units	mil.t	mil.t	mil.t	mil.t	Elec Heat	GWh TJ	GWh TJ	GWh TJ	GWh TJ	000 t	000 t	oil equiv., mil.t				
Afghanistan (LDC)	All-s	0.02	0.82	0.32	1.24	Elec	0	0	209	625	57	n/a	n/a	n/a	0.18	0.18(est)	Gas, coal. Oil not exploited.
	PE&HP	n/a	n/a	n/a	n/a	Heat	n/a	n/a	n/a	n/a							
	MI&C	n/a	n/a	n/a	n/a												
Albania (Annex B)	All-s	0.16	1.69	0.05	1.90	Elec	0	200	0	5,926	100	0	0.024	0.51	0.019	1.19	Oil, gas. *Heat output from fuels n/a.
	PE&HP	0	0.13	0	0.13	Heat				*1,189							
	MI&C	0	0.21	0.04	0.25												
Algeria	All-s	2.04	24.37	36.17	62.58	Elec	0	659	18,862	19,714	3,089	2,418	0.4	8.5	20.1	29.1	Gas, oil, coal.
	PE&HP	0	0.36	13.24	13.60	Heat	0	0	0	0	gas	g.64%					
	MI&C	1.41	0.13	4.88	6.42							c.36%					
Andorra	All-s	n/a	n/a	n/a	n/a	Elec				115*	n/a	n/a	n/a	n/a	n/a	n/a	*Mostly hydro. Imports elec.
	PE&HP	n/a	n/a	n/a	n/a	Heat	n/a	n/a	n/a	n/a							
	MI&C	n/a	n/a	n/a	n/a												
Angola (LDC)	All-s	0	2.11	0.33	2.44	Elec	0	103	0	1,028*	*149*	0	0	0.95	0.14	6.0	Oil, gas. *Mostly hydro.
	PE&HP	0	0.35	0	0.35	Heat	0	0	0	0							
	MI&C	0	0.02	0.33	0.35												
Antigua and Barbuda	All-s	n/a	n/a	n/a	n/a	Elec	n/a	n/a	n/a	95*	n/a	n/a	n/a	n/a	n/a	n/a	Imports fuel. *Mostly thermal.
	PE&HP	n/a	n/a	n/a	n/a	Heat	n/a	n/a	n/a	n/a							
	MI&C	n/a	n/a	n/a	n/a												
Argentina	All-s	2.48	68.93	56.88	128.29	Elec	1,520	3,800	33,706	69,759	3,189	404	1.0	21.0	29.6	56.2	Gas, oil, coal.
	PE&HP	1.63	2.07	14.68	18.37	Heat	0	0	0	0							
	MI&C	0.89	3.32	14.16	18.38												

Armenia	All-s	0.01	0.86	2.67	3.54	Elec	0	127	2,191	6,214	100	0	0.02	0.15	0.89	1.79	Imports fuel.
	PE&HP	0	0.35	0.99	1.35	Heat	0	506	3,256	3,762							
	MI&C	0	0	0.54	0.54												
Azerbaijan	All-s	0.01	21.12	13.68	34.81	Elec	0	83	15,467	17,088	100	33	0	8.8	6.70	15.7	Oil, gas, coal.
	PE&HP	0	0	4.41	4.41	Heat	0	0	20,891	20,891		ea.100%					
	MI&C	0	0.50	6.30	6.80												
Bahamas (WBA/h.i.)	All-s	n/a	n/a	n/a	n/a	Elec				985*	n/a	n/a	n/a	n/a	n/a	n/a	*Imports diesel and gas for e/g.
	PE&HP	n/a	n/a	n/a	n/a	Heat	n/a	n/a	n/a	n/a							
	MI&C	n/a	n/a	n/a	n/a												
Bahrain	All-s	0	1.39	11.80	13.19	Elec	0	*	4,883	4,883	0	394	0	1.28	5.27	6.55	Gas, oil.
	PE&HP	0	0	3.51	3.51	Heat	0	0	0	0							*Diesel, oil
	MI&C	0	0	6.06	6.06												also used.
Bangladesh (LDC)	All-s	0	6.61	13.61	20.22	Elec	0	741	9,994	11,474	*140*	254	0	2.90	6.80	9.80	Gas, oil.
	PE&HP	0	0.60	5.98	6.59	Heat	0	0	0	0		ea.100%					IEA reports
	MI&C	0	0.62	5.48	6.10												high crw %.
Barbados	All-s	n/a	n/a	n/a	n/a	Elec				603*	n/a	n/a	n/a	n/a	n/a	n/a	Oil, gas.
	PE&HP	n/a	n/a	n/a	n/a	Heat	n/a	n/a	n/a	n/a							*Gas, diesel
	MI&C	n/a	n/a	n/a	n/a												used.
Belize	All-s	0	0.41	0	0.41	Elec	0	148*	0	148	0	n/a				0.24**	Imports fuel.
	PE&HP	n/a	n/a	n/a	n/a	Heat	n/a	n/a	n/a	n/a							*Diesel used.
	MI&C	n/a	n/a	n/a	n/a												**Com.e, trad.f.
Benin (LDC)	All-s	0	0.26	0	0.26	Elec	0	47	*	47	*189*	0	0	0.14	0	1.92	Oil, gas.
	PE&HP	0	0.01	0	0.01	Heat	0	0	0	0							Diesel for e/g.
	MI&C	0	0.03	0	0.03												*Gas also used.
Bhutan (LDC)	All-s	0.06	0.11	n/a	0.24	Elec		7		1,717*	*70*	n/a				0.38**	Negligible.
	PE&HP	n/a	n/a	n/a	n/a	Heat	n/a	n/a	n/a	n/a							*Mostly hydro.
	MI&C	n/a	n/a	n/a	n/a												**Trad.f.
Bolivia	All-s	0	6.04	1.90	7.94	Elec	0	200	956	3,232	349	0	0	1.96	0.72	3.63	Gas, oil, coal.
	PE&HP	0	0.10	0.79	0.89	Heat	0	0	0	0							Small coal
	MI&C	0	0.25	0.61	0.85												deposits.
Bosnia & Herzegovina	All-s	1.41	2.72	0.49	4.62	Elec	783	0	0	2,203	75	77	0.35	0.89	0.21	1.75	Coal, lignite.
	PE&HP	1.07	0.35	0.80	2.22	Heat	966	0	0	966		p.100%					
	MI&C	0	0.18	0	0.18												

Table 5.1. *Continued*

Country	CO$_2$ emissions from source. Estimates are made for countries which do not have specific data. Total incl. cement manufacturing and gas flaring. 1995 (est)					Electricity and heat generation (hydro, nuclear and other energy types incl. in total)/ Comments 1996 (est), latest fig.					CO$_2$ from cement manuf. 1995	CO$_2$ from iron/ steel 1995 (est)	Primary energy consumption (hydro and nuclear incl. in in total). Part of fossil fuels used for thermal electric gen. 1997, latest fig.				Fossil fuel availability, in order of proved reserves 1997, latest fig.
	Fuel	Coal	Oil/ pet.	Gas	Total		Coal	Pet. pdts.	Gas	Total			Coal	Oil	Gas	Total	Fuels: proved recoverable basis/ Comments
	Units	mil.t	mil.t	mil.t	mil.t	Elec Heat	GWh TJ	GWh TJ	GWh TJ	GWh TJ	000 t	000 t	oil equiv., mil.t				
Botswana	All-s PE&HP MI&C	2.24 n/a n/a	n/a n/a n/a	n/a n/a n/a	2.24 n/a n/a	Elec Heat	925 n/a	n/a n/a	n/a n/a	1,017 n/a	0	n/a	n/a	n/a	n/a	n/a	Coal.
Brazil	All-s E&HP MI&C	45.87 4.71 39.28	232.45 3.79 34.44	9.17 0.01 6.14	287.48 8.52 79.86	Elec Heat	4,764 0	8,929 0	714 0	289,817 0	12,707	48,840 c.60% crw.25%, ea.7%, g.4%, p.3%	11.3	82.8	5.3	124.1	Oil, gas, coal. IEA reports high crw %.
Brunei (WBA/h.i.)	All-s PE&HP MI&C	0 0 0	1.32 0.02 0.08	6.52 1.51 0	7.83 1.53 0.08	Elec Heat	0 0	311 0	1,457 0	1,768 0	n/a	0	0	0.58	2.75	3.35	Gas, oil.
Burkina Faso (LDC)	All-s PE&HP MI&C	0.004 n/a n/a	0.956 n/a n/a	0 n/a n/a	0.96 n/a n/a	Elec Heat	 n/a	145* n/a	 n/a	220 n/a	0	n/a				2.57**	Imports fuel. *Mostly diesel. **Trad.f.
Burundi (LDC)	All-s PE&HP MI&C	0.018 n/a n/a	0.198 n/a n/a	0 n/a n/a	0.213 n/a n/a	Elec Heat	 n/a	2 n/a	 n/a	120* n/a	n/a	n/a				1.04**	Imports fuel. *Mostly hydro. **Trad.f.
Cambodia (LDC)	All-s PE&HP MI&C	0 n/a n/a	0.498 n/a n/a	n/a n/a n/a	0.498 n/a n/a	Elec Heat	n/a n/a	120* n/a	n/a n/a	194 n/a	0	n/a				1.74**	Imports fuel. *Mostly diesel. **Trad.f.
Cameroon	All-s PE&HP MI&C	0 0 0	3.64 0.07 0.39	0 0 0	3.64 0.07 0.39	Elec Heat	0 0	33 0	0 0	2,901* 0	259	0	0	0.38	0	5.0	Gas, oil. *Mostly crw.

Cape Verde	All-s	n/a	n/a	n/a	n/a	Elec				39*	n/a	n/a	n/a	n/a	n/a	n/a	Imports fuel.
(LDC)	PE&HP	n/a	n/a	n/a	n/a	Heat	n/a	n/a	n/a	n/a							*Mostly diesel.
	MI&C	n/a	n/a	n/a	n/a												
Central Af.	All-s	n/a	0.234	n/a	0.234	Elec	0	21	0	102*	234	n/a				0.85**	Imports fuel.
R (LDC)	PE&HP	n/a	n/a	n/a		Heat	n/a	n/a	n/a	n/a							*Mostly hydro.
	MI&C	n/a	n/a	n/a													**Trad.f.
Chad (LDC)	All-s	n/a	0.095	n/a	0.095	Elec	0	89*	0	89	0	n/a				0.95**	Imports fuel.
	PE&HP	n/a	n/a	n/a		Heat	n/a	n/a	n/a	n/a							*Mostly diesel.
	MI&C	n/a	n/a	n/a													**Trad.f.
Chile	All-s	9.36	29.50	3.23	42.09	Elec	6,820	2,350	303	28,027	1,495	1,867	2.3	10.9	1.6	16.4	Gas, coal, oil
	PE&HP	5.38	0.36	0.12	5.86	Heat				1,879*		c.69%					*Heat output
	MI&C	3.91	6.47	1.68	12.06							ea.21%, p.10%					from fuels n/a.
China	All-s	2,550	420.5	36.3	3,006.8	Elec	810,449	65,000	2,263	1,080,017	*222,049*	*406,678*	681.8	185.6	17.4	904.6	Coal, oil, gas.
	PE&HP	942.8	56.5	1.8	1,001.2	Heat	0	0	0	0		c.67%					
	MI&C	1,299.1	86.7	16.0	1,401.9							ea.20%, p.4%, g.1%					
China,	All-s	21.75	22.24	*	43.99	Elec	28,000	442	*	28,442	n/a	n/a	3.4	9.1	2.4	14.9	Imports fuel.
Hong Kong	PE&HP	21.74	0.49	0	22.24	Heat	0	0	0	0							*CCGT active.
	MI&C	0.05	6.04	0	6.09												
Colombia	All-s	17.24	38.24	9.59	65.06	Elec	3,080	241	5,706	44,605	4,796	1,426	2.9	12.9	4.8	23.2	Coal, oil, gas.
	PE&HP	4.96	0.23	3.98	9.17	Heat	0	0	0	0		ea.64%					
	MI&C	10.14	4.28	2.11	16.54							p.21%, c.7%					
Comoros	All-s	n/a	n/a	n/a	n/a	Elec	0	16*	0	16	n/a	n/a	n/a	n/a	n/a	n/a	Imports fuel.
(LDC)	PE&HP	n/a	n/a	n/a	n/a	Heat	n/a	n/a	n/a	n/a							*Mostly diesel.
	MI&C	n/a	n/a	n/a	n/a												
Congo,	All-s	0.99	3.95	0	4.94	Elec	0	130	0	6,261*	*50*	0	0.241	1.317	0	13.79**	Oil, coal, gas.
Dem.R.	PE&HP	0	0.15	0	0.15	Heat	0	0	0	0							*Mostly hydro.
(LDC) (Zaire)	MI&C	0.76	0.18	0	0.94												**Mostly crw.
Congo, Rep.	All-s	0	0.89	0.01	0.90	Elec	0	3	3	438*	50	0	0	0.331	0.003	1.20	Oil, gas.
	PE&HP	0	0	0.01	0.01	Heat	0	0	0	0							*Mostly hydro.
	MI&C	0	0.01	0	0.01												
Cook	All-s	n/a	n/a	n/a	n/a	Elec	n/a	n/a	n/a	n/a	n/a	n/a	n/a	n/a	n/a	n/a	Imports fuel.
Islands/NZ	PE&HP	n/a	n/a	n/a	n/a	Heat	n/a	n/a	n/a	n/a							
	MI&C	n/a	n/a	n/a	n/a												

Table 5.1. *Continued*

Country	CO_2 emissions from source. Estimates are made for countries which do not have specific data. Total incl. cement manufacturing and gas flaring. 1995 (est)					Electricity and heat generation (hydro, nuclear and other energy types incl. in total)/ Comments 1996 (est), latest fig.					CO_2 from cement manuf. 1995	CO_2 from iron/ steel 1995 (est)	Primary energy consumption (hydro and nuclear incl. in in total). Part of fossil fuels used for thermal electric gen. 1997, latest fig.				Fossil fuel availability, in order of proved reserves 1997, latest fig.
	Fuel	Coal	Oil/ pet.	Gas	Total		Coal	Pet. pdts.	Gas	Total			Coal	Oil	Gas	Total	Fuels: proved recoverable basis/
	Units	mil.t	mil.t	mil.t	mil.t	Elec Heat	GWh TJ	GWh TJ	GWh TJ	GWh TJ	000 t	000 t	oil equiv., mil.t				Comments
Costa Rica	All-s	0	4.81	0	4.81	Elec	0	680	0	4,895*	493	0	0	1.5	0	2.25	Imports fuel.
	PE&HP	0	0.61	0	0.61	Heat	0	0	0	0							*Mostly hydro.
	MI&C	0	0.69	0	0.69												
Cote d'Ivoire	All-s	0	3.88	0	3.88	Elec	0	1,143**	0	3,015	249	0	0	1.43	0	5.3*	*High crw %.
(Ivory Coast)	PE&HP	0	0.88	0	0.88	Heat	0	0	0	0							**Mostly diesel.
	MI&C	0	0.50	0	0.50												
Croatia	All-s	0.73	11.63	4.17	16.52	*Elec*	304	3,082	1,119	11,110	389	374	0.18	3.46	2.07	6.76	Coal, gas, oil.
(Annex B)	PE&HP	0.24	2.26	0.34	2.83	EP	281	1,865	560	9,306		ea.44%					
	MI&C	0.43	1.56	1.80	3.78	CHP	23	1,217	559	1,804		g.35%, c.11%, p.10%					
						Heat	0	11,528	2,475	14,003							
						CHP	0	9,840	1,388	11,228							
						HP	0	1,688	1,087	2,775							
Cuba	All-s	0.47	30.43	0.08	30.97	Elec	0	12,206	16	13,236	34	37	0.121	10.3	0.036	15.95	Oil, gas.
	PE&HP	0	8.59	0	8.59	Heat	0	0	0	0							
	MI&C	0.43	7.71	0.06	8.20												
Cyprus	All-s	0.05	5.85	0	5.89	Elec	0	2,592*	0	2,592	n/a	0	0.01	2.09	0	2.12	Imports fuel.
	PE&HP	0	2.06	0	2.06	Heat	0	0	0	0							*Mosty fuel oil.
	MI&C	0.05	1.17	1.22	1.22												
Djibouti	All-s	n/a	n/a	n/a	n/a	Elec	n/a	185*	n/a	185	n/a	n/a	n/a	n/a	n/a	n/a	Imports fuel.
(LDC)	PE&HP	n/a	n/a	n/a	n/a	Heat	n/a	n/a	n/a	n/a							*Mostly diesel.
	MI&C	n/a	n/a	n/a	n/a												

Dominican Republic	All-s	0.27	10.71	0	10.98	Elec	296	4,666	*	6,847	724	0	0.08	3.65	0	5.19	Imports fuel.
	PE&HP	0.27	2.55	0	2.81	Heat	0	0	0	0							*Gas also used.
	MI&C	0	1.05	0	1.05												
Ecuador	All-s	0	4.90	0	4.90	Elec	0	3,011	0	9,438	1,146	0	0	6.78	0	8.55	Oil, gas, coal.
	PE&HP	0	1.15	0	1.15	Heat	0	0	0	0							
	MI&C	0	0.94	0	0.94												
Egypt	All-s	2.33	63.94	24.31	90.58	Elec	0	21,352	25,399	57,561	7,973	653	1.0	25.9	10.6	38.5	Gas, oil, coal.
	PE&HP	0	14.32	15.03	29.34	Heat	0	0	0	0		c.100%					CCGT active.
	MI&C	2.90	18.84	4.22	25.95												
El Salvador	All-s	0	4.90	0	4.90	Elec	0	928	0	3,487*	436	3	0.001	1.449	0	4.058	Imports fuel.
	PE&HP	0	1.15	0	1.15	Heat						c.100%					*Mostly hydro.
	MI&C	0	0.94	0	0.94												
Equatorial Guinea (LDC)	All-s	n/a	0.13	n/a	0.13	Elec	0	18*	0	20	0	n/a				0.14**	Gas, oil.
	PE&HP	n/a	n/a	n/a	n/a	Heat	n/a	p.100%	n/a	n/a							*Mostly diesel.
	MI&C	n/a	n/a	n/a	n/a												**Trad.f.
Eritrea (LDC)	All-s	n/a	n/a	n/a	n/a	Elec	n/a	n/a	n/a	n/a	n/a	n/a	n/a	n/a	n/a	n/a	Imports fuel.
	PE&HP	n/a	n/a	n/a	n/a	Heat	n/a	n/a	n/a	n/a							
	MI&C	n/a	n/a	n/a	n/a												
Ethiopia (LDC)	All-s	0	3.0	0	3.0	Elec	0	103	0	1,316*	29	0	0	1.03	0	16.56**	Gas, oil.
	PE&HP	0	0.22	0	0.22	Heat	0	0	0	0							*Mosty hydro.
	MI&C	0	0.64	0	0.64												**Mostly crw.
Fiji	All-s	0.051	0.645	0	0.736	Elec	0	115*	0	544**	39	n/a				0.78***	Imports fuel.
	PE&HP	n/a	n/a	n/a	n/a	Heat	n/a	n/a	n/a	n/a							*d, **m.hydro.
	MI&C	n/a	n/a	n/a	n/a												***com.e, trad.f.
Gabon	All-s	0	1.51	0.16	1.67	Elec	0	125	104	740*	65	0	0	0.612	0.066	1.578**	Oil, gas.
	PE&HP	0	0.12	0.09	0.22	Heat	0	0	0	0							*Mostly hydro.
	MI&C	0	0.33	0	0.33												**High crw %.
Gambia (LDC)	All-s	0	0.216	0	0.216	Elec	0	74*	0	74	0	n/a				0.33**	Imports fuel.
	PE&HP	n/a	n/a	n/a	n/a	Heat	n/a	n/a	n/a	n/a							*Diesel used.
	MI&C	n/a	n/a	n/a	n/a												**Trad.f.
Georgia	All-s	0.43	0.14	2.85	3.43	Elec	0	75	1,000	7,195*	50	0	0.102	0.132	0.77	1.576	Gas, oil.
	PE&HP	0	0.06	0.99	1.04	Heat				19,950**							*Mostly hydro.
	MI&C	0	0	0.70	0.70												**Oil and gas.

Table 5.1. *Continued*

Country	CO_2 emissions from source. Estimates are made for countries which do not have specific data. Total incl. cement manufacturing and gas flaring. 1995 (est)					Electricity and heat generation (hydro, nuclear and other energy types incl. in total)/ Comments 1996 (est), latest fig.					CO_2 from cement manuf. 1995	CO_2 from iron/ steel 1995 (est)	Primary energy consumption (hydro and nuclear incl. in in total). Part of fossil fuels used for thermal electric gen. 1997, latest fig.				Fossil fuel availability, in order of proved reserves 1997, latest fig.
	Fuel	Coal	Oil/ pet.	Gas	Total		Coal	Pet. pdts.	Gas	Total			Coal	Oil	Gas	Total	Fuels: proved recoverable basis/ Comments
	Units	mil.t	mil.t	mil.t	mil.t	Elec Heat	GWh TJ	GWh TJ	GWh TJ	GWh TJ	000 t	000 t	oil equiv., mil.t				
Ghana	All-s	0.01	3.10	0	3.10	Elec	0	7	0	5,980*	698	0	0.002	1.446	0	6.657	Gas, oil.
	PE&HP	0	0.15	0	0.15	Heat	0	0	0	0							*Mostly hydro.
	MI&C	0	0.19	0	0.19												
Grenada	All-s	n/a	n/a	n/a	n/a	Elec				62*	n/a	n/a	n/a	n/a	n/a	n/a	Imports fuel.
	PE&HP	n/a	n/a	n/a	n/a	Heat	n/a	n/a	n/a	n/a							*Research reqd.
	MI&C	n/a	n/a	n/a	n/a												
Guatemala	All-s	0	6.01	0	6.01	Elec	0	887	0	4,298*	778	0	0	1.96	0	5.22	Oil, gas.
	PE&HP	0	0.52	0	0.52	Heat	0	0	0	0							*Mostly hydro.
	MI&C	0	1.07	0	1.07												
Guinea (LDC)	All-s	0	1.08	0	1.08	Elec	0	352	0	543	0	n/a				1.35*	Imports fuel.
	PE&HP	n/a	n/a	n/a	n/a	Heat	n/a	n/a	n/a	n/a							*Mostly trad.f
	MI&C	n/a	n/a	n/a	n/a												
Guinea-Bissau (LDC)	All-s	0	0.23	0	0.23	Elec	0	43*	0	43	0	n/a				0.16**	Imports fuel, *d.
	PE&HP	n/a	n/a	n/a	n/a	Heat	n/a	n/a	n/a	n/a							**Com.e, trad.f.
	MI&C	n/a	n/a	n/a	n/a												Hydro potential.
Guyana	All-s	n/a	0.93	0	0.93	Elec	0	313*	0	318	0	n/a				0.47**	Imports fuel.
	PE&HP	n/a	n/a	n/a	n/a	Heat	n/a	n/a	n/a	n/a							*Oil, trad.f.
	MI&C	n/a	n/a	n/a	n/a												**Mostly com.e.
Haiti (LDC)	All-s	0	0.97	0	0.97	Elec	0	333*	0	625	25	n/a	0	0.37	0	1.97**	Imports fuel.
	PE&HP	0	0.16	0	0.16	Heat	0	0	0	0							*Diesel used.
	MI&C	0	0.14	0	0.14												**Mostly crw.

Honduras	All-s	0	3.52	0	3.53	Elec	0	14	0	2,815*	326	3	0.001	1.162	0	2.925**	Imports fuel.
	PE&HP	0	0.01	0	0.01	Heat	0	0	0	0		c.100%					*Mostly hydro.
	MI&C	0	1.10	0	1.10												**Mostly crw.
India	All-s	535.25	231.73	36.02	803.00	Elec	318,357	12,000	27,189	435,075	*34,881*	*61,726*	146.4	83.1	22.0	260.3	Coal, oil, gas.
	PE&HP	337.83	8.04	12.61	358.48	Heat	0	0	0	0		c.80%					IEA reports
	MI&C	191.41	46.30	17.98	255.69							ea.16%, p.4%					high crw %.
Indonesia	All-s	23.60	124.75	78.69	227.04	Elec	173,44	17,018	21,447	67,062	*9,717*	*6,042*	6.4	46.3	29.5	83.0	Coal, gas, oil.
	PE&HP	15.91	8.21	10.85	34.97	Heat	0	0	0	0		g.56%					IEA reports
	MI&C	4.61	22.75	10.67	38.03							p.44%					high crw %.
Iran	All-s	4.30	163.75	64.94	232.99	Elec	0	33,741	49,734	90,851	*8,122*	*1,934*	1.5	60.8	38.6	101.5	Oil, gas, coal.
	PE&HP	0	20.09	23.56	43.65	Heat	0	0	0	0							CCGT active.
	MI&C	4.36	26.12	23.61	54.09												
Israel	All-s	16.15	34.71	0.05	50.91	Elec	22,397	10,038	0	32,497	1,744	254	4.95	10.81	0.012	16.185	Imports fuel.
	PE&HP	16.15	8.17	0.01	24.33	Heat											Small oil and
	MI&C	0.03	4.05	0.04	4.12												gas deposits.
Jamaica	All-s	0.13	9.11	0	9.24	Elec	0	5,631	0	6,038	261	0	0.039	3.131	0	3.718	Imports fuel.
	PE&HP	0	2.02	0	2.07	Heat	0	0	0	0							
	MI&C	0.13	0.58	0	0.71												
Jordan	All-s	0	12.18	0.44	12.62	Elec	0	5,306	729	6,058	1,993	11	0	4.307	0	4.487	Imports fuel.
	PE&HP	0	4.03	0.44	4.47	Heat	0	0	0	0							Gas and oil
	MI&C	0	1.54	0	1.54												deposits.
Kazakhstan	All-s	130.61	31.16	23.81	185.58	Elec	42,250	4,277	0	58,657	897	9,889	24.0	11.6	8.8	45.3	Coal, gas, oil.
	PE&HP	89.23	4.34	6.14	99.72	Heat				256*		ea.83%					*CHP operative.
	MI&C	41.38	0	0	41.38							c.17%					
Kenya	All-s	0.23	6.88	0	7.11	Elec	0	355	0	4,040*	747	0	0.055	1.965	0	13.279	Imports fuel.
	PE&HP	0	0.24	0	0.24	Heat	0	0	0	0							*Mostly hydro.
	MI&C	0.23	1.10	0	1.34												Geothermal/ solar.
Kiribati (LDC)	All-s	n/a	n/a	n/a	n/a	Elec			7*	n/a	n/a	n/a	n/a	n/a	n/a	n/a	Imports fuel.
	PE&HP	n/a	n/a	n/a	n/a	Heat	n/a	n/a	n/a	n/a							*Fuel sources
	MI&C	n/a	n/a	n/a	n/a												to investigate.
Korea, Dem. P's. R. (N)	All-s	82.13	4.98	0	87.11	Elec	12,519	0	0	35,036*	1,334	6,894	19.54	1.51	0	24.0	Coal.
	PE&HP	19.10	0	0	19.10	Heat	0	0	0	0		c.100%					Imports fuel.
	MI&C	62.37	0.75	0	63.12												*Mostly hydro.

Table 5.1. *Continued*

Country	CO_2 emissions from source. Estimates are made for countries which do not have specific data. Total incl. cement manufacturing and gas flaring. 1995 (est)					Electricity and heat generation (hydro, nuclear and other energy types incl. in total)/ Comments 1996 (est), latest fig.					CO_2 from cement manuf. 1995	CO_2 from iron/ steel 1995 (est)	Primary energy consumption (hydro and nuclear incl. in in total). Part of fossil fuels used for thermal electric gen. 1997, latest fig.				Fossil fuel availability, in order of proved reserves 1997, latest fig.
	Fuel	Coal	Oil/ pet.	Gas	Total		Coal	Pet. pdts.	Gas	Total			Coal	Oil	Gas	Total	Fuels: proved recoverable basis/ Comments
	Units	mil.t	mil.t	mil.t	mil.t	Elec Heat	GWh TJ	GWh TJ	GWh TJ	GWh TJ	000 t	000 t	oil equiv., mil.t				
Korea,	All-s	109.74	221.86	21.50	353.10	*Elec*	78,941	40,683	27,097	223,069	27,472	32,030	34.0	105.9	14.8	175.1	Coal.
Rep. (S)	PE&HP	38.39	28.72	10.82	77.94	EP	78,941	40,159	14,710	210,158		ea.45%					Imports fuel.
	MI&C	58.44	53.27	0	111.71	CHP	0	524	12,387	12,911		c.40%, p.15%					CCGT active.
						Heat	0	9,039	20,740	29,779							
						CHP	0	0	18,461	18,461							
						HP	0	9,039	2,279	11,318							
Kuwait	All-s	0	32.36	9.65	42.0	Elec	0	5,522	19,953	25,475	997	0	0	8.1	8.5	16.6	Oil, gas.
(WBA/h.i.)	PE&HP	0	3.41	9.65	13.05	Heat	0	0	0	0							
	MI&C	0	5.03	0	5.03												
Kyrgyz	All-s	1.76	0.97	1.65	4.38	Elec	910	593	0	13,758*	149	0	0.592	0.599	0.883	2.952	Coal, oil, gas.
Republic	PE&HP	0.67	0	0.95	1.62	Heat				19,162**							*Mostly hydro.
	MI&C	1.08	0	0	1.08												CHP active.
Laos, P. Dem.	All-s	0.004	0.308	n/a	0.312	Elec	0	43	0	908*	0	n/a				1.16**	Imports fuel.
R. (LDC)	PE&HP	n/a	n/a	n/a	n/a	Heat	n/a	n/a	n/a	n/a							*Mostly hydro.
	MI&C																**Trad. f.
Lebanon	All-s	0.46	12.94	0	13.40	Elec	0	6,167	0	6,965	1,495	0	0.132	4.359	0	4.747	Imports fuel.
	PE&HP	0	3.68	0	3.68	Heat	0	0	0	0							
	MI&C	0.46	2.34	0	2.80												
Lesotho	All-s	n/a	n/a	n/a	n/a	Elec				1.21*	0	n/a	n/a	n/a	n/a	n/a	Imports fuel.
(LDC)	PE&HP	n/a	n/a	n/a	n/a	Heat	n/a	n/a	n/a	n/a							and electricity.
	MI&C	n/a	n/a	n/a	n/a												*hydro, diesel.

Liberia (LDC)	All-s	n/a	n/a	n/a	n/a	Elec				486*	0	n/a				**1.36	Imports fuel.
	PE&HP	n/a	n/a	n/a	n/a	Heat											*Diesel, gas.
	MI&C	n/a	n/a	n/a	n/a												**Trad.f.
Liech.	All-s	6.67	2.42	0	9.09	Elec	5,601	38	0	6,489	0	1,279	1.632	1.512	0	3.421	Coal.
(Ann.B)? data	PE&HP	5.67	0.23	0	5.89	Heat	669	5,011	0	6,208		ea.76%					
Macedonia	MI&C	0.54	0.41	0	0.95							c.19%, p.4%					
Madagascar	All-s	0.044	1.048	0	1.125	Elec	0	258	0	611*	*30*	n/a				2.90**	Imports fuel.
(LDC)	PE&HP	n/a	n/a	n/a	n/a	Heat	n/a	n/a	n/a	n/a							*Diesel, hydro.
	MI&C	n/a	n/a	n/a	n/a												**Trad.f.
Malawi (LDC)	All-s	n/a	n/a	n/a	n/a	Elec				803*	*69*	n/a				2.74**	Imports fuel.
	PE&HP	n/a	n/a	n/a	n/a	Heat	n/a	n/a	n/a	n/a							*Mostly hydro.
	MI&C	n/a	n/a	n/a	n/a												**Trad.f.
Malaysia	All-s	6.29	60.82	25.17	92.27	Elec	3,285	6,321	36,615	51,407	*5,315*	578	1.6	20.2	16.0	38.5	Gas, oil, coal.
	PE&HP	3.71	6.76	14.99	25.47	Heat	0	0	0	0							CCGT and
	MI&C	2.80	15.25	59.40	20.98												CCGT-Cogen active.
Maldives	All-s	n/a	n/a	n/a	n/a	Elec	0	40*	0	40	n/a	n/a	n/a	n/a	n/a	n/a	Imports fuel.
(LDC)	PE&HP	n/a	n/a	n/a	n/a	Heat	n/a	n/a	n/a	n/a							*Diesel used.
	MI&C	n/a	n/a	n/a	n/a												*High risk, s.l.r.*
Mali (LDC)	All-s	0	454	0	465	Elec	0	65	0	290*	10	n/a				1.60**	Imports fuel.
	PE&HP	n/a	n/a	n/a	n/a	Heat	n/a	n/a	n/a	n/a							*Mostly hydro.
	MI&C	n/a	n/a	n/a	n/a												**Trad.f.
Malta	All-s	0.12	2.50	0	2.62	Elec	0	1,658	0	1,658	0	0	0	894	0	894	Imports fuel.
	PE&HP	0.12	1.51	0	1.63	Heat	n/a	n/a	n/a	n/a							Oil, gas used
	MI&C	0	0.07	0	0.07												for e/g.
Marshall	All-s	n/a	n/a	n/a	n/a	Elec	n/a	n/a	n/a	n/a	n/a	n/a	n/a	n/a	n/a	n/a	Imports fuel.
Islands	PE&HP	n/a	n/a	n/a	n/a	Heat	n/a	n/a	n/a	n/a							Diesel used for
	MI&C	n/a	n/a	n/a	n/a												e/g.
Mauritania	All-s	0.015	2.865	0	3.067	Elec	0	124*	0	152	*187*	n/a				0.95**	Imports fuel.
(LDC)	PE&HP	n/a	n/a	n/a	n/a	Heat											*Diesel used.
	MI&C	n/a	n/a	n/a	n/a												**Com.e.
Mauritius	All-s	0.176	1.315	0	1.491	Elec				1,120*	0	n/a				0.81**	Imports fuel.
	PE&HP	n/a	n/a	n/a	n/a	Heat	n/a	n/a	n/a	n/a							*Mostly
	MI&C	n/a	n/a	n/a	n/a												thermal. **Com.e, trad.f.

Table 5.1. *Continued*

Country	CO_2 emissions from source. Estimates are made for countries which do not have specific data. Total incl. cement manufacturing and gas flaring. 1995 (est)					Electricity and heat generation (hydro, nuclear and other energy types incl. in total)/ Comments 1996 (est), latest fig.					CO_2 from cement manuf. 1995	CO_2 from iron/ steel 1995 (est)	Primary energy consumption (hydro and nuclear incl. in in total). Part of fossil fuels used for thermal electric gen. 1997, latest fig.				Fossil fuel availability, in order of proved reserves 1997, latest fig.
	Fuel	Coal	Oil/ pet.	Gas	Total		Coal	Pet. pdts.	Gas	Total			Coal	Oil	Gas	Total	Fuels: proved recoverable basis/ Comments
	Units	mil.t	mil.t	mil.t	mil.t	Elec Heat	GWh TJ	GWh TJ	GWh TJ	GWh TJ	000 t	000 t	oil equiv., mil.t				
Mexico	All-s	21.32	246.84	59.40	327.56	Elec	17,648	81,382	18,442	162,526	*11,945*	22,010	6.5	77.6	29.5	118.8	Oil, gas, coal.
	PE&HP	12.53	54.12	9.31	75.96	Heat	0	0	0	0		g.36%					CCGT active.
	MI&C	8.37	32.05	35.88	76.30							ea.28%, c.26%					Geothermal.
Micronesia	All-s	n/a	n/a	n/a	n/a	Elec	n/a	n/a	n/a	n/a	n/a	n/a	n/a	n/a	n/a	n/a	Imports fuel.
	PE&HP	n/a	n/a	n/a	n/a	Heat	n/a	n/a	n/a	n/a							Diesel used for
	MI&C	n/a	n/a	n/a	n/a												e/g.
Moldova	All-s	2.30	3.17	5.78	11.25	*Elec*	1,360	384	4,012	6,122	25	518	0.513	0.970	2.927	4.601	Imports fuel.
	PE&HP	1.84	1.04	2.63	5.51	EP	1,339	211	3,010	4,926		ea.100%					CHP active.
	MI&C	0.05	0.10	0.84	0.99	CHP	21	173	1,002	1,196							
						Heat	0	2,030	12,568	14,598							
						CHP	0	1,741	9,315	11,056							
						HP	0	289	3,253	3,542							
Monaco (Annex B)	All-s	n/a	n/a	n/a	n/a	Elec	n/a	n/a	n/a	n/a	n/a	n/a	n/a	n/a	n/a	n/a	Negligible.
	PE&HP	n/a	n/a	n/a	n/a	Heat	n/a	n/a	na/	n/a							Imports elec.
	MI&C	n/a	n/a	n/a	n/a												from France.
Mongolia	All-s	n/a	n/a	n/a	n/a	Elec				2,629*	54	n/a				2.42**	Oil, coal.
	PE&HP	n/a	n/a	n/a	n/a	Heat	n/a	n/a	n/a	n/a							*Coal, diesel.
	MI&C	n/a	n/a	n/a	n/a												**Com.e.
Morocco	All-s	7.81	17.81	0.04	25.67	Elec	5,550	4,868	0	12,356	3,239	0	2.08	6.15	0.015	8.82	Imports fuel.
	PE&HP	5.35	3.68	0	9.03	Heat	0	0	0	0							Small gas and
	MI&C	1.87	1.65	0.04	3.56												oil depsits.
Mozambique (LDC)	All-s	0.13	1.36	0.31	1.80	Elec	74	391*	82	597	10	n/a	0.036	0.498	0.295	7.813**	Imports fuel.
	PE&HP	0.07	0.14	0.31	0.52	Heat	0	0	0	0							*Diesel used.
	MI&C	0.07	0	0	0.07												**High crw %.

Myanmar	All-s	0.03	1.64	3.60	5.28	Elec	5	678	1,943	4,256*	*258*	0	0.009	1.338	1.322	12.767**	Gas, oil, coal.
(LDC)	PE&HP	0	0.23	0.85	1.08	Heat	0	0	0	0							*High hydro %.
	MI&C	0.03	0.38	2.75	3.16												**High crw %.
Namibia	All-s	n/a	n/a	n/a	n/a	Elec				1,257*	n/a	n/a	n/a	n/a	n/a	n/a	Imports fuel.
	PE&HP	n/a	n/a	n/a	n/a	Heat	n/a	n/a	n/a	n/a							*Mostly hydro.
	MI&C	n/a	n/a	n/a	n/a												
Nauru	All-s	n/a	n/a	n/a	n/a	Elec				30*	n/a	n/a	n/a	n/a	n/a	n/a	Imports fuel.
	PE&HP	n/a	n/a	n/a	n/a	Heat	n/a	n/a	n/a	n/a							*Diesel used.
	MI&C	n/a	n/a	n/a	n/a												
Nepal (LDC)	All-s	0.28	1.58	0	1.86	Elec	0	117	0	1,191	*110*	0	0.071	0.531	0	6.974**	Imports fuel.
	PE&HP	0	0.02	0	0.02	Heat	0	0	0	0							*Mostly hydro.
	MI&C	0.24	0.33	0	0.57												*High crw %.
Nicaragua	All-s	0	2.63	0	2.63	Elec	0	1,156*	0	1,907	174	0	0	0.896	0	2.391**	Imports fuel.
	PE&HP	0	0.86	0	0.86	Heat	0	0	0	0							*Diesel used.
	MI&C	0	0.32	0	0.32												**High crw %.
Niger (LDC)	All-s	0.458	0.645	0	1.118	Elec	0	175*	0	175	15	n/a				1.64**	Coal, imp.f.
	PE&HP	n/a	n/a	n/a	n/a	Heat	n/a	n/a	n/a	n/a							*Diesel used.
	MI&C	n/a	n/a	n/a	n/a												*m.trad.e.
Nigeria	All-s	0.06	40.59	10.28	50.93	Elec	0	3,919	5,572	14,991	1,296	187	0.086	12.939	4.456	82.669*	Gas, oil, coal.
	PE&HP	0	0.88	3.68	4.56	Heat	0	0	0	0		g.100%					*High crw %.
	MI&C	0.06	1.80	1.73	3.59												
Niue/NZ	All-s	n/a	n/a	n/a	n/a	Elec	n/a	n/a	n/a	n/a	n/a	n/a	n/a	n/a	n/a	n/a	Imports fuel.
	PE&HP	n/a	n/a	n/a	n/a	Heat	n/a	n/a	n/a	n/a							
	MI&C	n/a	n/a	n/a	n/a												
Oman	All-s	0	5.81	4.62	10.42	Elec	0	1,237	5,565	6,802	698	0	0	2.753	2.096	4.848	Oil, gas.
	PE&HP	0	1.16	4.09	5.25	Heat	0	0	0	0							CCGT active.
	MI&C	0	0.02	0.12	0.13												
Pakistan	All-s	8.08	45.68	27.58	81.34	Elec	440	17,547	15,280	56,956	4,278	1,666	2.3	16.5	14.3	35.3	Coal, gas, oil.
	PE&HP	0.07	13.07	9.00	22.14	Heat	0	0	0	0		c.100%					
	MI&C	7.42	5.78	11.07	24.27												
Palau	All-s	n/a	n/a	n/a	n/a	Elec				22*	n/a	n/a	n/a	n/a	n/a	n/a	Imports fuel.
	PE&HP	n/a	n/a	n/a	n/a	Heat	n/a	n/a	n/a	n/a							*Do research.
	MI&C	n/a	n/a	n/a	n/a												

Table 5.1. *Continued*

Country	CO_2 emissions from source. Estimates are made for countries which do not have specific data. Total incl. cement manufacturing and gas flaring. 1995 (est)					Electricity and heat generation (hydro, nuclear and other energy types incl. in total)/ Comments 1996 (est), latest fig.					CO_2 from cement manuf. 1995	CO_2 from iron/ steel 1995 (est)	Primary energy consumption (hydro and nuclear incl. in in total). Part of fossil fuels used for thermal electric gen. 1997, latest fig.				Fossil fuel availability, in order of proved reserves 1997, latest fig.
	Fuel	Coal	Oil/ pet.	Gas	Total		Coal	Pet. pdts.	Gas	Total			Coal	Oil	Gas	Total	Fuels: proved recoverable basis/ Comments
	Units	mil.t	mil.t	mil.t	mil.t	Elec Heat	GWh TJ	GWh TJ	GWh TJ	GWh TJ	000 t	000 t	oil equiv., mil.t				
Palestine (W.Bank, Gaza)	All-s																British Gas is exploring gasfields onshore Gaza
	PE&HP																
	MI&C																
Panama	All-s	0.13	4.65	0	4.78	Elec	0	1,464	0	3,915*	174	0	0.034	1.495	0	2.28	Imports fuel. *High hydro %.
	PE&HP	0	1.07	0	1.07	Heat	0	0	0	0							
	MI&C	0.13	0.88	0	1.01												
Papua New Guinea	All-s	0.004	2.323	0.158	2.481	Elec				1,790*	0	n/a				2.29**	Gas, oil. *Mostly thermal. **High trad.f.%.
	PE&HP	n/a	n/a	n/a	n/a	Heat	n/a	n/a	n/a	n/a							
	MI&C	n/a	n/a	n/a	n/a												
Paraguay	All-s	0	3.49	0	3.49	Elec	0	61	0	48,200	284	0	0	0.160	0.912	4.285**	Imports fuel. **High hydro and crw %. Elec. exp.
	PE&HP	0	0	0	0	Heat	0	0	0	0							
	MI&C	0	0.34	0	0.34												
Peru	All-s	1.68	22.89	1.92	26.49	Elec	0	3,612	230	17,280*	1,046	140 c.100%	0.306	7.440	0.805	13.933**	Coal, gas, oil. *Mostly hydro. **High crw %.
	PE&HP	0	0.73	0	0.73	Heat	0	0	0	0							
	MI&C	1.15	2.25	0	3.40												
Philippines	All-s	4.59	45.57	0	50.16	Elec	4,855	18,173	20	36,656	*4,883*	*2,114* p.59% c.20%, ea.18%	2.3	18.2	0.009	21.0*	Coal, gas, oil. Net fuel importer. *incl. gt., crw.
	PE&HP	1.67	9.32	0	11.00	Heat	0	0	0	0							
	MI&C	2.23	8.38	0	10.61												
Qatar (WBA/h.i.)	All-s	0	2.92	16.45	19.37	Elec	0	0	6,575	6,575	0	782 ea.100%	0	1.017	7.715	8.733	Gas, oil. CCGT active.
	PE&HP	0	0	10.12	10.12	Heat	0	0	0	0							
	MI&C	0	0.01	3.02	3.04												

Rwanda	All-s	0	0.484	0	0.491	Elec	0	4	0	164*	3	n/a				1.43**	Imports fuel.
(LDC)	PE&HP	n/a	n/a	n/a	n/a	Heat	n/a	n/a	n/a	n/a							*Mostly hydro.
	MI&C	n/a	n/a	n/a	n/a												**Trad.f.
Saint Kitts &	All-s	n/a	n/a	n/a	n/a	Elec				40*	n/a	n/a	n/a	n/a	n/a	n/a	Imports fuel.
Nevis	PE&HP	n/a	n/a	n/a	n/a	Heat	n/a	n/a	n/a	n/a							*Mostly
	MI&C	n/a	n/a	n/a	n/a												thermal.
Saint Lucia	All-s	n/a	n/a	n/a	n/a	Elec				182*	n/a	n/a	n/a	n/a	n/a	n/a	Imports fuel.
	PE&HP	n/a	n/a	n/a	n/a	Heat	n/a	n/a	n/a	n/a							*Mostly diesel.
	MI&C	n/a	n/a	n/a	n/a												
Saint Vincent	All-s	n/a	n/a	n/a	n/a	Elec				51*	n/a	n/a	n/a	n/a	n/a	n/a	Imports fuel.
& Grenadines	PE&HP	n/a	n/a	n/a	n/a	Heat	n/a	n/a	n/a	n/a							*Mostly hydro.
	MI&C	n/a	n/a	n/a	n/a												
Samoa (LDC)	All-s	n/a	n/a	n/a	n/a	Elec				48*	n/a	n/a	n/a	n/a	n/a	n/a	Imports fuel.
	PE&HP	n/a	n/a	n/a	n/a	Heat	n/a	n/a	n/a	n/a							*Mostly
	MI&C	n/a	n/a	n/a	n/a												biomass.
San Marino	All-s	n/a	n/a	n/a	n/a	Elec				*	n/a	n/a	n/a	n/a	n/a	n/a	Imports fuel.
	PE&HP	n/a	n/a	n/a	n/a	Heat	n/a	n/a	n/a	n/a							*Imports elec.
	MI&C	n/a	n/a	n/a	n/a												from Italy.
Sao Tome &	All-s	n/a	n/a	n/a	n/a	Elec				16*	n/a	n/a	n/a	n/a	n/a	n/a	Imports fuel.
Principe	PE&HP	n/a	n/a	n/a	n/a	Heat	n/a	n/a	n/a	n/a							*Mostly diesel.
	MI&C	n/a	n/a	n/a	n/a												
Saudi Arabia	All-s	0	150.08	76.99	227.06	Elec	0	55,807	42,012	97,819	7,973	0	0	55.7	39.5	95.2	Oil, gas.
	PE&HP	0	41.2	15.22	56.42	Heat	n/a	n/a	n/a	n/a							CCGT active.
	MI&C	0	24.79	0	24.79												
Senegal	All-s	0	2.48	0.11	2.58	Elec	0	1,034	140	1,174	294	0	0	1.023	0.036	2.588*	Imports fuel.
	PE&HP	0	0.83	0	0.83	Heat	0	0	0	0							*High crw %.
	MI&C	0	0.46	0.11	0.56												
Seychelles	All-s	n/a	n/a	n/a	n/a	Elec				109*	n/a	n/a	n/a	n/a	n/a	n/a	Imports fuel.
	PE&HP	n/a	n/a	n/a	n/a	Heat	n/a	n/a	n/a	n/a							*Diesel used.
	MI&C	n/a	n/a	n/a	n/a												
Sierra Leone	All-s	n/a	0.443	n/a	0.443	Elec	0	241*	0	241	0	n/a				0.86**	Imports fuel.
(LDC)	PE&HP	n/a	n/a	n/a	n/a	Heat	0	0	0	0							*Mostly diesel.
	MI&C	n/a	n/a	n/a	n/a												*m.trad.f.

Table 5.1. *Continued*

Country	CO$_2$ emissions from source. Estimates are made for countries which do not have specific data. Total incl. cement manufacturing and gas flaring. 1995 (est)					Electricity and heat generation (hydro, nuclear and other energy types incl. in total)/ Comments 1996 (est), latest fig.					CO$_2$ from cement manuf. 1995	CO$_2$ from iron/ steel 1995 (est)	Primary energy consumption (hydro and nuclear incl. in in total). Part of fossil fuels used for thermal electric gen. 1997, latest fig.				Fossil fuel availability, in order of proved reserves 1997, latest fig.
	Fuel	Coal	Oil/ pet.	Gas	Total		Coal	Pet. pdts.	Gas	Total			Coal	Oil	Gas	Total	Fuels: proved recoverable basis/ Comments
	Units	mil.t	mil.t	mil.t	mil.t	Elec Heat	GWh TJ	GWh TJ	GWh TJ	GWh TJ	000 t	000 t	oil equiv., mil.t				
Singapore	All-s	0.13	55.33	3.26	58.71	Elec	0	18,964	4,494	24,100	947	761	0	29.1	1.4	30.4	Imports fuel.
(WBA/h.i.)	PE&HP	0	17.42	3.26	20.68	Heat	0	0	0	0		ea.100%					
	MI&C	0.10	1.32	0	1.42												
Slovenia	All-s	4.96	6.89	1.56	13.41	*Elec*	4,189	346	0	12,770*	498	666	1.161	2.726	0.656	6.167*	Imports fuel.
(Annex B)	PE&HP	3.34	0.06	0.15	3.55	EP	532	94	0	8,861		ea.45%					*incl. hydro
	MI&C	0.33	0.61	0.88	1.83	CHP	3,657	252	0	3,909		g.42%					and nuclear.
						Heat	5,831	786	2,989	9,702		c.10%					
						CHP	5,821	175	0	5,996							
						HP	10	611	2,989	3,706							
Solomon	All-s	0	0.121	0	0.121	Elec	0	29*	0	29	0	n/a				0.12*	Imports fuel.
Islands (LDC)	PE&HP	n/a	n/a	n/a	n/a	Heat	0	0	0	0							*Diesel used.
	MI&C	n/a	n/a	n/a	n/a												**trad.f, com.e.
Somalia	All-s				0.012*	Elec	0	272	0	272*	12	n/a	n/a	n/a	n/a	n/a	Gas deposits.
(LDC)	PE&HP	n/a	n/a	n/a	n/a	Heat	n/a	n/a	n/a	n/a							Imports fuel.
	MI&C	n/a	n/a	n/a	n/a												*Cement manuf.
South Africa	All-s	285.64	31.23	4.01	320.88	Elec	184,952	0	0	198,346	4,520	*20,197*	83.1	21.0	0	107.7	Coal, gas, oil.
	PE&HP	146.20	0.31	0	146.51	Heat				300		c.73%					
	MI&C	53.94	3.96	0	57.89							ea.22%, p.3%, g.2%					

Sri Lanka	All-s	0.01	6.23	0	6.25	Elec	0	1,278	0	4,530*	448	0	0.001	2.587	0	6.792	Imports fuel.
	PE&HP	0	0.32	0	0.32	Heat	0	0	0	0							*High hydro %.
	MI&C	0	0.53	0	0.53												
Sudan (LDC)	All-s	0	5.01	0	5.01	Elec	0	989	0	2,063*	*125*	0	0	1.406	0	10.787**	Gas, oil.
	PE&HP	0	0.71	0	0.71	Heat	0	0	0	0							Imports fuel.
	MI&C	0	1.17	0	1.17												*High hydro %.
																	**Mostly crw.
Suriname	All-s	0	2.151	0	2.151	Elec	0	335	0	1,614*	25	n/a				0.59**	Imports fuel.
	PE&HP	n/a	n/a	n/a	n/a	Heat	n/a	n/a	n/a	n/a							*Mostly hydro.
	MI&C	n/a	n/a	n/a	n/a												**Com.e.
Swaziland	All-s	0.454	0	0	0.454	Elec				126*	n/a	n/a	n/a	n/a	n/a	n/a	Coal deposits.
	PE&HP	n/a	n/a	n/a	n/a	Heat	n/a	n/a	n/a	n/a							Imports fuel.
	MI&C	n/a	n/a	n/a	n/a												*Mostly hydro.
Syrian Arab Rep.	All-s	0.01	34.14	5.22	39.37	Elec	0	4,950	5,107	17,001*	2,990	3.34	0.001	11.702	2.235	14.541*	Oil, gas.
	PE&HP	0	7.84	2.86	10.70	Heat	0	0	0	0		c.100%					*High hydro %.
	MI&C	0.01	3.24	1.92	5.17												
Tajikstan	All-s	0.05	3.56	1.71	5.32	Elec	0	175	0	15,000*	50	0	0.051	1.228	0.931	3.513*	Gas, oil.
	PE&HP	0	0	0.80	0.80	Heat	0	0	0	0							Imports fuel.
	MI&C	0	0	0	0												**High hydro %.
Tanzania, Un. Rep (LDC)	All-s	0.02	2.46	0	2.47	Elec	0	243	0	1,991*	*399*	0	0.003	0.726	0	13.798*	Coal, gas.
	PE&HP	0	0.25	0	0.25	Heat	0	0	0	0							Imports fuel.
	MI&C	0.02	0.24	0	0.26												*High crw %.
Thailand	All-s	27.65	105.13	22.76	155.55	Elec	17,507	25,604	36,749	87,467	*13,205*	*3,811*	9.0	38.5	12.0	60.1	Coal, gas, oil.
	PE&HP	14.18	17.34	17.47	48.99	Heat	0	0	0	0		ea.79%					(coal: lignite).
	MI&C	14.61	14.62	1.83	31.06							p.21%.					
Togo (LDC)	All-s	0	0.572	0	0.744	Elec	0	87*	0	93	174	n/a				0.71**	Imports fuel.
	PE&HP	n/a	n/a	n/a	n/a	Heat	n/a	n/a	n/a	n/a							*Diesel used.
	MI&C	n/a	n/a	n/a	n/a												**Trad.f, com.e.
Trinidad & Tobago	All-s	0	3.19	11.55	11.74	Elec	0	0	4,508	4,541	229	1,135	0	1.445	6.413	7.887*	Gas, oil.
	PE&HP	0	0	3.20	3.20	Heat	0	0	0	0		g.100%					CCGT active.
	MI&C	0	0.24	6.51	6.75												

Table 5.1. *Continued*

Country	CO_2 emissions from source. Estimates are made for countries which do not have specific data. Total incl. cement manufacturing and gas flaring. 1995 (est)					Electricity and heat generation (hydro, nuclear and other energy types incl. in total)/ Comments 1996 (est), latest fig.					CO_2 from cement manuf. 1995	CO_2 from iron/ steel 1995 (est)	Primary energy consumption (hydro and nuclear incl. in in total). Part of fossil fuels used for thermal electric gen. 1997, latest fig.				Fossil fuel availability, in order of proved reserves 1997, latest fig.
	Fuel	Coal	Oil/ pet.	Gas	Total		Coal	Pet. pdts.	Gas	Total			Coal	Oil	Gas	Total	Fuels: proved recoverable basis/ Comments
	Units	mil.t	mil.t	mil.t	mil.t	Elec Heat	GWh TJ	GWh TJ	GWh TJ	GWh TJ	000 t	000 t	oil equiv., mil.t				
Tunisia	All-s	0.31	10.01	4.50	14.81	Elec	0	1,066*	6,400	7,533	2,143	397	0.069	3.513	1.948	6.676	Gas, oil.
	PE&HP	0	0.38	3.47	3.85	Heat	0	0	0	0		c.58%					*Diesel used.
	MI&C	0.31	2.36	0.85	3.51							ea.42%					
Turkey	All-s	66.66	80.76	13.08	160.50	Elec	30,413	6,540	17,174	94,862	*16,520*	*12,358*	0.3	30.8	9.9	64.3	Coal, oil, gas.
	PE&HP	30.44	3.01	6.02	39.47	EP	29,341	4,114	16,058	90,072		ea.60%					*CHP active.
	MI&C	21.32	14.03	3.37	38.73	CHP	1,072	2,426	1,116	4,790		c.23%, p.16%					Geothermal.
Turkmenistan	All-s	*	8.20	26.18	34.39	Elec	*	0	10,095	10,100	199	0	0	4.0	7.0	11.0	Gas, oil.
	PE&HP	0	0	9.12	9.12	Heat	0	0	0	0							Coal also used,
	MI&C	0	0	0	0												n/m by sources.
Tuvalu (LDC)	All-s	n/a	n/a	n/a	n/a	Elec	0	3	0	3	n/a	n/a	n/a	n/a	n/a	n/a	Imports fuel.
	PE&HP	n/a	n/a	n/a	n/a	Heat	n/a	n/a	n/a	n/a							Solar. *High*
	MI&C	n/a	n/a	n/a	n/a												*risk, s.l.r.*
Uganda (LDC)	All-s	0	978	0	1,044	Elec	0	7	0	792*	65	n/a				3.9**	Imports fuel.
	PE&HP	n/a	n/a	n/a	n/a	Heat	n/a	n/a	n/a	n/a							*Mostly hydro.
	MI&C	n/a	n/a	n/a	n/a												**m.trad.f.
United Arab Emirates	All-s	0	31.50	44.08	75.57	Elec	0	3,378	16,359	19,737	2,990	0	0	18.0	27.0	45.0	Oil, gas.
	PE&HP	0	5.0	8.78	13.78	Heat	0	0	0	0							CCGT active.
	MI&C	0	1.37	3.44	4.80												
Uruguay	All-s	0.01	4.70	0	4.70	Elec	0	838	0	6,670*	299	0	0	1.937	0	2.955	Imports fuel.
	PE&HP	0	0.31	0	0.31	Heat	0	0	0	0							*Mostly hydro.
	MI&C	0.01	0.62	0	0.63												

Uzbekistan	All-s	5.07	26.37	84.16	115.60	*Elec*	1,827	5,069	31,998	45,419	1,744	0	1.9	6.8	37.5	46.8	Coal, gas, oil.
	PE&HP	3.34	4.97	31.90	40.20	EP	1,238	2,320	16,929	27,012							
	MI&C	0	0	0	0	CHP	589	2,749	15,069	18,407							
						Heat	3,939	12,429	91,507	107,875							
						CHP	3,874	10,594	40,678	55,146							
						Heat	65	1,835	50,829	52,729							
Vanuatu (LDC)	All-s	n/a	n/a	n/a	n/a	Elec				29*	n/a	n/a	n/a	n/a	n/a	n/a	Imports fuel.
	PE&HP	n/a	n/a	n/a	n/a	Heat	n/a	n/a	n/a	n/a							*do research.
	MI&C	n/a	n/a	n/a	n/a												
Venezuela	All-s	0.01	54.56	58.62	113.19	Elec	0	1,862	19,666	75,372	3,438	12,845	0.2	19.7	27.8	52.6	Oil, gas, coal.
	PE&HP	0	2.21	10.86	13.07	Heat	0	0	0	0		g.71%					*High hydro %.
	MI&C	0.01	7.64	22.18	29.84							ea.27%, p.2%					
Vietnam	All-s	8.29	12.39	1.10	21.78	Elec	2,376	1,460	13,105	16,945*	3,737	0	3.403	6.261	3.118	33.750**	Gas, coal, oil.
	PE&HP	1.81	0.57	0.55	2.92	Heat	0	0	0	0							*High hydro
	MI&C	4.94	4.40	0.55	9.89												% 95.
																	**High crw %.
Yemen (LDC)	All-s	0	8.70	0	8.70	Elec	0	2,440	0	2,440	498	0	0	2.858	0	2.936	Oil, gas.
	PE&HP	0	1.00	0	1.00	Heat	0	0	0	0							
	MI&C	0	0.45	0	0.45												
Zambia (LDC)	All-s	0.88	1.59	0	2.47	Elec	40	2	0	7,796*	*149*	0	0.102	0.564	0	5.790**	Imports fuel.
	PE&HP	0	0	0	0	Heat	0	0	0	0							*Mostly hydro.
	MI&C	0.78	0.37	0	1.14												**High crw %.
Zimbabwe	All-s	13.30	2.67	0	15.97	Elec	5,559	0	0	7,876*	498	1,533	3.434	1.504	0	10.442**	Coal.
	PE&HP	6.96	0.09	0	7.05	Heat	0	0	0	0							*Incl. hydro.
	MI&C	4.62	0.30	0	4.92												**High crw %.

Table 5.1. *Continued*

Table 5.1 note

Notes: All these countries are independent entities, but some have "free" associations with countries from which they have sought independence. The list includes countries which may join UNFCCC and participate in CDM or JI projects. Certain cells or rows are left blank because the relevant data is neither available nor provided separately by the source/sources. In the country column, LDC countries are *underlined.* In the columns providing quantitative volumes of carbon dioxide emissions from the cement and iron & steel industries, the amounts of emissions from major country emitters are *underlined.* Liechstenstein, a non-Annex I country, is not included due to inavailability of figures for this country at the time of data compilation. For some countries which have separate entries for electricity and heat generation, these utility sectors are *underlined.*

Non-Annex I and non-Annex I/Annex B host countries and emissions indicators

Notes /symbols /acronyms:	Annex B	Annex B country listing in the Kyoto Protocol.
	LDC/UNCTAD	Least Developed Country classification by UNCTAD.
	WBA/h.i.	High income economy according to World Bank Atlas 1999.
	All-s	All sectors, public and private, contributing to carbon dioxide emissions.
	EP	Electricity plant.
	HP	Heat plant.
	PE&HP	Public electricity and heat production.
	MI&C	Manufacturing industry and construction.
	TPES	Total primary energy supply.
	c	Refers to coal used in iron and steel works.
	com.e	Commercial energy – solid, liquid and gaseous fuels; and non-combustible energy sources (nuclear, hydro, renewables).
	crw	Combustible renewables and waste fuels.
	ea	Refers to electric arc furnaces used in iron and steel works.
	elec	Electricity.
	e/g	Electricity generation.
	est	Estimate.
	exp.	Exporter/exports.
	d	Diesel fuel.
	f	Fuel.
	g	Refers to natural gas used iron and steel works.

	gas	Refers to natural gas in the last column. The density of North Sea gas, 0.7 kg/cubic metre, is used to calculate reserves of natural gas in mass terms so that it can be ranked comparatively against other fossil fuels. For CDM host countries, the density of the natural gas sourced domestically or country of export should be utilised.
	gt	Geothermal.
	imp.f	Imports fuel.
	incl.	Includes.
	mil.t	Million tonnes.
	m.hydro	Mostly hydro.
	m.trad.f.	Mostly traditional fuels.
	n/a	Information not available at time of data compilation.
	nil	Zero or insignificant quantity.
	n/m	Not mentioned by source publications.
	pet, p	Petroleum products.
	s.l.r.	Sea level rise.
	sub-bit.	Sub-bituminous coal.
	t	Tonne (metric).
	trad.f.	Traditional fuels – fuelwood, charcoal, bagasse, animal and vegetal wastes.
	* ** ***	The single, double and triple star asterisks refer to proportion of the major contributing source to the individual or total energy amount as indicated by the asterisk/s in columns 9 to 17; such proportions quantified in column 18.
Units	GWh	Giga Watt-hours, 10^9 Watt-hours or 10^6 kWh.
	TJ	Tera Joules; 1TJ = 10^{12} Joules or 1 trillion joules (UK definition of a trillion).

Approximate mass relationships: 1,000 cubic metres of natural gas (North Sea) = 0.7 tonne of coal = 1 tonne of oil

Approximate heat energy relationships: 1 billion cubic metres of natural gas = 0.9 million tonnes of oil equivalent

Sources: International Energy Agency's *Energy Statistics and Balances of non-OECD Countries 1995–96*, 1998 edition.
World Resources 1998–99, World Resources Institute 1998.
International Energy Agency's *Energy Balances of OECD Countries*, 1995–96, 1998 edition.

Table 5.1. *Continued*

Table 5.1 note continued

Notes to table sub-columns

Electricity and heat production

Aggregate figures for electricity and heat production from different fossil fuel types were not available for all countries at time of data compilation, but for certain countries the total amounts from one or both sources are provided.

Iron and steel works

The conversion factor for calculating carbon dioxide from iron & steel works is 3.34 (1 t oil equiv. = 1.5 t coal = 3.34 t CO_2). IPCC's default value of 2.23 is used for conversion of coal mass units (if coke is combusted in the furnace) to carbon dioxide mass units – based on oil equivalent heat units from IEA's *Energy Statistics and Balances for non-OECD Countries 1995–96*. The percentages of fossil fuel types used for iron/steel works, where this data is available, are shown below the CO_2 total figure for this industry; c = coal, g = gas, o = oil. See Comments (4) for further explanation.

Cement manufacturing

Estimates of fossil fuel types used for cement manufacturing is shown below the CO_2 total from this industry; in the same format as for iron and steel works. A fuel which provides more than 90% of the energy supplied is shown without a percentage.

Fossil fuel availability

The last column refers to coal, oil and natural gas. "Import fuels" refer to importing of these fuels into the economy.

Comments on country factors, sources of statistical data and methods of carbon dioxide calculations

1. CO_2 emissions from fuel combustion. Source: Carbon Dioxide Emissions from Fuel Combustion – 1997 edition, International Energy Agency (data for 1995 year).

Comments: There may exist a critical size of CDM investment in value terms so that expected returns – which include CO_2 – may be realised. For example, larger reductions of CO_2 per unit cost of equipment may only be achieved in plants operating above a certain capacity, compared to lower reductions in smaller installations. Pollution control equipment twice the size of a standard unit may cost a third more and perform at a higher efficiency. If there are a limited number of opportunities in a small country, then a CDM investor may not be able to replicate successful projects to increase the volume of returns. Small and medium-size economies including small island states should be included in a global CDM portfolio, as global warming is non-country discriminate. For some countries, total emissions figures are provided where industry sector contributions are not available – these figures include emissions from the domestic sector as well as non-manufacturing business and commercial entities. For certain countries, emissions from certain fuels are not recorded in the source – perhaps due to their small contributons to the total. Sub-figures, expressed in two decimals places, may not add up to total figures due to rounding.

2. Fossil fuel contribution to electricity and heat generated. Sources: International Energy Agency's *Energy Statistics and Balances of non-OECD Countries 1995–96. World Resources 1998–89*, World Resources Institute.

Comments: Countries with high coal utilisation for power generation could consider technologies to lower carbon dioxide emissions; e.g. Integrated Gasification Combined Cycle (IGCC) using coal, petcoke or other petroleum products; burn lower carbon content fuels like natural gas (which is price sensitive); or switch to Combined Heat and Power schemes which enables the plant to generate its own electricity and heat. For some countries, where the percentage contribution by different fossil fuels to electricity generated is not known at the time of data compilation, e.g. Bahamas, the total amount in GWh is shown. The types of fossil fuels used for power generation for those countries in which detailed information is not available are provided in the last column. In a number of cases, e.g. Hong Kong, Special Administrative Region of China, electricity generated by clean power technology (Combined Cycle Gas Turbine or CCGT plants) are not accounted for in IEA's publication. More concise electricity and heat data are provided for those countries which have adopted energy efficiency measures; Croatia is exemplary.

3. CO_2 emissions from cement manufacturing. Source: *World Resources 1998–99*, World Resources Institute.

Comments: Selected data by the World Resources Institute, perhaps limited by available information on global CO_2 emissions. Emissions from countries which are not covered by WRI are estimated in this current study, based on figures for the construction industry in International Energy Agency's *Energy Statistics and Balances of non-OECD countries 1995–96*. Such emissions figures for some, but not all countries covered by both publications appear to exhibit general consistency. The estimates hence should be treated with caution.

4. CO_2 emissions from iron and steel works. Source: International Energy Agency's *Energy Statistics and Balances of non-OECD Countries 1995–96.*

Comments: Heat energy in tonnes of oil equivalent are converted into carbon dioxide volumes by a factor of 3.34, as explained in the notes section. There are sharp discrepancies between the coal consumption figures expressed in tonnes of oil equivalent for the iron and steel industry quoted in the energy balance statistical section (III) and the hard coal & brown coal statistical section (II) of the source. The higher energy figures in section II may be due to coal transformation (steel coke formation) and distribution losses not accounted for in section III. On comparison, the lower figures from section III were chosen to maintain consistency with electricity data in the table quoted from the same section.

CO_2 from electric arc furnaces are calculated by multiplying the amount of energy in tonnes of oil equivalent consumed with the reciprocal of 33% (electrical efficiency) and then by 3.34 to obtain emissions values in mass terms. A problem in calculation arises when the electricity is supplied from thermal, hydro, nuclear, combustible renewables and waste, as separate data for the respective energy contributions from these types of plants to the furnace are not available from the source publication, e.g. Brazil. In countries where both fossil and non-fossil energy resources are available, estimates of upstream CO_2 emissions produced by fossil fuel combustion in thermal plants are provided based on the relative proportion of thermal electricity to total electricity. In utilising data from the source publication, energy from combustible renewables and waste used to generate electricity for electric arc furnaces should be distinguished from energy from the same source used to provide heat in the blast furnace.

Only very broad estimates of upstream CO_2 emissions for electric arc furnaces are hence provided for countries where such separate data is not currently available, based on the assumption that the electricity is supplied from a proportionate ratio of thermal to total electricity plants. Corrections will have to be made to these estimates when the exact amounts of electricity supplied from non-fossil fuel resources are known.

Table 5.1. *Continued*

Table 5.1 note continued

5. Primary energy consumption statistics. Sources: *BP Statistical Review of World Energy 1998*, International Energy Agency's *Energy Statistics & Balances of non-OECD Countries 1995–96*.

Comments: BP's publication provides data on developed and developing countries in which primary energy, based on commercially traded fuels, form a significant part of the economy. The amounts of coal, gas and oil consumed are utilised for thermal power generation, process industries, transport, commercial and domestic sectors.

The company's Review, however, does not provide primary energy consumption figures for other developing and least developed countries; estimates for these countries are made from IEA's *Energy Statistics and Balances of non-OECD Countries 1995–96*. Primary energy consumption figures for these countries are sourced from IEA's TPES (Total Primary Energy Supply) figures – these are the amounts available for consumption by the electricity and heat generation sectors, petroleum refineries, fuel processing plants, industry, transport, agricultural, commercial and domestic customers.

IEA's TPES and BP's primary energy consumption figures were found to be inconsistent, as BP excludes combustible renewables and wastes in its report – on the basis that external or publicly available data on these resources are unreliable. However, IEA publishes detailed figures on these alternative fuels. For the purposes of comparison in the current study, BP's figures – which are more recent (1997) than IEA's (1996) – are quoted.

6. Fossil fuel availability, proved recoverable basis. Source: *Survey of Energy Resources 1998*, World Energy Council.

Comments: CDM project planners can identify the kinds of fossil fuels available for a change of one type of fuel to another, and the amounts of these fuel reserves, for maximising CO_2 reductions. The choice of fuels will depend on the market price of competing fuels over the life of the project and the cost of appropriate technology. For coal, WEC's *Survey of Energy Resources 1998* provides figures only for countries with more than 1 million tonnes of the fuel.

Table 5.2. Global emissions of methane and nitrous oxide: Non-Annex I Countries. Comparative survey of selected countries

Country	Year	Methane emissions from anthropogenic sources (000 tonnes)						Nitrous oxide (000) tonnes
		Fossil fuel extraction	Fuel combustion	Agriculture Livestock	Agriculture other	Waste	Total methane	
Bangladesh	1990	80	190	520	473	74	1,335	3
Bolivia	1990	14	115	429	29	n/a	597	1
Brazil	1994	0	n/a	n/a	n/a	n/a	n/a	n/a
China	1990	5,650	50	8,940	18,400	790	33,830	1,100
Cote d'Ivoire	1990	0	127	50	78	370	626	1
Ecuador	1990	21	20	281	108	64	555	1
Egypt	1990	52	13	424	n/a	n/a	489	n/a
Ethiopia	1992	15	340	974	171	28	1,597	5
Gambia	1993	n/a	2	13	8	7	30	0
Indonesia	1990	527	50	864	2,039	n/a	3,746	2,769
Kazakhstan	1990	841	n/a	939	n/a	1,763	3,555	7
Mauritius	1994	0	2	1	0	3	6	0
Mongolia	1990	9	5	301	n/a	15	329	0
Nepal	1990	84	*	370	542	n/a	996	1
Oman	1992	0	n/a	n/a	n/a	1,518	1,518	n/a
Peru	1990	22	158	383	298	130	1,433	7
Philippines	1990	8	220	315	559	138	1,290	8
Venezuela	1990	1,827	12	853	98	221	3,170	454

* Included in the fuel extraction figure.
Source: World Resources 1998–99, World Resources Institute 1998

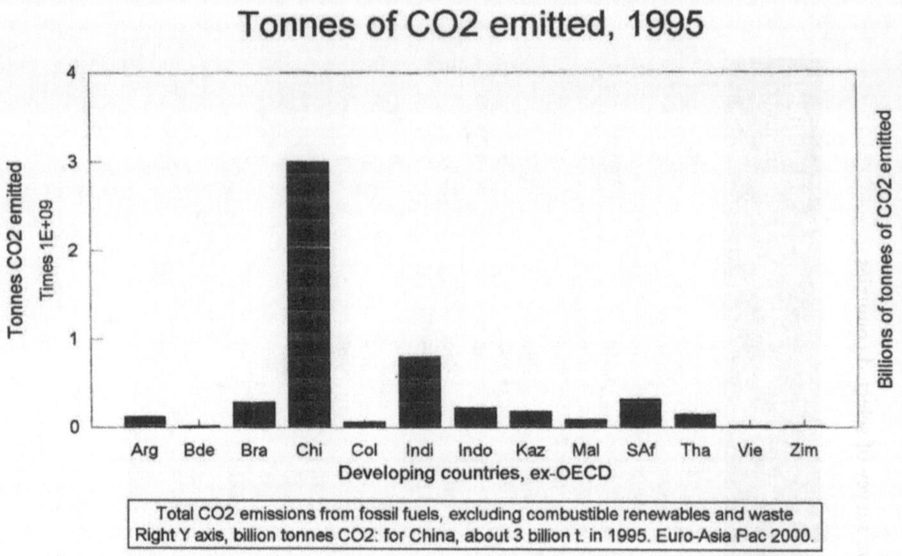

Fig. 5.1. Tonnes of CO_2 emitted, 1995. The most attractive markets in the developing world for the Clean Development Mechanism are compared in terms of the total amount of carbon dioxide emissions that can be reduced. Six of the top ten countries - which are Asian - accounted for 80% of emissions in this grouping. China's emissions of 3 bn. tonnes in 1995, is expected to exceed 3.5 bn. tonnes in 2000

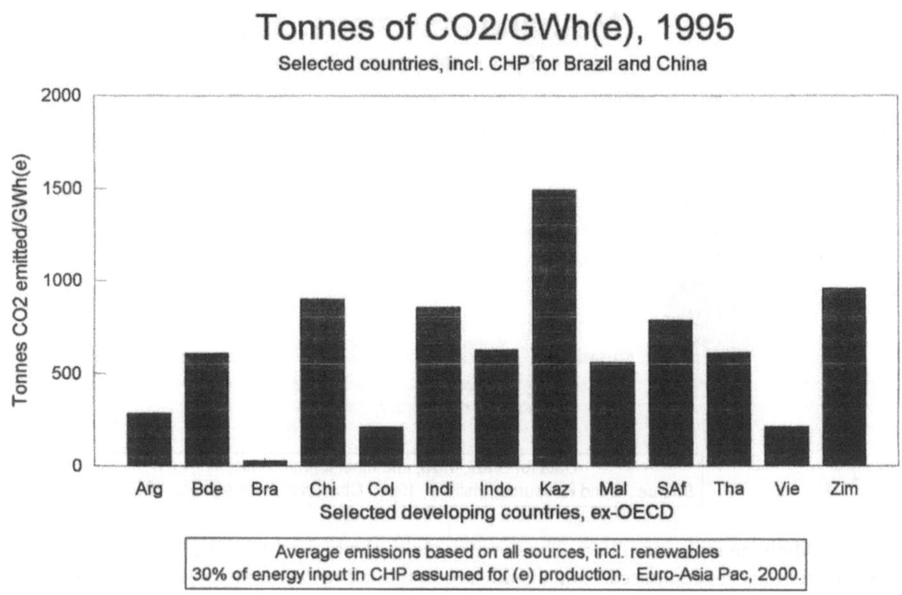

Fig. 5.2. Tonnes of CO_2 per GWh_e, 1995. The largest markets for carbon dioxide eduction in terms of average emission levels per unit output of electricity are shown. On the basis of total electricity output from all energy sources (fossil fuels and renewables), carbon dioxide tonnage emitted per electric gigawatt-hour (GWh_e) is highest in Kazakhstan, followed by Zimbabwe, China and India. Brazil has the lowest emission levels/GWh_e in the top thirteen countries surveyed, primarily due to large hydroelectric resources

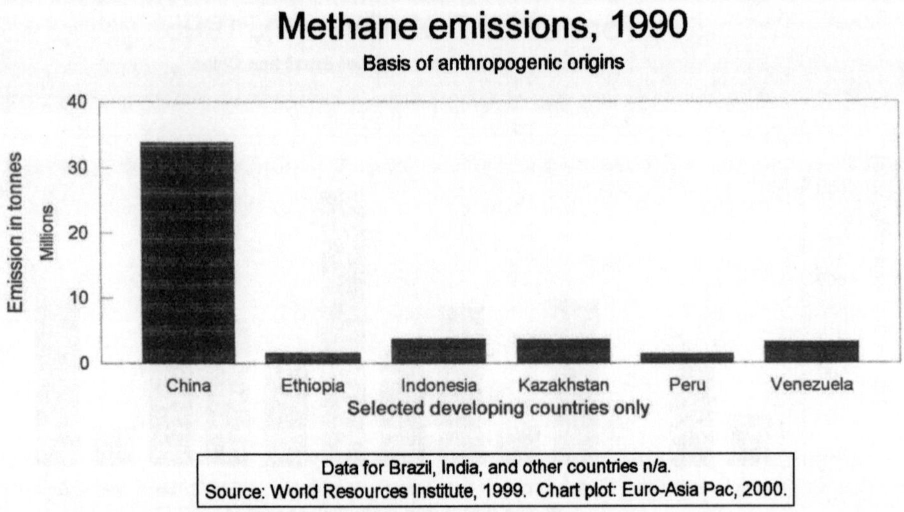

Fig. 5.3. Methane emissions, 1995. The countries with the highest potential for methane emissions reductions are compared. China is the largest market with emissions exceeding 30 million tonnes in 1995. The World Resources Institute – the data source – however, did not provide figures for India, which emit substantial quantities of coal-bed methane and consume high levels of fertilisers. Indonesia and Kazakhstan offer ample opportunities for methane emissions abatement

6
Clean Development Mechanism Project Prospects

This chapter, in the form of an easy-reference table, provides information on multi-sector industries which consume fossil-fuel energy and identifies key power plant sources burning coal, lignite and oil for electricity generation. The largest thermal power stations are listed under the name of the holding state or private power company. Smaller capacity plants and power facilities in the interior of the country are not included.

CDM project profitability check list for financiers and investors

1. Is domestic economy growing?
   Which industries? What is future market demand, local, regional and global?
2. Power projects: Power Purchase Agreements with state or private buyers. If PPA private buyers, what is the shape of these industries – are they going to expand in domestic and export markets?
3. Profitability of energy intensive industries (large consumers of energy and electricity) – How are these sectors likely to perform in the future? Will they affect investment decision of CDM projects?

Projects planned for 2001 and beyond

Global carbon funds may invest or co-finance – pending approval of their shareholders – new projects in developing countries for the purchase of equipment or technology which ultimately result in reduction of carbon dioxide emissions, such as:

- Coal gasification systems for power generation.
- Combined Cycle Gas Turbine systems – with increased thermal efficiency.
- IGCC, Integrated Gasification Combined Cycle (using coal gasification for electricity generation via gas and steam turbines).
- CHP, Combined Heat and Power or Co-generation.
- Landfill gas and Waste-to-Energy plants.

Alternatively, such funds may only approve renewable energy technologies to replace plants emitting carbon dioxide, and reject clean coal or thermal technologies in their mandate.

6.1 Basis of Country, Power Plant and Industry Sector Survey

Table 6.1. Country-by-country survey

Country	Industry sectors	Location/region/remarks
Afghanistan	Coal	Kabul. Mines derelict.
LDC	Oil/gas	Angot, Hadia-Gugerdag, Mazar-i-Sharif, Yetym-Tag (derelict fields).
	Textiles	Kabul, Herat.
Albania	Chem/petchem	Ballsh, Cerrik, Stalin, Tirana, Vlore.
	Food processing	Tirana.
	Smelting	Bajram Curri.
	Textiles	Tirana.
	Oil/gas	Devoli-Kucova, Deviaka, Dumre, Fier, Patos; offshore south-west of the Albania/Adriatic Sea-Ionian Sea.
Algeria	Chem/petchem	Algiers, Annaba, Arzew, Oran, Skikda.
	Heavy ind.	Algiers, Annaba. > Audit.
	Oil/gas	Ghadames-Illizi-In Sallah, Hassi Berkina, Hassi Messaoud, Hassi R'Mel Qoubba (excess flaring, fugitive gas losses).
	Iron & steel	Annaba.
	Vehicle manuf.	Oran (new plants).
Angola	Cement manuf.	Luanda.
LDC	Chem/petchem	Luanda, Cabinda.
	Iron & steel	Luanda.
	Oil/gas	Benfica, Bengo, Cabinda, Dalia, East Lombo, Essungo, Galinda, Garissol, Luanda, Mavango, Tobias.
Antigua and Barbuda	Chem/petchem	St John's.
Argentina	Coal	West of Rio Gallegos.
	Chem/petchem	Bahia Blanca, Buenos Aires, Comodoro Rivadavia, Mendoza, Neuquen, Resistencia, Rio Grande.
	Heavy eng.	Cordoba.
	Iron & steel	Buenos Aires.
	Oil/gas	Campo Duran, Cape Horn, Comodoro Rivadavia, Malvinas/N. Falkland, Mendoza, Neuquen, Rio Grande, San Jorge Basin, San Salvador de Jujuy.
	Tim/paper/pulp	Buenos Aires, Rosario.
	Vehicle manuf.	Buenos Aires, Cordoba, Rosario, San Miguel de Tucuman, Santa Fe.
	Carbonaceous or hydrocarbon-fired power stations	
		AGUA Y ENERGIA ELECTRICA
	Coal-oil-gas	San Nicholas I, at Buenos Aires, 350 MW.
	90% coal	AES Corporation.
		SEGBA S.A.
	Coal-oil-gas 90% coal	Puerto Nuevo – Buenos Aires, 340 MW.
Armenia	Chem/petchem	Yerevan.
	Smelting	Alaverdi (copper).
	Textiles	Yerevan.

Table 6.1. *Continued*

Country	Industry sectors	Location/region/remarks
Azerbaijan	Chem/petchem	Baku.
	Iron & steel	Baku.
	Oil/gas	Baku, SE Baku-offshore: Azeri Megastructure - Chirag-Guneshli-Azeri fields.
Bahrain	Oil/gas	Awali, Manama.
	Smelting	Bahrain city.
Bangladesh	Engineering	Chittagong, Dhaka.
	Cotton process.	Bogra, Chittagong, Dhaka, Nawabganj.
	Iron & steel	Chittagong.
	Chem/petchem	Chittagong.
	Oil/gas	Bay of Bengal.
	Jute processing	Chittagong, Dhaka, Khulna, Mymensingh.
	Textiles	Dhaka.
Barbados	Chem/petChem	Bridgetown.
Belize	Food processing	Belize City.
	Textiles	Belize City.
	Oil/gas	Corozal Basin/Chetumal Bay.
Benin	Chem/petchem.	Porto Novo.
LDC	Oil/gas	Seme gas field.
Bhutan LDC	Mineral processing.	Thimphu.
Bolivia	Chem/petchem	Cochabamba, Sucre, Santa Cruz.
	Oil/gas	Camiri, Sanandita, Santa Cruz, Subandino Sur, Tuichi-Pilcamayo, Tarija.
	Textiles	Santa Cruz, La Paz, Oruro.
	Tim/paper/pulp	Riberalta, Santa Cruz.
Bosnia &	Coal	Banovici, Konjic, Kladanj.
Herzegovina	Chem/petchem	Banja Luka, Lukavica, Maglaj, Tuzla.
	Iron & steel	Zenica.
	Textiles	Bihac, Bosanski Novi, Sarajevo, Travnik.
	Tim/paper/pulp	Banja Luka, Bugojno, Maglaj, Olovo, Sarajevo, Vitez, Zavidovici.
	Vehicle manuf.	Hrasnica, Maglaj, Sarajevo.
	Carbonaceous or hydrocarbon-fired power stations	
	Lignite	Gacko, at Gacko, 300 MW.
	Lignite	Kakanj, at Zenica, 568 MW.
	Lignite	Tuzla, at Tuzla, 779 MW.
	Lignite	Ugljevik II, at Ugljevik, 300 MW.
Botswana	Mineral proc.	Gaborone
Brazil	Chem/petchem	Belo Horizonte, Curitiba, Fortaleza, Maceio, Manaus, Porto Alegre, Recife, Rio de Janeiro, Rio Grande, Salvador, Sao Paulo.
	Iron & steel	Belo Horizonte, Curitiba, Maceio, Manaus, Maraba, Porto Allegre, Recife, Rio de Janeiro, Salvador, Sao Paulo.
	Oil/gas	Offshore Natal and Fortaleza - Agulha, Potiguar, Xareu; offshore Recife, Rio de Janeiro - offshore Campos - Enchova, MarlimSul, Pargo.
	Tim/paper/pulp	Belem, Manaus, Porto Alegre.

Table 6.1. *Continued*

Country	Industry sectors	Location/region/remarks
Brazil	Vehicle manuf.	Belo Horizonte, Curitiba, Rio de Janeiro, Porto Alegre, Recife, Salvador, Sao Paulo.
	Carbonaceous or hydrocarbon-fired power stations	
		COMPANHIA ESTUDUAL DE ENERGIA ELECTRICA
	Coal-oil	Candiota II, at Bage, 452 MW. 2 × 66 MW (1974). 2 × 160 MW (1987). Particulate collectors on all units.
		ELETROSUL
	Coal	Charqueadas, at Sao Jeronimo, 72 MW 3 × 18 MW, 1962. 1 × 18 MW, 1965. Particulate collectors, all units.
	Coal	Jorge Lacerda I, at Sao Catarina, 100 MW. 1 × 50 MW, 1965. 1 × 50 MW, 1967.
	Coal	Jorge Lacerda II, at Sao Catarina, 132 MW. 1 × 66 MW, 1974.
	Coal	Jorge Lacerda III, at Sao Catarina, 250 MW. 2 × 125 MW, 1980.
	Coal	Jorge Lacerda IV, at Sao Catarina, 350 MW. 1997. Particulate collector.
	Power stations in 1998, 1999, 2000	
	Coal	COMPANHIA ESTUDUAL DE ENERGIA ELETRICA Candiota III-2, at Bage, 350 MW. 1 × 350 MW. Planned for 2000.
	Coal	ELETROSUL Jacui I, at Sao Jeronimo, 350 MW. Planned for 1998. Particulate collector.
	Power stations in 2001 or after 2001	
		COMPANHIA ESTUDUAL DE ENERGIA ELETRICA
	Coal	Candiota III-1, at Bage, 350 MW. Planned for 2002.
	Coal	Candiota III-3, at Bage, 350 MW. Planned for 2002.
Brunei	Chem/petchem	Bandar Seri Begawan.
	Oil/gas	Baram, Fairley, Salbiah, Seria, Sri Ampa.
Burkina Faso LDC	Food processing	Bobo-Dioulasso, Ouagadougou.
	Textiles	Bobo-Dioulasso, Ouagadougou.
Burundi LDC	Mineral proc.	Bujumbura.
Cambodia LDC	Fertilisers	Batdambang, Kampot.
	Chem/petchem	Sihanoukville.
	Textiles	Phnom Penh.
	Tim/paper/pulp	Phnom Penh.

Table 6.1. *Continued*

Country	Industry sectors	Location/region/remarks
Cameroon	Chem/petchem	Pointe Limboh.
	Food processing	Douala.
	Oil/gas	Bavo, Betika, Douala, Kribi, Kole, Kombo Centre.
	Tim/paper/pulp	Douala.
Central African Rep LDC	Tim/paper/pulp	Bangui.
Chad LDC	Oil/gas	Doba, in the south (new fields).
Chile	Chem/petchem	Antofagasta, Concepcion, Concon/Valparaiso, Iquique, Punta Arenas.
	Coal	Concepcion.
	Heavy ind.	Santiago, Talcahuano, Vina del Mar.
	Iron & steel	Concepcion.
	Oil/gas	Arauco Basin, Straits of Magellan, Tamarugal Basin.
	Tim/paper/pulp	Concepcion, Punta Arenas, Valdivia.
	Vehicle manuf.	Arica.
	Carbonaceous or hydrocarbon-fired power stations	
		CHILGENER
	Coal	Guacolda, at Huasco, 300 MW. 2 × 150 MW, 1995, 1996. Particulate collector.
	Coal	Renca, at Santiago, 100 MW. 2 × 50 MW, 1962. Particulate collector.
	Coal	New Tocopilla, at Tocopilla, 125 MW. 1995. Particulate collector.
	Coal	New Tocopilla, at Tocopilla, 125 MW. 1997. Particulate collector.
	Coal	Laguna Verde, at Valparaiso, 55 MW. 1 × 32 MW, 1 × 22 MW. 1939.
	Coal	Las Ventanas, at Valparaiso, 338 MW. 1 × 120 MW, 1964. 1 × 218 MW, 1977. Particulate collector.
		CODELCO.
	Bit. and sub-bit. coal	Tocopilla, at Tocopilla, 565 MW.
		EDELNOR
	Coal	Mejillones, at Mejillones, 150 MW. 1996.
		ENDESA
	Coal	Bocamina, at Concepcion, 125 MW. 1970. No particulate, SOx or NOx controls.

Table 6.1. *Continued*

Country	Industry sectors	Location/region/remarks
Chile	*Power stations in 1998, 1999, 2000*	
		EDELNOR
	Coal	Mejillones (2), at Mejillones, 150 MW. Planned for 1998.
		ENDESA
	Coal	Punta Patache, at Iquique, 150 MW. Planned for 1998.
China	Cement	Guangzhou, Urumqi.
	Coal	Heilongjiang Province – Dalian, Changchun, Harbin; Provinces of Anhui, Henan, Shanxi, *Shaanxi, Sichuan.*
	Chem/petchem	Anda, Anshan, Beijing, Chengdu, Chongqing, Fushun, Guangzhou, Harbin, Karamay, Kunmimg, Lanzhou, Luda, Maoming, Nanjing, Shanghai, Shengli, Shenyang, Taiyuan, Tianjin, Urumqi, Wuhan, Xian.
	Iron & steel	Anshan, Beijing, Changchun, Chongqing, Guangzhou, Hong Kong, Kunming, Nanjing, Shanghai, Shenyang, Taiyuan, Tianjin, Urumqi, Wuhan, Xian.
	Oil/gas	Daqing, Lanzhou, Urumqi; Provinces of Anhui, Shanxi and Sichuan; Da Hinggan Ling Province – north-east China; Tarim Basin – north-west China; Bohai Gulf, East China Sea, Pearl River Mouth, South China Sea, South Yellow Sea.
	Tim/paper/pulp	Chongqing, Guangzhou.
	Vehicle manuf.	Beijing, Chongqing, Kunming, Nanjing, Wuhan.
	Carbonaceous or hydrocarbon-fired power stations	
		CHINA LIGHT & POWER
	Coal-oil	Castle Peak A, 1–4, at New Territories, 1400 MW. 4 × 350 MW. 1982. Low NOx burners. Particulate collector.
	Coal-oil	Castle Peak B, 1–4, at New Territories, 2708 MW. 1986. 4 × 677 MW. Low NOx burners. Particulate collector.
		FUJIAN, MINISTRY OF ELECTRIC POWER
	Coal	Huaneng Fuzhou I, at Fuzhou, 700 MW. 2 × 350 MW.
	Coal	Xiamen, at Xiamen, 50 MW.
	Coal	Yongan, at Yongan, 350 MW.
	Coal	Zhangping, at Zhangping, 200 MW.

Table 6.1. *Continued*

Country	Industry sectors	Location/region/remarks
China		FUJIAN TRANSATLANTIC RESOURCES CORPORATION
	Coal	Shoawn, 250 MW. 2 × 125 MW. 1977. IPP.
		FUJIAN, XIAMEN MUNICIPAL PEOPLE'S GOVERNMENT
	Coal	Songyu, at Xiamen, 600 MW. 2 × 300 MW. 1997.
		GUANGDONG, CHEUNG KONG HOLDINGS LTD
	Coal	Gaolan Island, at Gaolan Island, 1320 MW. 2 × 660 MW. 1996.
		GUANGDONG, GUANGDONG ELECTRIC CORPORATION
	Coal	Zuhai, at Zuhai, 660 MW. 1 × 660 MW. 1995. Bright World Enterprises/Zuhai Special Economic Zone.
		GUANGDONG, HUANENG INTERNATIONAL POWER DEVELOPMENT CORPORATION
	Coal	Huaneng Shantou, 300 MW. 2 × 150 MW. Unit 1, 1995. Unit 2, under construction.
		GUANGDONG, MINISTRY OF ELECTRIC POWER
	Coal	Nanhai, 400 MW. 2 × 200 MW. 1995, 1996.
	Coal	Nantang, 171 MW. 1 × 121 MW. 1 × 50 MW. 1995.
	Coal	Guangzhou, at Guangzhou, 150 MW.
	Coal	Huangpu, at Guangzhou, 1100 MW.
	Coal	Shajiao A, at Guangzhou, 900 MW.
	Coal	Shajiao B, at Guangzhou, 700 MW. 2 × 350 MW. 1987. Joint-venture with Hopewell Holdings.
	Coal	Shajiao C, at Guangzhou, 1980 MW. 3 × 660 MW. Joint-venture with Hopewell Holdings. 1996. BOT transfer in 20 years.
	Coal	Yunfu, at Hekou, 250 MW.

Table 6.1. *Continued*

Country	Industry sectors	Location/region/remarks
China	Coal	Meixian, at Meizhou, 100 MW.
	Coal	Nansha, at Nanshan, 660 MW.
		Joint-venture with New World Development, Guangzhou Economic Construction
		Corporation, Pearl River Power Corporation.
		1995.
	Coal	Shaoguan, at Qujiang, 224 MW.
		4 × 60 MW (gross).
	Coal	Shaoguan B, at Shaguan, 240 MW.
		4 × 60 MW.
		1995.
	Coal	Zhongshan, at Zhongshan, 50 MW.
		GUANGDONG, SHENZHEN CITY ELECTRICITY AUTHORITY
	Coal	Mawan (1), at Shenzhen, 300 MW.
		Joint-venture with Hong Kong companies.
		GUANGDONG, SOUTH CHINA UNITED POWER DEVELOPMENT CORPORATION
	Coal	Shenzhen, at Shenzhen, 300 MW.
		1994.
		GUANGXI, MINISTRY OF ELECTRIC POWER
	Coal	Laibin, at Laibin, 250 MW.
	Coal	Tiangdong, at Tiangdong, 74 MW.
		Heshan, at Yishan, 505 MW.
		HAINAN, MINISTRY OF ELECTRIC POWER
	Coal	Haikou City, at Haikou, 100 MW.
	Coal	Machun, at Haikou, 250 MW.
		HONG KONG ELECTRIC COMPANY LTD
	Coal-oil	Lamma Island (1–7), at Lamma Island, 2150 MW.
	99% coal	3 × 250 MW.
		4 × 350 MW.
		1982.
		Sulphur dioxide control on units 6, 7.
		Wet scrubber.
		Low NOx burners, overfire air.
		Particulate collector.
		Unit 7, 1995.
	Coal-oil	Lamma Island (8), on Lamma Island, 350 MW.
		1997.
	Coal	MINISTRY OF POWER
		150 MW, PFBC.
		Planned.
		SHANGHAI, HUANENG INTERNATIONAL POWER DEVELOPMENT CORPORATION
	Coal	Huaneng Shidongku 3 & 4, at Shanghai, 1200 MW.
		2 × 600 MW.
		Unit 3 operating since 1994.
		Unit 4 planned.

Table 6.1. *Continued*

Country	Industry sectors	Location/region/remarks
China		SHANGHAI, MINISTRY OF ELECTRIC POWER
	Coal	Baoshan Iron Corporation, at Shanghai.
		Unit 1, 700 MW.
		Unit 2, 83 MW.
	Coal	Congming, at Shanghai, 140 MW.
	Coal	Gaoqiao, at Shanghai, 228 MW.
		Co-generation plant.
	Coal	Jinshan, at Shanghai, 277 MW.
	Coal	Minhang, at Shanghai, 818 MW.
	Coal	Shidongkou, at Shanghai, 2400 MW.
		2 × 1200 MW.
		Unit 2 operating since 1992.
		First super-critical plant.
		Best heat rate of any coal-fired station in China.
	Coal	Shinan, at Shanghai, 139 MW.
	Coal	Wujing, at Shanghai, 950 MW.
		Co-generation plant.
	Coal	Yangshupu, at Shanghai, 216 MW.
	Coal	Zgabei, at Shanghai, 455 MW.
		ZHEJIANG, FORMOSA CHEMICALS & FIBRES
	Coal	Changhua, at Changhua, 70 MW.
		Low NOx burners, overfire air.
		ZHEJIANG, MINISTRY OF ELECTRIC POWER
	Coal	Banshan, at Hangzhou, 237 MW.
	Coal	Taizhou, at Linhai, 750 MW.
	Coal	Meixi, at Meixi, 75 MW.
	Coal	Bailun, at Ningbo, 600 MW.
	Coal	Dinghai, at Ningbo, 66 MW.
	Coal	Juhua, at Quzhou, 124 MW.
		Co-generation plant.
	Coal	Hunzhou, at Wenzhou, 250 MW.
	Coal	Zhenhai, at Zhenhai, 1050 MW.
	Power stations in	*1998, 1999, 2000*
		FUJIAN, AES CHIGEN
	Coal	Quan Zhou, at Quan Zhou, 600 MW.
		2 × 300 MW.
		Planned for 1998.
		IPP.
		FUJIAN, HUANENG INTERNATIONAL POWER DEVELOPMENT CORPORATION
	Coal	Huaneng Fuzhou (3–4), at Fuzhou, 1320 MW.
		2 × 660 MW.
		Planned for 1998.
		FUJIAN, LIPPO GROUP
	Coal	Meizhouwan, 700 MW.
		2 × 350 MW.
		Planned for 1999.
		IPP.

Table 6.1. *Continued*

Country	Industry sectors	Location/region/remarks
China		FUJIAN, MINISTRY OF ELECTRIC POWER
	Coal	Huaneng Fuzhou, at Fuzhou, 700 MW. 2 × 350 MW. Planned for 1998 and 1999.
		GUANGDONG, GUANGDONG ELECTRIC POWER CORPORATION
	Coal	Zhuhai (2), at Zhuhai, 660 MW. Joint-venture with Zhuhai Special Economic Zone/Bright World Enterprises.
		GUANGDONG, GUANGDONG POWER BUREAU
	Coal	Taishan, at Taishan, 1320 MW. 2 × 660 MW. Planned for 1998.
	Coal	Taishan City, at Tongyu Bay, 1320 MW. Planned for 2000. Joint-venture with Marubeni Energy Power Group.
		GUANGDONG, SHENZHEN CITY ELECTRICITY AUTHORITY
	Coal	Mawan (2–3), at Shenzhen, 600 MW. Joint-venture with Hong Kong companies.
		GUANGXI, NEW WORLD DEVELOPMENT COMPANY LTD
	Coal	Beihai Nansha, at Beihai, 700 MW. 2 × 350 MW. Planned for 2000.
		SHANGHAI, MINISTRY OF ELECTRIC POWER
	Coal	Jinshan, at Shanghai, 125 MW.
		SHANGHAI, SHANGHAI ELECTRIC POWER BUREAU
	Coal	Yangshupu, at Shanghai, 250 MW. 2 × 125 MW. East China Electric Power Corporation.
		ZHEJIANG, ZHEJIANG POWER BUREAU
	Coal	Beilungang, at Beilungang, 1800 MW. 3 × 600 MW. Planned for 1999/2000.
	Coal	Ningbo (3–4), at Ningbo, 1200 MW. 2 × 600 MW. Planned for 1998–2000.
	Power stations in 2001 or after 2001	
		GUANGDONG, CONSOLIDATED ELECTRIC POWER ASIA
	Coal	Shenzhen Eastern, 1320 MW. 2 × 660 MW. Planned for 2003. IPP.

Table 6.1. *Continued*

Country	Industry sectors	Location/region/remarks
China		GUANGDONG, GUANGDONG GENERAL POWER COMPANY
	Coal	Swatow, at Swatow, 1200 MW.
		2 × 600 MW.
		Planned for 2003.
		IPP.
		GUANGDONG, HUANENG INTERNATIONAL POWER DEVELOPMENT CORPORATION
	Coal	Huaneng Shantou (3–4), 700 MW.
		2 × 350 MW.
		Planned for 2001.
		GUANGDONG, SHENZHEN CITY ELECTRICITY AUTHORITY
	Coal	Mawan (4–6), at Shenzhen, 900 MW.
		Joint-venture with Hong Kong companies.
		GUANGXI, MINISTRY OF ELECTRIC POWER
	Coal	Laibin B (1–2), 720 MW.
		2 × 360 MW.
		BOT, joint-venture with Alstom and EdF.
		SHANGHAI, MINISTRY OF ELECTRIC POWER
	Coal	Waigaoqiao, at Shanghai, 1200 MW.
		4 × 300 MW.
		Proposed super-critical plant.
		Low NOx burners, ESP.
		FGD to be included.
		ZHEJIANG, ZHEJIANG POWER BUREAU
	Coal	Ningpo (5), at Ningpo, 600 MW.
		Planned for 2001.
		ZHEJIANG, ZHEJIANG PROVINCE POWER DEVELOPMENT CORPORATION
	Coal	Jiaxing II, at Jiaxing, 2400 MW.
		4 × 600 MW.
		Joint-venture with National Power, GEC, East China Power Corporation.
		Planned for 2002.
	Lignite-fired power stations	
		GUANGDONG, ZHUHAI ELECTRICITY BOARD
	Lignite	Zhuhai Steam, at Zhuhai, 1320 MW.
		2 × 660 MW. Planned for 2000.
Colombia	Coal	Cali
	Chem/petchem	Barranquilla, Barrancabermeja, Bogota, Cali, Cartagena, El Guamo, Medellin, Tibu.
	Iron & steel	Barranquilla, Medellin.
	Oil/gas	Arauca, Cano Limon, La Cira, Orito.
	Tim/paper/pulp	Cali, Ibague, Medillin.
	Vehicle manuf.	Bogota, Ibague, Medellin
Congo Dem.	Chem/petchem	Kinshasa.
Rep. (Zaire)	Oil/gas	Boma, Yanga-Sendji, Takula.
LDC	Iron & steel	Kinshasa.
	Tim/paper/pulp	Kinshasa, Kisangani, Lubumbashi.

Table 6.1. *Continued*

Country	Industry sectors	Location/region/remarks
Congo, Rep.	Food processing	Brazzaville, Pointe-Noire.
	Chem/petchem	Pointe-Noire.
	Oil/gas	Djambala, Hautemer, Kitina, Moho, Yambo.
Costa Rica	Chem/petchem	Limon.
	Smelting	Boruca (bauxite).
Cote d'Ivoire (Ivory Coast)	Chem/petchem	Abidjan.
	Iron & steel	Abidjan.
	Oil/gas	Belia, Belier, Espoir, Foxtrot, Lion, Panthere.
	Tim/paper/pulp	Abidjan.
Croatia	Chem/petchem	Lendava, Rijeka, Sisak, Slavonski Brod, Zagreb, Zapresic.
	Oil/gas	Dugo Selo, Maribor, Molve, Obod, Petisovci.
	Textiles	Zagreb.
Cuba	Chem/petchem	Cabaiguan, Cienfuegos, Santiago de Cuba.
	Manufacturing	Havana.
	Textiles	Havana.
	Oil/gas	Archipeilago de Camaguey, Cardenas Bay.
Djibouti LDC	Chem/petchem	Djibouti.
Dominican Rep	Chem/petchem	Bonao, Nagua.
		DOMINICAN ELECTRIC CORPORATION
	Coal	Barahona, at Barahona, 42 MW. 1994. Particulate collector.
	Coal	Itabo, at Santo Domingo, 250 MW. Unit 1, 1984. Unit 2, 1988. Particulate collector.
	Power stations in	*2001 or after 2001*
		DOMINICAN ELECTRIC CORPORATION
	Coal	Itapo (3–4), at Santo Domingo, 250 MW. 2 × 125 MW. Planned.
Ecuador	Chem/petchem	Esmeraldas, La Libertad.
	Food processing	Guayaquil, Quito.
	Iron & steel	Guayaquil.
	Oil/gas	Caribe, Lago Agrio, Santa Elena fields, Shushufindi.
	Textiles	Quito.
	Tim/paper/pulp	Guayaquil.
Egypt	Cement	Alexandria, Helwan.
	Chem/petchem	Alexandria, Aswan, Asyut, Cairo, Helwan, Musturud, Suez, Tanta.
	Heavy ind.	Alexandria, Asyut, Cairo, Suez.
	Oil/gas	North Egypt – Abu Madi, Abu Qir, Baltim, Wakar; Red Sea – Geisum, Morgan; Sinai; West Egypt – A Gharadig, Alamein, Um, Barka.
	Iron & steel	Alexandria, Cairo, Helwan.
	Textiles	Alexandria, Aswan, Asyut, Cairo, El Minya, Port Said.
	Vehicle manuf.	Cairo, Helwan.

Table 6.1. *Continued*

Country	Industry sectors	Location/region/remarks
Egypt	*Power stations in 2001 and after 2001*	
	Coal	Zafarana, in Suez Canal Region, 1200 MW. 4 × 300 MW. Planned.
El Salvador	Textiles	San Miguel, San Salvador.
Equatorial Guinea LDC	Oil/gas	New fields discovered, onshore/offshore: Alba and Zafiro.
Eritrea LDC	Cement	Asmara.
	Chem/petchem	Assab.
	Textiles	Asmara.
Ethiopia LDC	Cement	Dire Dawa.
	Chem/petchem	Addis Ababa, Assab.
	Food processing	Addis Ababa, Dire Dawa, Gonder, Nazret.
	Oil/gas	Ogaden region.
	Textiles	Addis Ababa, Dire Dawa, Nazret.
	Tim/paper/pulp	Addis Ababa.
Fiji	Brewing	Suva.
	Food processing	Suva.
Gabon	Chem/petchem	Port Gentil.
	Food processing	Libreville, Port Gentil.
	Iron & steel	Belinga.
	Oil/gas	Anguille, Astrid Marin, Grondin, Port Gentil, Torpille.
	Textiles	Libreville.
	Tim/paper/pulp	Libreville, Port Gentil.
Gambia LDC	Mineral proc.	Banjul.
Georgia	Chem/petchem	Batumi, Tbilisi.
	Food processing	Tbilisi.
	Oil/gas	Offshore Black Sea, Block 12/east Georgia, Block 10/northeast Georgia.
	Textiles	Tbilisi.
Ghana	Chem/petchem	Accra, Tema.
	Food manuf.	Accra.
	Oil/gas	Cape Three Block, Salt Pond, Tano.
	Textiles	Accra.
	Tim/paper/pulp	Accra.
	Vehicle manuf.	Accra.
Guatemala	Chem/petchem	Escuintla.
	Food processing	Guatemala city.
	Oil/gas	Rubelsanto, West Chinaja, Xan.
	Textiles	Guatemala city.
	Tim/paper/pulp	Guatemala city.

Table 6.1. *Continued*

Country	Industry sectors	Location/region/remarks
Guatemala	*Carbonaceous or hydrocarbon-fired power stations*	
	Power stations in 1998, 1999, 2000	
		EMPRESA ELECTRICA DE GUATEMALA
	Coal	San Jose, 120 MW.
		Planned for 2000.
		BOO, Coastal Power, Teco Power Systems.
Guinea LDC	Food processing	Conakry.
Guyana	Food processing	Georgetown.
	Oil/gas	Karanambo.
	Tim/paper/pulp	Georgetown.
Haiti LDC	Sugar proc.	Port-Au-Prince.
Honduras	Chem/petchem	Puerto Cortes, San Pedro Sula.
	Textiles	Tegucigalpa, San Pedro Sula.
	Tim/paper/pulp	San Pedro Sula.
India	Coal	Jamshedpur, West Bengal - Dhanbad, Bhagalpur.
	Chem/petchem	Ahmadabad, Anklesvar, Bangalore, Barauni, Baroda, Bombay, Calcutta, Delhi, Digboi, Duar Gauhati, Madras, Nowgong.
	Heavy ind.	Bangalore, Bombay, Calcutta/Haldia, Delhi, Hyderabad, Indore, Kanpur, Madras, Nagpur.
	Iron & steel	Ahmadabad, Calcutta, Jamshedpur, South-east India.
	Oil/gas	Bombay High/Bassein.
	Textiles	Ahmadabad, Bangalore, Bombay, Calcutta, Delhi, Hyderabad, Madras, Madurai, Nagpur, Kanpur.
	Tim/paper/pulp	Bombay.
	Vehicle manuf.	Bombay, Jamshedpur, Madras.
	Carbonaceous or hydrocarbon-fired power stations	
		ANDHRA PRADESH, ANDHRA PRADESH STATE ELECTRICITY BOARD
	Coal	Ramagundam, at Karimnagar, 63 MW.
	Coal-lignite	Vijayawada 1–6, 1260 MW. 6 × 210 MW.
	Coal	Rayalaseema, at Muddanur, 420 MW.
		2 × 210 MW. 1994.
	Coal-lignite	Kothagudem, at Paloncha, 670 MW.
		3 × 60 MW.
		2 × 105 MW.
		2 × 110 MW.
		ANDHRA PRADESH, NATIONAL THERMAL POWER CORPORATION LTD
	Coal-lignite	Ramagundam STPS, at Karimnagar, 2100 MW.
		3 × 200 MW.
		3 × 500 MW.
		GUJARAT, AHMADABAD ELECTRIC COMPANY

Table 6.1. *Continued*

Country	Industry sectors	Location/region/remarks
India	Coal	Ahmadabad, at Ahmadabad, 509 MW. 3 × 110 MW. 1 × 36 MW. 1 × 44 MW. 1 × 99 MW.
		GUJARAT, GUJARAT ELECTRICITY BOARD
	Coal	Dhuvaran, near coast, 534 MW.
	Coal	Gandhinagar, at Ghandinagar, near coast, 660 MW. 2 × 120 MW. 2 × 210 MW.
	Coal	Sikka, at Jamnagar, near coast, 440 MW.
	Coal	Wanakbori, at Sevalia, 1260 MW. 6 × 210 MW.
	Coal	Ukai, at Ukai Dam, near coast, 850 MW. 2 × 120 MW. 2 × 200 MW. 1 × 210 MW.
		KARNATAKA, KARNATAKA ELECTRICITY BOARD
	Coal	Raichur, at Raichur, 630 MW. 3 × 210 MW.
		MAHARASHTRA, BOMBAY SUBURBAN ELECTRICITY SUPPLY
	Coal	Dahanu 1-2, at North Bombay, 500 MW. 2 × 250 MW. Unit 1, 1995.
		MAHARASHTRA, MAHARASHTRA STATE ELECTRICITY BOARD
	Coal	Bhadrawati, 1082 MW. 1997. Joint-venture: MetChem, Mitsui, Ispat.
	Coal	Bhusawal, at Bhusawal, 478 MW. 1 × 58 MW. 2 × 210 MW.
	Coal	Chandrapur, at Durgapur, 1840 MW. 4 × 210 MW. 2 × 500 MW.
	Coal	Nasik, at Eklahore, 910 MW. 2 × 140 MW. 3 × 210 MW.
	Coal	Khaperkheda, at Khaperkheda, 420 MW. 2 × 210 MW.
	Coal	Koradi, at Koradi, 1080 MW. 4 × 115 MW. 2 × 210 MW. 1 × 210 MW.
	Coal	Parli, at Parli Vaijnath, 690 MW. 2 × 30 MW. 3 × 210 MW.

Table 6.1. *Continued*

Country	Industry sectors	Location/region/remarks
India	Coal	Paras, at Vydyut Nagar, 58 MW. MAHARASHTRA, TATA ELECTRIC COMPANY
	Coal	Trombay, at Bombay, 1337 MW. 1992. Multi-fuel capacity. Wet scrubber. TAMIL NADU, TAMIL NADU ELECTRICITY BOARD
	Coal-oil	North Madras (2–3), at Cuddalore, 420 MW. 2 × 210 MW. 1995.
	Coal	Ennore, at Madras, 450 MW. 2 × 60 MW. 3 × 110 MW.
	Coal	Mettur, at Mettur, 840 MW. 4 × 210 MW.
	Coal	Tuticorin, at Tuticorin, 1050 MW. 5 × 210 MW. WEST BENGAL, CALCUTTA ELECTRIC SUPPLY CORPORATION
	Coal	Titagarth, at Titagarth, 240 MW. 2 × 60 MW (only?) WEST BENGAL, DAMODAR VALLEY CORPORATION
	Coal	Durgapur, at Durgapur, 350 MW. 1 × 140 MW. 1 × 210 MW. WEST BENGAL, WEST BENGAL STATE ELECTRICITY BOARD
	Coal	Bandel, at Calcutta, 530 MW. 4 × 80 MW.
	Coal	Budge-Budge, at Calcutta, 250 MW. 1 × 250 MW. 1996.
	Coal	Calcutta & South Generating, at Calcutta, 341 MW. 2 × 68 MW. 1 × 75 MW. 1 × 130 MW.
	Coal	DPL, at Durgapur, 390 MW. 2 × 30 MW. 2 × 75 MW. 1 × 70 MW. 1 × 110 MW.
	Coal	Farakka 1–4, at Farakka, 1130 MW. 3 × 210 MW. 1 × 500 MW.
	Coal	Farakka II, at Farakka, 1130 MW. Proposed in 8th Five Year Plan.
	Coal	Kolaghat, at Medinipurt, 1050 MW. 5 × 210 MW.

Table 6.1. *Continued*

Country	Industry sectors	Location/region/remarks
India	Coal	Santaldih, at Purilya, 480 MW. 4 × 120 MW.
	Power stations in 1998, 1999, 2000	
		WEST BENGAL, WEST BENGAL STATE ELECTRICITY BOARD
	Coal	Budge-Budge (2–3), at Calcutta, 500 MW. 2 × 250 MW. Planned for 1998.
	Power stations in 2001 or after 2001	
	Coal	ANDHRA PRADESH, NATIONAL THERMAL POWER CORPORATION LTD Simhadri, at Vishakahpatnam, 1000 MW. 2 × 500 MW.
		GUJARAT, GUJARAT ELECTRICITY BOARD
	Coal	Ghandi Nagar, at Ghandi Nagar, 550 MW. Under construction.
		KARNATAKA, KARNATAKA ELECTRICITY BOARD
	Coal	Chamalapura, at Mysore, 1000 MW. 4 × 250 MW. Mysore Power Generation Pte Ltd.
	Coal	Mangalore, at Nandikur, 1000 MW. 2 × 500 MW. Planned for 2001. Dakhina Kannada District. Cogentrix Energy (USA), China Light & Power. Developer: Nagarjuna Power Corporation Ltd.
	Coal	Raichur, at Raichur, 210 MW. Under construction.
		TAMIL NADU, TAMIL NADU ELECTRICITY BOARD
	Coal	Cuddalore, at Cuddalore, 1320 MW. 2 × 660 MW. Coastal location. Developer: Cuddalore Power Company.
	Coal	North (Chennai) - Madras Stage II, at Cuddalore, 1050 MW. 2 × 525 MW. IPP. Developer: Videocon Power Ltd.
	Coal	Tuticorin IV, at Tuticorin, 500 MW. Under construction.
	Lignite-fired power stations	
		TAMIL NADU, JAYAMKONDAM LIGNITE POWER CORPORATION
	Lignite	Jayamkondam, at Jayamkondam, 500 MW. Under negotiation.

Table 6.1. *Continued*

Country	Industry sectors	Location/region/remarks
India		TAMIL NADU, NEYVELI LIGNITE CORPORATION
	Lignite	Neyveli, at Neyveli, 1050 MW. 5 × 210 MW. Units 1–3 in operation. Units 4, 5 under negotiation.
		TAMIL NADU, TAMIL NADU ELECTRICITY BOARD
	Lignite	Zero Unit, at Neyveli, 210 MW. Proposed in 8th 5 Year Plan.
Indonesia	Coal	Adaro.
	Chem/petchem	Balikpapan, Cilacap, Dumai, Medan, Palembang/Sungei Gurong, Pangkalan Berandan, Sungei Pakning, Surabaya/Wonokromo.
	Heavy ind.	Jakarta, Surabaya.
	Iron & steel	Jakarta.
	Oil/gas	Balikpapan, Banjarmasin, Jakarta, Medan, Natuna field, Palembang, Sorong; North Sumatra – Arun, Rantau; Central Sumarta – Duri, Minas, Puncak, Beruk; South Sumatra – Benakat, Pendopo, Limau; North-Java – Ardjuna, Jatibarang, Sindang; South and south-east Kalimantan – Badak, Bungkap, Handil, Tandjung. New gas fields: Timor Gap Zone/Timor Sea, Natuna field.
	Textiles	Bandung, Jakarta, Medan, Surabaya.
	Vehicle manuf.	Jakarta.
	Carbonaceous or hydrocarbon-fired power stations	
		JAVA (EAST), PERUSAHAAN UMUM LISTRIK NEGARA
	Coal	Paiton 1–2, at Surabaya, 800 MW. 2 × 400 MW. 1994. Low NOx burners. Particulate collector.
		JAVA (WEST), PERUSAHAAN UMUM LISTRIK NEGARA
	Coal	Suralaya 1–4, at Banten, 1600 MW. 4 × 440 MW. 1984, 1989.
	Coal	Suralaya 5–6, at Banten, 1200 MW. 2 × 600 MW. 1997.
	Coal	Suralaya 7, at Banten, 600 MW. 1997.
		KALIMANTAN (WEST), PRIVATE SECTOR
	Coal	Pontianak, at Pontianak, 75 MW. 3 × 25 MW. 1997. PPA in process with government.

Table 6.1. *Continued*

Country	Industry sectors	Location/region/remarks
Indonesia		SUMATRA (SOUTH), PERUSAHAAN UMUM LISTRIK NEGARA
	Coal	Bukit Assam 1–2, at Tanjong Enim, 130 MW. 2 × 65 MW. 1986, 1988. Particulate collector.
	Coal	Bukit Assam 3–4, at Tanjong Enim, 130 MW. 2 × 65 MW. 1994. Particulate collector.
		SUMATRA (WEST), PERUSAHAAN UMUM LISTRIK NEGARA
	Coal	Ombilin, at Sawah Lunto, 200 MW. 2 × 100 MW. 1996.
	Power stations in	*1998, 1999, 2000*
		JAVA (CENTRAL), PRIVATE SECTOR
	Coal	Tanjung Jati B, at Jepara, 1320 MW. 2 × 660 MW. Planned for 2000. BOO, Hopewell Holdings/ PT CEPA. National Power, Tomen, PT Bakrie Power, PT Maharani Paramitra.
		JAVA (EAST), PRIVATE SECTOR
	Coal	Paiton I (1–2), at Surabaya, 1230 MW. 2 × 615 MW. Planned for 1998, 1999. Mission Energy, Mitsui, PT Batu Hitam, General Electric.
	Coal	Paiton II (1–2), at Surabaya, 1220 MW. 2 × 610 MW. Planned for 1999, 2000. BOO, Siemens, Powergen, Bimantara, PT Java Power.
		JAVA (WEST), PRIVATE SECTOR
	Coal	Cilegon I, at Banten, 400 MW. 40% Powergen. Planned for 1999. PT Daya Listrik Pratuma. BOO. Project delayed in September 1997.
	Coal	Cilegon II, at Banten, 450 MW. Planned for 2000. PT Bukit Sunur Wijawa, Morrison Knudsen, Powergen.
		SULAWESI (NORTH), PRIVATE SECTOR
	Coal	Amurang, 55 MW. Planned for 2000. Transmega Ekacipta Primarindo.

Table 6.1. *Continued*

Country	Industry sectors	Location/region/remarks
Indonesia		SUMATRA (NORTH), PRIVATE SECTOR
	Coal	Sibolga, at Sibolga, 200 MW. Planned for 1999. Transmega Ekacipta Primarindo.
		SUMATRA (SOUTH), PRIVATE SECTOR
	Coal	Tarahan, at Lampung, 200 MW. 2 × 200 MW. Planned for 1999.
	Power stations in 2001 or after 2001	
		JAVA (CENTRAL), PRIVATE SECTOR.
	Coal	Cilacap, at Cilacap, 400 MW. PT Citra Kartika Daya. Planned for 2001. Postponed.
	Coal	Tanjung Jati A, at Jepara, 1200 MW. 2 × 600 MW. Planned for 2002. Bakrie Power Corporation, Maharani Paramitra, Tomen Power Corporation. PPA signed in 1997.
	Coal	Tanjong Jati C, at Jepara, 1320 MW. 2 × 660 MW. Under construction. Planned for 2003. PPA signed in 1997.
		JAVA (EAST), PERUSAHAAN UMUM LISTRIK NEGARA
	Coal	Paiton (3–4), at Surabaya, 800 MW. 2 × 400 MW. Unit 3, 2003. Unit 4, 2004.
		JAVA (EAST), PRIVATE SECTOR
	Coal	T. Awar-Awar, at Surabaya, 1320 MW. 2 × 660 MW. Planned for 2003. Indopasifik Electomatra GHD, Black & Veatch.
		JAVA (WEST), PERUSAHAAN UMUM LISTRIK NEGARA
	Coal	Serang, at Serang, 400 MW. Joint-venture with Ansaldo Energia, Sachen Holding, Fenergia Cipta. Planned for 2001. PPA. Postponed.
		KALIMANTAN (SOUTH), PRIVATE SECTOR
	Coal	Tanjung, 200 MW. Envirodrindo Energitama. Planned for 2001.

Table 6.1. *Continued*

Country	Industry sectors	Location/region/remarks
Indonesia		KALIMANTAN (WEST), PRIVATE SECTOR
	Coal	Pontianak (4), at Pontianak, 25 MW. PPA under negotiation.
		SULAWESI (NORTH), PRIVATE SECTOR
	Coal	Amurang (2), 55 MW. Unit 2, 2002. Transmega Ekacipta Corporation, Primarindo Finance Corporation. PPA signed.
		SULAWESI (SOUTH), PRIVATE SECTOR
	Coal	Takalar, 130 MW. 2 × 65 MW. Planned for 2003. Okha Hutama Satrya.
	Coal	Ujung Pandang, 130 MW. 2 × 65 MW. Mutiara Situju Corporation. Planned for 2001.
		SUMATRA (NORTH), PERUSAHAAN UMUM LISTRIK NEGARA
	Coal	Padang Sidempuan, at Padang Sidempuan, 400 MW. Design Power (NZ), PT Daya Mandri Eres Lamas.
		SUMATRA (NORTH), PRIVATE SECTOR
	Coal	Sibolga B, at Sibolga, 200 MW. 2 × 200 MW. Unit 1, 2001. Unit 2, 2002. BOO. Sumatra Energy Company.
Iran	Cement	Esfahan, Mashad, Shiraz, Tehran.
	Chem/petchem	Ahvaz, Bakhtaran, Bandar-e-Abbas, Esfahan, Kharg Island, Shiraz, Tabriz, Tehran.
	Iron & steel	Esfahan.
	Oil/gas	S.W. Iran – fields aorund Dezfal, Abadan, Ganaweh, Kangan, South Parsfield.
	Textiles	Ahvaz, Bakhtaran, Bandar-e-Abbas, Esfahan, Kerman, Mashad, Qom, Rasht, Shiraz, Tabriz, Tehran.
	Vehicle manuf.	Tehran.
Israel	Chem/petchem	Haifa, Tel Aviv-Yafo.
	Oil/gas	Kochav-Heletz-Bror fields.
	Textiles	Jerusalem, Tel Aviv-Yafo.
	Vehicle manuf.	Tel Aviv-Yafo.
	Carbonaceous or hydrocarbon-fired power stations	
		ISRAEL ELECTRIC CORPORATION
	Coal-oil	Rutenberg, at Ashkalon, 1100 MW. 2 × 575 MW 1990. Particulate collector.

Table 6.1. *Continued*

Country	Industry sectors	Location/region/remarks
Israel	Coal-oil	Maor David 1–4, at Hadera, 1400 MW. 4 × 360 MW. 1981. Particulate collector.
	Coal	Maor David, at Hadera, 1100 MW. 2 × 550 MW. 1996. Particulate collector.
	Power stations in 1998, 1999, 2000	
		ISRAEL ELECTRIC CORPORATION
	Coal-oil	Rutenberg (3–4), at Ashkalon, 1100 MW. 2 × 550 MW. Planned for 1999, 2000.
	Power stations in 2001 or after 2001	
	Coal	ISRAEL ELECTRIC COPRORATION Haifa D, at Haifa, 1140 MW. 2 × 575 MW. Unit 1, 2003. Unit 2, 2004.
Jamaica	Chem/petchem	Kingston.
	Smelting	Lucca (bauxite), Montego Bay.
Jordan	Chem/petchem	Amman, Zerka.
	Oil/gas	New fields discovered.
Kazakhstan	Cement	Karaganda.
	Coal	Karaganda.
	Chem/petchem	Aktyubinsk, Atyrau, Karaganda, Shymkent.
	Iron & steel	Karaganda.
	Oil/gas	Akshabulak/south-east Kazakhstan, Aktyubinsk, Atyrau, Karachaganak/west Kazakhstan, North-east Caspian Sea.
Kenya	Cement	Mombasa, Nairobi, Nakuru.
	Chem/petchem	Eldoret, Mombasa, Nakuru, Thika.
	Oil/gas	Great Rift Valley, Lake Turkuna.
	Iron & steel	Mombasa.
	Vehicle manuf.	Nairobi, Mombasa, Thika.
Korea, Dem People's R. (North)	Chem/petchem	Pyongyang, Unggi.
	Coal	Pyongyang.
	Iron & steel	Chongjin, Pyongyang.
	Textiles	Pyongyang.
Korea, Rep (South)	Chem/petchem	Inchon, Kwangju, Pohang, Pusan, Seoul, Ulsan, Yosu.
	Iron & steel	Changwon, Inchon, Kunsan, Pohang, Pusan, Seoul.
	Tim/paper/pulp	Pusan.
	Vehicle manuf.	Kwangju, Seoul, Pusan, Ulsan.
	Carbonaceous or hydrocarbon-fired power stations	
		KOREA ELECTRIC POWER COMPANY
	Anthracite	Kunsan, at Cholla Puk Do, 75 MW. Particulate collector.

Table 6.1. *Continued*

Country	Industry sectors	Location/region/remarks
Korea, Rep (South)	Coal	Po Ryong, at Chungchong Nam Do, 3000 MW. 6 × 500 MW. 1983, 1993, 1994. Unit 3, low NOx burners. Particulate collector.
	Coal	Souchon, at Chungchong Nam Do, 400 MW. 2 × 200 MW. 1983. Particulate collector.
	Coal	Taean 1–2, at Chungchong Nam Do, 1000 MW. 2 × 500 MW. 1995. Low NOx burners. Particulate collector.
	Coal	Taean 3–4, at Chungchong Nam Do, 1000 MW. 1997.
	Anthracite	Yong Dong, at Kangwon Do, 325 MW. 1 × 125 MW. 1 × 200 MW. Particulate collector.
	Anthracite	Yongwol, at Kwangwon Do, 100 MW. 2 × 50 MW. Particulate collector.
	Coal	Hadong (3–4), at Kyongsang Nam Do, 1000 MW. 1997.
	Coal	Samchonpo 1–3, at Kyongsang Nam Do, 2240 MW. 3 × 560 MW. 1983, 1984, 1994. Unit 3, low NOx burners. Particulate collector.
	Coal	Samchonpo, at Kyongsang Nam Do, 560 MW. 1994.
	Coal	Samchonpo, at Kyongsang Nam Do, 1000 MW. 1997.
	Anthracite	Pusan, at Pusan, 120 MW. 2 × 60 MW. Particulate collector.
	Power stations in 1998, 1999, 2000	
		KOREA ELECTRIC POWER COMPANY
	Coal	Dangjin, at Chungchong Nam Do, 1000 MW. Planned for 2000. Low NOx burners.
	Coal-anthracite	Donghae, at Kangwon, 200 MW. Circulating Fluidised Bed planned for 1998.
	Coal	Hadong, at Kyongsang Nam Do, 1000 MW. 2 × 500 MW. Low NOx burners. Particulate collector. Planned for 2000.

Table 6.1. *Continued*

Country	Industry sectors	Location/region/remarks
Korea, Rep (South)	Coal	New Coal I (1), at Kyongsang Nam Do, 500 MW. Planned for 1999.
	Coal	New Coal I (2), at Kyongsang Nam Do, 500 MW. Planned for 1999.
	Coal	New Coal II (1), at Kyongsang Nam Do, 500 MW. Planned for 2000.
	Coal	New Coal II (2), at Kyongsang Nam Do, 500 MW. Planned for 2000.
	Power stations in 2001 or after 2001	
		KOREA ELECTRIC POWER COMPANY
	Coal	Dangjin (3–4), at Chungchong Nam Do, 1000 MW. Planned for 2001. Low NOx burners.
	Coal	Yongheungdo, at Kyongki Do, 1600 MW. Planned for 2001.
	Coal	Hadong (5–6), at Kyongsang Nam Do, 1000 MW. 2 × 500 MW. Particulate collector. Planned for 2002.
Kuwait	Chem/petchem	Kuwait, Mina Abdullah, Mina Al-Zour, Mina Al-Ahmadi, Mina Saud, Ras Khafji, Shuaiba.
	Oil/gas	Magwa, Raudhatain.
Kyrgyz Rep	Food processing	Bishkek.
	Oil/gas	Fergana fields.
	Textiles	Bishkek.
Laos, People's Dem. Rep, LDC	Oil/gas	New fields discovered.
	Textiles	Vientiane.
	Tim/paper/pulp	Vientiane.
Lebanon	Chem/petchem	Sidon, Tripoli.
	Textiles	Beirut.
Liberia LDC	Chem/petchem	Monrovia.
	Food processing	Monrovia.
	Iron & steel	Mount Nimba (not exploited).
	Textiles	Monrovia.
Libya	Cement	Benghazi.
	Chem/petchem	Al Burayqah, Benghazi, Marsa-el-Brega, Ras Lanuf, Sider, Tripoli.
	Iron & steel	Misratah, Tripoli.
	Oil/gas	Al Burayqah, Dahra-Hofra, Murzuk, Sarir, Surt Basin, Zelten.
Macedonia	Chem/petchem	Skopje.
	Food processing	Skopje.
	Carbonaceous or hydrocarbon-fired power stations	
		ELEKTROSTOPANSTVO NA MAKEDONIJA
	Lignite	Bitola, at Bitola, 630 MW. In operation.
	Lignite	Oslomej, at Kicevo, 120 MW. In operation.

Table 6.1. *Continued*

Country	Industry sectors	Location/region/remarks
Madagascar	Chem/petchem	Anatananarivo
LDC	Food processing	Anatananarivo
Malaysia	Cement	Kuala Lumpur.
	Chem/petchem	Labuan, Lutong and Miri (East Malaysia), Port Dickson.
	Iron & steel	Kuala Lumpur.
	Oil/gas	Baram, Baronia, Kinarut, Miri, Samarang Kechil, Temana (East Malaysia);
		Bintang-Jerneh-Pilong-Tapis-Pulai-Anding-Angsi-Besar-Duyong-Udang fields (West Malaysia, offshore Kerteh).
	Tim/paper/pulp	Kuala Lumpur.
	Vehicle manuf.	Kuala Lumpur.
	Carbonaceous or hydrocarbon-fired power stations	
		SARAWAK ELECTRICITY SUPPLY CORPORATION
	Coal	Kuching, at Sejinkat Sarawak, 100 MW.
		1997.
		TENAGA NASIONAL BERHAD
	Coal-oil-gas	Port Kelang (3–4), at Pelabuhan Kelang, 600 MW.
		2 × 300 MW.
		1988.
		Low NOx burners.
		Particulate collector.
	Power stations in 1998, 1999, 2000	
		TENAGA NASIONAL BERHAD
	Coal-oil-gas	Port Kelang 3 (5–6), 1000 MW.
		Planned for 1998.
	Coal	Jainmanjung, at Penang Island, 1000 MW.
		Planned for 1999, 2000.
	Power stations in 2002 or after 2001	
		PRIVATE SECTOR
	Coal	Yan Kedah, 1000 MW.
		Planned for 2002.
		TENAGA NASIONAL BERHAD
	Coal	Lumut Perak, 2000 MW.
		Planned for 2001/2003.
		IPP.
		Developer: TNB Janamanjung Sendirian Berhad.
Malawi	Coal	Rumphi.
LDC	Food processing	Blantyre.
Mali	Food processing	Bamako
LDC	Textiles	Bamako
Mauritania	Food processing	Nouakchott.
LDC	Iron & steel	Zouerat.
	Textiles	Nouakchott.
Mauritius	Textiles	Port Louis (export processing zones).
Mexico	Chem/petchem	Ascapotzalco, Cadeyreta, Guadalajara, Irapuato, Mexico City, Minatitlan, Monterey, Reynosa, Salamanca, Salina Cruz, Tampico.

Table 6.1. *Continued*

Country	Industry sectors	Location/region/remarks
Mexico	Iron & steel	Manzanillo, Monterey, Piedras Negras.
	Oil/gas	Ebano, Minatitlan, Ojiacaque, Poza Rica, Reynosa, Tampico; Gulf of Mexico – Akal, Bacab, Bay of Campeche-Cantarell, Ebano-Panuco fields.
	Vehicle manuf.	Mexico City, Puebla.
	Carbonaceous or hydrocarbon-fired power stations	
		COMMISION FEDERAL DE ELECTRICIDAD
	Coal	Carbon II, 1–2, at Piedras Negras, 700 MW. 2 × 350 MW.
	Coal	Carbon II-3, at Piedgras Negras, 350 MW. 1995.
	Coal	Carbon II-4, at Piedgras Negras, 350 MW. 1996.
	Coal	Rio Escondido, at Piedgras Negras, 1200 MW. 4 × 300 MW.
Moldova	Food processing	Chisinau.
	Textiles	Chisinau.
	Carbonaceous or hydrocarbon-fired power stations	
	Thermal plant	Plant name: Moldova Moldavskaya.
Mongolia	Food processing	Ulan Bator.
	Textiles	Ulan Bator.
	Oil/gas	New fields discovered in south-east: Tsagaan Els, Zunnbayan.
Morocco	Chem/petchem	Casablanca, Safi.
	Fertilisers	Boudnib, Youssoufia.
	Heavy ind.	Casablanca.
	Vehicle manuf.	Casablanca.
	Carbonaceous or hydrocarbon-fired power stations	
		OFFICE NATIONAL DE L'ELECTRICITE
	Anthracite	Roches Noires, at Casablanca, 60 MW. 1968.
	Coal-oil	Jorf Lasfar, at El Jadida, 660 MW. 2 × 330 MW. ESP (electrostatic precipitators). 1994.
	Anthracite-oil	Jerada, at Jerada, 165 MW. 3 × 55 MW. 1971. Electrostatic cyclone precipitators.
	Coal	Mohammedia (3–4), at Mohammedia, 300 MW. 2 × 150 MW. 1994, 1995. ESP (electrostatic precipitators).
	Power stations in 1998, 1999, 2000	
		OFFICE NATIONAL DE L'ELECTRICITE
	Coal	Jorf Lasfar (3–6), at El Jadida, 1320 MW. 4 × 330 MW.

Table 6.1. *Continued*

Country	Industry sectors	Location/region/remarks
Morocco		CMS-ABB joint-venture.
		BOT.
		Planned for 1999, 2000.
Mozambique	Chem/petchem	Maputo.
LDC	Coal	Tete.
	Textiles	Beira, Maputo.
	Carbonaceous or hydrocarbon-fired power stations	
	Coal	Maputo, at Maputo, 57.5 MW.
		1961.
		Particulate collector.
Myanmar	Chem/petchem	Chauk, Syriam, Yangon.
LDC	Iron & steel	Yangon.
	Oil/gas	Arakan fields, Gulf of Martaban: Yadana-Yetagun fields.
	Tim/paper/pulp	Moulmein, Yangon.
Namibia	Smelting	Windhoek (lead, copper)
	Carbonaceous or hydrocarbon-fired power stations	
		SWA WATER AND ELECTRICITY CORPORATION
	Coal	Van Eck, at Windhoek, 120 MW.
		4 × 30 MW.
Nicaragua	Chem/petchem	Managua.
	Textiles	Leon, Managua.
	Tim/paper/pulp	Bluefields.
Niger	Food manufact.	Niamey.
LDC	Textiles	Niamey.
Nigeria	Cement	Kano, Shagamu.
	Chem/petchem	Alesa-Eleme, Kano, Kaduna, Lagos, Port Harcourt.
	Iron & steel	Kaduna.
	Tim/paper/pulp	Lagos.
	Oil/gas	Lagos; Niger River Estuary, fields onshore and offshore from Port Harcourt to Wari; Anambra-Benue-Chad-Dahomey Basins, Asasa, Bonga.
Oman	Cement	Muscat.
	Chem/petchem	Mina Al Fahal, Muscat.
	Oil/gas	Birba, Fahud, Jazal, Jebel-Aswad, Marmul, Natih, Qarn Alam-Wafra-Qaharir fields, Rima; Saih Niyad, Shaheer; West Bukha Structure in Gulf of Hormuz.
Pakistan	Chem/petchem	Karachi, Hyderabad, Lahore, Rawalpindi.
	Iron & steel	Karachi, Lahore.
	Oil/gas	Indus River Estuary, Sukkur-Sui fields; fields south of Islamabad.
	Textiles	Faisalabad, Lahore, Quetta.
	Tim/paper/pulp	Karachi.
	Vehicle manuf.	Karachi.

Table 6.1. *Continued*

Country	Industry sectors	Location/region/remarks
Pakistan	*Power stations in 1998, 1999, 2000*	
		CONSOLIDATED ELECTRIC POWER ASIA LTD
	Coal-lignite	Keti Bandar, at Keti Bandar, 1320 MW. 2 × 660 MW. CEPA of Hong Kong. Planned for 1998, 2000.
	Lignite-fired power stations	
		PAKISTAN WATER AND POWER DEVELOPMENT AUTHORITY
	Lignite	Khanot, also known at Lakhra (1–3), at Lakhra, 150 MW. 3 × 50 MW. Environmental standards to be applied.
	Lignite	Lakhra (4–6), at Lakhra, 150 MW. 3 × 50 MW.
		POWER GENERATION SERVICE
	Lignite	Lakhra Private, at Lakhra, 200 MW.
Palestine:	Oil/gas	Gurim-Kidod-Zohar fields West Bank, Gaza
Panama	Chem/petchem	Colon, Panama City.
	Smelting	Cerro Colorado (copper).
	Tim/paper/pulp	Colon.
Papua New Guinea	Oil/gas	New fields discovered on New Guinea: hinterland of Kikori, Kutubu-Uramu-Elvala fields.
	Smelting	Panguna/Bougainville Island and Ok Tedi (copper).
Paraguay	Chem/petchem	Asuncion-Villa Elisa.
	Textiles	Asuncion.
		Large hydro-power capacity at Iguazu Falls, Itaipu.
Peru	Chem/petchem	Lima, Pucallpa, Talara.
	Iron & steel	Chimbote.
	Textiles	Arequipa, Ica, Lima.
	Tim/paper/pulp	Iquitos.
	Oil/gas	Arica, Brea-Parinas fields, Camesea-Ucayali Basin, Capahuari.
	Vehicle manuf.	Lima.
	Power stations in 1998, 1999, 2000	
		ENERSUR
	Coal	250 MW. Planned for 2000, 2001. Hitachi, Tractebel. Coal and gas firing systems.
Philippines	Chem/petchem	Batangas, Manila, Subic Bay.
	Oil/gas	Offshore Palawan Island: Galoc-Nida, Malampaya-Camago.
	Smelting	Baguio, Cebu, Lupon.
	Textiles	Cebu, Manila.
	Tim/paper/pulp	Davao, Manila.

Table 6.1. *Continued*

Country	Industry sectors	Location/region/remarks
Philippines	*Carbonaceous or hydrocarbon-fired power stations*	
		NATIONAL POWER CORPORATION
	Bit., and sub-bit. coal	Calaca, at Batangas, 600 MW. 2 × 300 MW. 1984, 1995. Particulate collector.
	Coal	Naga I, at Visayas, 50 MW. 1981. Particulate collector.
	Coal-oil	Naga II, at Visayas, 52 MW. 1986. Particulate collector.
	Coal	Toleda Atlas, at Visayas, 55 MW. 1993. Fluidised Gas Desulphuriser.
		PRIVATE SECTOR
	Coal	Pagbilao, at Quezon City. 2 × 350 MW. 1996. Hopewell Power, BOT, CEPA contract for 29 years. Particulate collector.
	Power stations in 1998, 1999, 2000	
		NATIONAL POWER CORPORATION
	Coal	Masinloc, at Zambales, 600 MW. 2 × 300 MW. Planned for 1998.
		PRIVATE SECTOR
	Coal	Sual, at Pangasinan, 1,320 MW. 2 × 660 MW. CEPA-Hopewell Power. Planned for 1998, 1999.
	Coal	Mauban, at Quezon City, 500 MW. PF, dry flue gas scrubber, ESP. Bechtel, Ogden, Philippine Board of Investments. Planned for 2000.
	Coal	Panay, at Visayas, 100 MW. 2 × 50 MW. BOO. Planned for 2000.
		PRIVATE SECTOR
	Coal	Pinamucan, at Batangas, 600 MW. 2 × 300 MW. Approved by Philippine Board of Investments. Magellan Utilities, Devil Corp, CMS Generation.
	Coal	Mindanao I, at Mindanao, 200 MW. Contract being negotiated – Hurbin Consortium. Planned for 2001. Mindanao II, at Mindanao, 150 MW. BOO. Planned for 2005.
	Coal	Pampanga, at Pampanga 100 MW. Aziz Pacific Energy Corporation – registered with Board of Investments.

Table 6.1. *Continued*

Country	Industry sectors	Location/region/remarks
Philippines	*Lignite-fired power stations*	
		PRIVATE SECTOR
	Lignite-oil	Calaca III, at Batangas, 300 MW. Mitsubishi Corporation.
Qatar	Chem/petchem	Umm Bab, Umm Said.
	Oil/gas	Umm Bab-Dukhan, Mahzam, North fields.
Rwanda LDC	Food processing	Kigali.
	Oil/gas	New fields at Lake Kivu.
	Textiles	Kigali.
Saudi Arabia	Cement	Al Hufuf, Jeddah, Riyadh, Yanbu Al Bahr.
	Chem/petchem	Al Jubayl, Jeddah, Rabigh, Ras Al Khafji, Ras Tanurah, Riyadh, Yanbu Al Bahr.
	Iron & steel	Jeddah, Yanbu Al Bahr.
	Oil/gas	Abqaiq, Ad Damman, Al Hufuf, Al Jubayl, Berri, Fareidoon, Ghawar, Harad, Manifa, Qatif, Ras Tanurah, Safaniya, Yanbu Al Bahr.
Senegal	Chem/petchem	Dakar.
	Food procesing	Dakar.
	Oil/gas	New fields discovered off Casamance estuary.
	Vehicle manuf.	Dakar.
Seychelles	Oil/gas	New fields discovered.
Sierra Leone LDC	Chem/petchem	Freetown. > Civil war crisis.
	Food processing	Freetown.
	Textiles	Freetown.
Singapore	Chem/petchem	Pulau Bukom, West Singapore.
	Textiles	West Singapore.
	Tim/paper/pulp	West Singapore.
Solomon Islands LDC	Smelting	Rennell Island (copper).
Somalia LDC	Chem/petchem	Mogadishu (unexploited mineral resources).
South Africa	Cement	Capetown.
	Chem/petchem	Capetown, Durban, Johannesburg.
	Heavy ind.	Johannesburg.
	Iron & steel	Pretoria.
	Oil/gas	Bredasdorp Basin-Oribi.
	Tim/paper/pulp	Capetown, Port Elizabeth.
	Vehicle manuf.	Capetown, Johannesburg, Port Elizabeth, Pretoria.
Sri Lanka	Food processing	Colombo.
	Textiles	Colombo.
Sudan	Cement	Khartoum.
	Chem/petchem	Khartoum, Port Sudan.
	Food processing	Khartoum.
	Oil/gas	Sharaf-Unity-Adar fields.
	Textiles	Khartoum.
Suriname	Food processing	Paramaribo.
	Smelting	Paramaribo (bauxite).
	Tim/paper/pulp	Paramaribo.

Table 6.1. *Continued*

Country	Industry sectors	Location/region/remarks
Swaziland	Tim/paper/pulp	Mbanane.
Syrian Arab Rep.	Cement	Adra, Damascus, Halab, Hamah, Tartus.
	Chem/petchem	Baniyas, Damascus, Dayr az Zawr, Homs.
	Oil/gas	Al Sukhne, Deir al-Zor, Ghouna, Palmyra, Tayyim.
Tajikstan	Food processing	Dushanbe.
	Oil/gas	Leninskoye, Mazar-i-Sharif (unexploited).
	Textiles	Dushanbe.
Tanzania,	Chem/petchem	Dar Es Salaam.
United Rep, LDC		Oil/gas Pemba Island, Songo Songo (offshore).
	Tim/paper/pulp	Dar Es Salaam, Dodoma.
Thailand	Chem/petchem	Chon Buri, Nakhon Ratchasima, Sri Racha.
	Oil/gas	Gulf of Thailand – Erawan, Moragot, Pailin, Platong, Satun, Thai-Malay jv Mali-Amarit-Wira fields.
	Textiles	Bangkok, Khon Kaen, Samut Prakan.
	Vehicle manuf.	Bangkok, Samut Prakan, Thon Buri.
	Power stations in 2001 or after 2001	
		ELECTRICITY GENERATING AUTHORITY OF THAILAND
	Coal-oil	New Thermal (3–4), 2000 MW.
		2 × 1000 MW.
		Unit 3, 2007.
		Unit 4, 2008.
	Coal-oil	Tap Sae Kae, at Rayong, 1400 MW.
		2 × 700 MW.
		Unit 1, 2002.
		Unit 2, 2003.
		BLCP Ltd – Banpu Plc and Loxley Plc.
		SAHAVINIYA STEEL COMPANY
	Coal	Bang Saphan, at Prachuap Kiri Khan, 1,400 MW.
		2 × 700 MW.
		Unit 1, 2001.
		Unit 2, 2002.
		Low NOx burners, FGD.
		Union Power Development Corporation.
	Lignite-fired power stations	
		ELECTRICITY GENERATING AUTHORITY OF THAILAND.
	Lignite	Mae Moh (1–3), at Muanghampang, 225 MW.
		3 × 75 MW.
		1978, 1979, 1981.
	Lignite	Mae Moh (12–13), at Muanghampang, 600 MW.
		2 × 300 MW.
		World Bank, Australian International Development.
		Assistance Bureau. Commissioned in 1995.
		Japan EXIM, US EXIM, EDC financed FGD project.

Table 6.1. *Continued*

Country	Industry sectors	Location/region/remarks
Thailand	Lignite	Mae Moh (4–7), at Muanghampang, 600 MW. 4 × 150 MW. 1984. FGD retrofitting to be completed by 1999.
	Lignite	Mae Moh (8–11), at Muanghampang, 1800 MW. FGD retrofitting planned.
Togo	Chem/petchem	Lome
LDC	Fertilisers	Lome
Trinidad	Chem/petchem	Point-a-Pierre, Point-Fortin, Port-of-Spain.
and Tobago	Oil/gas	Onshore/offshore – Dolphin, Forest Reserve, Immortelle, Samaan, Soldado, Parrylands-Teak-Radix.
Tunisia	Cement	Tunis.
	Chem/petchem	Bizerte, Gabes, Gafsa, Kasserine, Sfax, Tunis.
	Fertilisers	Gafsa, Metlaoul.
	Iron & steel	Tunis.
	Oil/gas	Ashtart, Bir Aouin, Douleb, El Borma, Ezzaoula, Sidi el Kilani; Gulf of Gabes – El Riban-Miskar-El Biban.
	Vehicle manuf.	Kairouan, Sousse, Tunis.
Turkey	Cement	Adana, Ankara, Antalya, Bursa, Diyarbakir, Erzurum, Gaziantep, Isparta, Konya, Istanbul, Samsun, Sivas.
	Chem/petchem	Adana, Ankara, Batman, Diyarbakir, Iskenderun, Istanbul, Izmir, Izmit, Kirikkale, Mersin.
	Iron & steel	Kirikkale, Izmir, Sivas.
	Oil/gas	Batman, Diyanbakir, Garzan-Raman-Batiraman-Kayakoy-Bulgardag fields.
	Vehicle manuf.	Bursa, Istanbul, Izmir.
	Carbonaceous or hydrocarbon-fired power stations	
		TURKIYE ELEKTRIK KURUMU
	Coal	Y. Catalagzi, at Catalagzi, 300 MW. ESP (electrostatic precipitators).
	Power stations in 2001 or after 2001	
		BOO PROJECTS
	Coal	Filyos, at Filyos, 1000 MW. 2 × 500 MW. ESP, FGD. Planned for 2006.
	Coal	Iskenderun, at Iskenderun, 1000 MW. 2 × 500 MW. Planned for 2003. ESP, FGD.
	Coal	Other power stations of 1000 MW capacities. Planned for 2007–2010.
		TURKIYE ELEKTRIK KURUMU
	Coal	Amasram, at Amasra, 600 MW. 2 × 300 MW. ESP, FGD. Planned for 2009/2010.

Table 6.1. *Continued*

Country	Industry sectors	Location/region/remarks
Turkey	Coal	Catalagzi, at Zonguldak, 600 MW. 2 × 300 MW. ESP, FGD. Planned for 2009/2010
	Lignite-fired power stations	
		BOT PROJECTS
	Lignite	Orta, at Cankiri-Orta, 250 MW. Planned for 2001. FGD, ESP.
	Lignite	A. Elbistan A (5–6), at Elbistan, 680 MW. 2 × 340 MW. Planned for 1998. FGD, ESP.
	Lignite	Adana-Tufanbeyli, at Adana, 600 MW. 2 × 300 MW. Planned for 1999/2000. FGD, ESP.
	Lignite	Cayirhan A, B, at Beypazari. Total of 1,800 MW. Current and future capacity. FGD, ESP in some units.
	Lignite	Bingol-Karliova, at Bingol-Karliova, 690 MW.
	Lignite	Keles, at Bursa Keles, 150 MW. Planned for 2007. FGD, ESP.
	Lignite	Can, at Canakkale-Can, 300 MW. Planned for 2004/2005. FGD, ESP.
	Lignite	A. Elbistan A (1–4), B, C, D, at Elbistan. Total of 5,440 MW. FGD, ESP. Current and future capacities.
	Lignite	Kemerkoy, at Kemerkoy, 630 MW. 3 × 210 MW. 1994. FGD.
	Lignite	Beysehir, at Konya-Beysehir, 340 MW. Planned for 2009. FGD, ESP.
	Lignite	Seyitomer, at Seyitomer, 600 MW. 4 × 150 MW. ESP.
	Lignite	Silopi, at Silopi, 100 MW.
	Lignite	Kangal, at Sivas-Kangal. Total of 600 MW. Some units with FGD and ESP. Current and future capacities.

Table 6.1. *Continued*

Country	Industry sectors	Location/region/remarks
Turkey	Lignite	Soma A, B, C at Soma. Total of 1634 MW. Some units with FGD and ESP. Current and future capacities .
	Lignite	Tuncbilek A, B, C, at Tuncbilek. Total of 729 MW. Some units with ESP.
	Lignite	Yatagan, at Yatagan, 630 MW. 3 × 210 MW. FGD contract signed.
	Lignite	Yenikoy, at Yenikoy, 420 MW. 2 × 210 MW. ESP.
Turkmenistan	Chem/petchem	Chardzhev, Krasnovodsk.
	Oil/gas	Bagadzha, Dauletabad-Donmez, Kotur-Tepe, Mary, Maiskoye-Bagadzha, Nebit Dag-Cheleken, Turkmenbashi.
	Textiles	Ashgabat, Chardzhev, Mary.
Uganda LDC	Chem/petchem	Kampala.
	Food processing	Kampala.
	Textiles	Kampala.
United Arab Emirates	Chem/petchem	Abu Dhabi, Bab, Jabal Dhanna, Ruwais, Umm Al Nar.
	Oil/gas	Abu Dhabi, Bab, Bu Hasa, Margham, Umm Shaif, Zakum, Zubbaya.
Uruguay	Chem/petchem	Montevideo.
	Heavy ind.	Montevideo.
	Textiles	Montevideo, Paysandu.
Uzbekistan	Chem/petchem	Bukhara, Kokand, Namangan, Samarkand, Tashkent.
	Oil/gas	Achak-Kirpichli, Farghona, Isbakent-Andizhan, Kishtuvan-Metedzhan, Namangan.
	Textiles	Bukhara, Farghona, Kokand, Nukus, Samarkand, Tashkent, Urganch.
Vanuatu LDC	Tim/paper/pulp	Port-Vila.
Venezuela	Chem/petchem	Amuay Las Piedras, Bajo Grande, Caracas, El Cardon, Maracaibo, Puerto La Cruz, San Roque, Valencia.
	Coal	Guanare River – coal deposits.
	Iron & steel	Ciudad de Guayana.
	Tim/paper/pulp	Ciudad de Guayana, Valencia.
	Oil/gas	Ciudad de Guayana, Jusepin-Santa Barbara, La Salina/Lake Maracaibo, Machete-Zuata-Hamaca-Cerro Negro fields, Quiriquire, Santa Rosa-La Ceibita-Zorro-Dacion fields.
	Vehicle manuf.	Caracas, Valencia.

Table 6.1. *Continued*

Country	Industry sectors	Location/region/remarks
Vietnam	Cement	Hai Phong.
	Chem/petchem	Hanoi, Da Nang, Ho Chi Minh.
	Iron & steel	Bien Hoa, Da Nang, Quang Nam.
	Oil/gas	Haiphong, Ho Chi Minh; South China Sea: Bach Ho-Rong, Cuu Long, Dragon, Dua, LanTay Lan Do, White Tiger fields.
	Textiles	Ho Chi Minh.
	Tim/paper/pulp	Ho Chi Minh.
	Vehicle manuf.	Hanoi, Ho Chi Minh.
Yemen LDC	Cement	Aden, Amran, Bajil.
	Chem/pethem	Aden, Marib.
	Oil/gas	Alif, Amal-Iyad, Marib.
	Textiles	Al Mukalia, Bajil, Sana, Taizz.
Zambia LDC	Cement	Lusaka.
	Chem/petchem	Lusaka, Ndola.
	Iron & steel	Lusaka.
	Tim/paper/pulp	Lusaka.
	Vehicle manuf.	Lusaka.
Zimbabwe	Cement	Bulawayo.
	Coal	Hwange, Kwekwe.
	Chem/petchem	Harare, Kwekwe.
	Iron & steel	Bulawayo, Harare, Kwekwe.
	Vehicle manuf.	Bulawayo, Mutare.

Notes

CDM host "developing countries" listing: "developing" countries according to World Resources Institute classification and UNFCCC non-Annex I listing – the interpretation is subjective to GDP/capita, particularly OPEC countries; OECD members Mexico, Turkey and South Korea; and highly industrialised nation states.

Capital or largest city is quoted when exact locations of industries are not known for certain countries.

Oil and gas sites may include the names of fields.

Two place names separated by a hyphen (/): the first place is located in a part of a larger or more widely known and identifiable geographical region.

Place names separated by hyphen (-): a stretch of particular region (e.g. oil and gas deposits) which extend across specific locations, or places which are close to each other and operate in the same industry sector. Place names include names of fields in the energy sector.

UNFCCC has not specifically identified or agreed on vehicle manufacturing plants as industries qualifying for greenhouse gases reductions, although the author has proposed the admissibility of this industry for CDM projects.

>: Refers to recent developments in this location or recommended remedial activity.

( ): Refers to current conditions in the location or project conditions. For the coal industry, it refers to project conditions.

"Audit": Environmental audit recommended for this sector.

Locations that are highlighted with underlines or italics represent the areas of the highest concentration of the particular industry in the country.

Table 6.1. *Continued*

For some thermal power stations, details of dates of commissioning, construction, operation and power output capacity were not available at time of data compilation.
The listing includes Libya and Myanmar – foriegn investors and financiers are advised to consult their governments for full/partial partial limitation of trade with these countries.

Abbreviations:

Chem/pet chem	Chemical and petrochemical industries.
Mineral proc.	Mineral processing.
Tim/paper/pulp	Timber, paper and pulp industries.

Power stations using coal: coal-importing regions only
Source: Coal Information 1998, OECD/International Energy Agency.
The names and geographical locations of existing and planned power stations in regions that import coal are listed. The names of power stations are alphabetically arranged in sub-heading format under the relevant government authority or type of private entity structure for each country. Energy consumption activity in coal-importing regions are indicative of current and future demand for world steam coal, as most export-oriented industries are established in or near these regions. In China and India, special economic zones with incentives in coal-importing regions – especially along or near the coast – offer potentially attractive investment returns on power projects compared to other areas in these countries.
The listing excludes those regions which are self-sufficient in coal; for China, as an example, it does not include power stations in the Heilongjiang Province where coal is abundant. The names of the local authority and/or owner of power station projects are also provided, where known.
Data on power stations was based on the source; current at April 1998.

Nomenclature for types of power stations burning coal:

AFBC	Atmospheric Fluidised Bed Combustion
PF	Advanced Pulverised Fuel Power Plant
PFBC	Pressurised Fluidised Bed Combustion
CGCC	Coal-Gasification Combined Cycle

Advanced power generating stations which operate under supercritical or ultra-super critical steam conditions are capable of producing thermal efficiencies more than 45%.

Type of power station financing or ownership

BOO	Build-own-operate
BOOT	Build-own-operate-transfer
BOT	Build-operate-transfer
IPP	Independent Power Producer

Other nomenclature

ESP	Electrostatic precipitator
FGD	Flue gas desulphurisation
NOx	Emissions of oxides of nitrogen
Power-station unit ( ):	( ) refers to phase or extension of project, e.g. (2) refers to second turbine

Clean Development Mechanism project prospects in selected countries

At end of September 2000, a projects portfolio covering eleven non-Annex I countries had been developed by Euro-Asia Pac Corp Limited, an environmental consultancy in London. Figures 6.1 and 6.2 show the average internal rates of return of these projects, and the augmentation on these rates by contributions from the project source and ten secondary trades of the original contract. Initial offer prices of emission credits in terms of carbon dioxide equivalents are indicated in Figure 6.3.

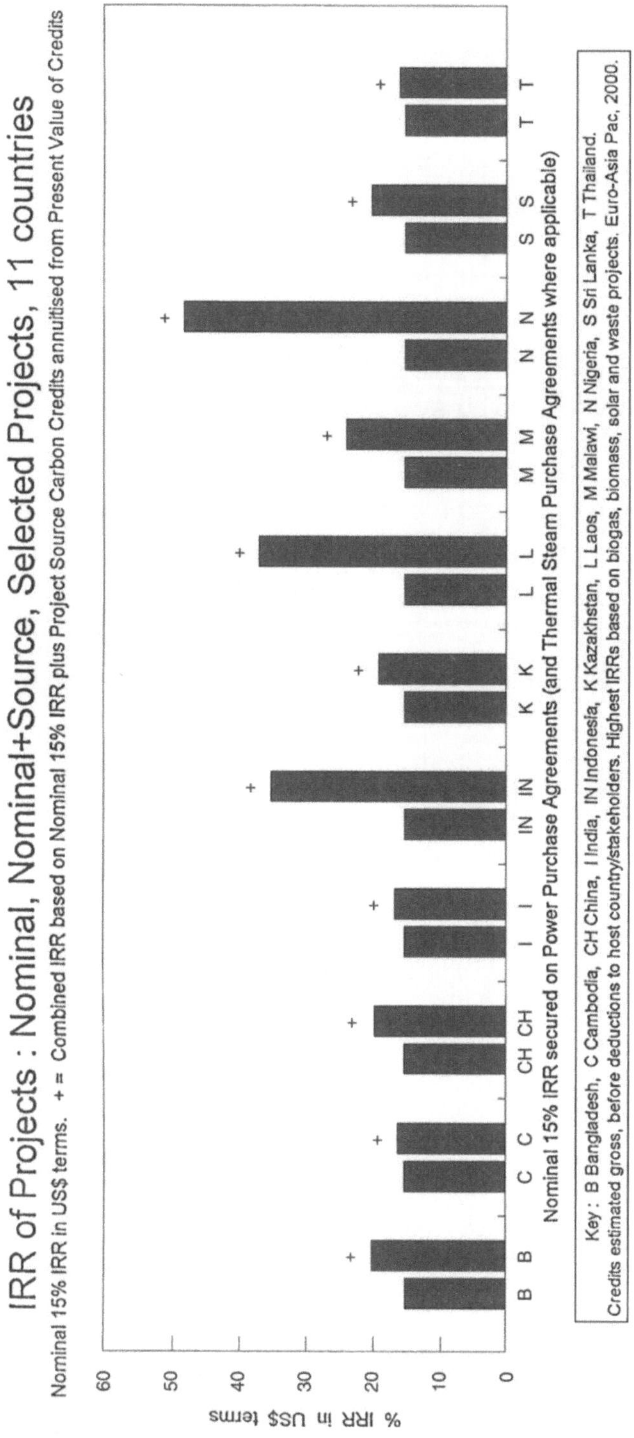

Fig. 6.1. IRR of projects: nominal, nominal + source, selected projects, 11 countries

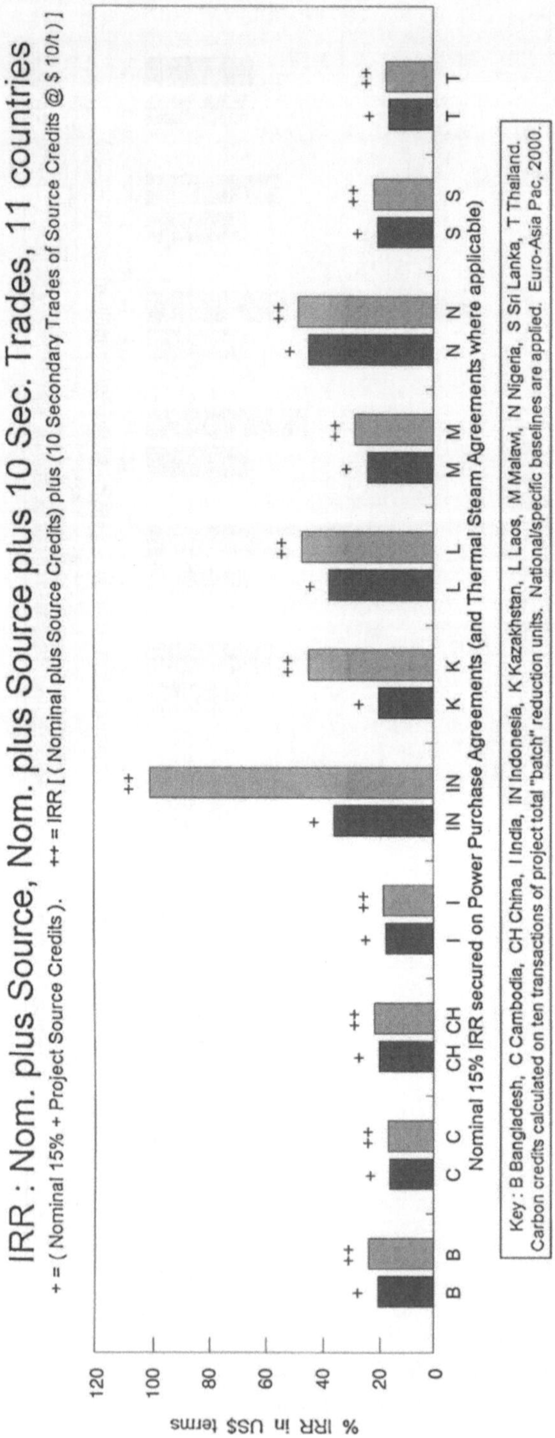

Fig. 6.2. IRR: nominal plus source, nominal plus source plus ten secondary trades, 11 countries

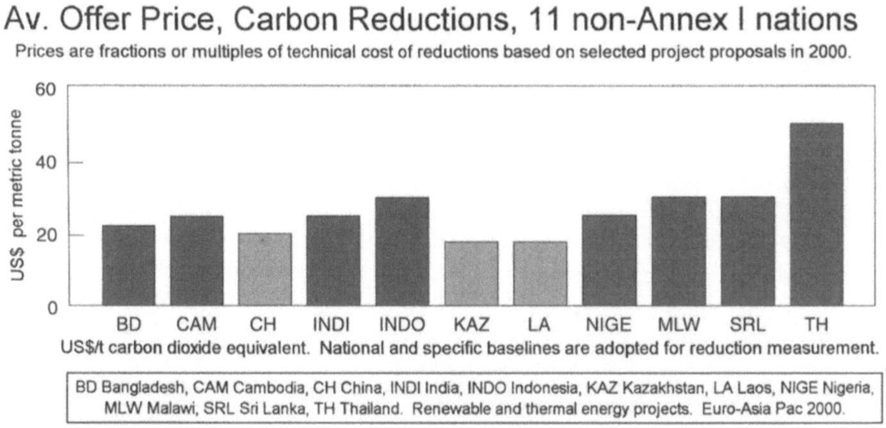

Fig. 6.3. Average offer price, carbon reductions, 11 non-Annex countries

Offer prices of carbon credits are based on discounts of 20%–60% of technical costs of abatement for projects. Laos and Kazakhstan present the best opportunities in this portfolio.

7
Project Parameters and Equitable Treatment

This chapter deals with the nature of baselines, upon which reductions of greenhouse gas emissions, or leakages, i.e. additional escape of emissions, may be measured; environmental and financial additionality; and equitable treatment of project proceeds and benefits under the Kyoto Protocol. At the 12th Meeting of Subsidiary Bodies in June 2000, held in Bonn, an "Adaptation Fund", and mandatory percentage contributions of the settled value of each and every carbon contract to the year 2012 towards this fund, were proposed. The author would like to express his appreciation to the United Nations Environmental Programme for adopting the generic concept of the LDC/IRISK (Least Developed Countries/Islands-at-Risk) Climate Change Fund in the June 2000 meeting. In the last part of the chapter, the criteria for selecting the project discount rate, and the effect of price support to projects, are examined.

7.1
Baselines

At mid-2000, a number of projects aimed at trading emissions – mostly bilateral corporate projects and transactions between "North" countries – had taken place in the European Union, a Baltic country, USA, Canada, Japan, New Zealand and Australia. Some "North-South" projects have also been initiated in Malaysia, South Africa, Central America and China. The definition of the most appropriate, or legal, baseline acceptable to UNFCCC is perhaps one of the difficult issues facing the investment community.

7.1.1
"Temporal" Baselines – Reference Years

Base year for measuring greenhouse gas emission reductions

Measuring emission reductions of carbon dioxide is a straightforward task – which can be made against a known quantity in a reference or base year. The five non-carbon dioxide gases can be measured in terms of carbon dioxide equivalents – for example, reducing methane emissions by half a tonne is equivalent to reducing about 10 tonnes of carbon dioxide. Hence, an Annex I country has the option of targeting carbon dioxide and other greenhouse gases to meet its carbon dioxide equivalent reduction commitments.

At the Kyoto Protocol, it was decided that three "core" greenhouse gases – carbon dioxide, methane and nitrous oxide, which constituted nearly 99% of greenhouse gases in Annex 1 countries – should be the prime targets for reduction.

The base year for measuring emission reductions of the three "core" greenhouse gases is 1990 for Annex I countries, except for Bulgaria (1988), Hungary (1985–87), Poland (1988) and Romania (1989). The base years for Croatia, Slovenia and Ukraine have not been agreed.

As for measuring reductions of the three trace industrial gases – hydrofluorocarbons, perfluorocarbons and sulphur hexafluoride, the Protocol has allowed a degree of flexibility for all Annex I countries – 1995 may be chosen as the base year. However, it is likely that these countries may adopt any year between 1990 and 1995 in which emissions of the trace gases were at their highest levels.

7.1.2
"Physical" Baselines – National vs. Specific

National baselines

For CDM and JI, national baselines are expected to be applied to renewable energy projects, while specific baselines are likely to be applied to individual projects involving thermal plant conversion or replacement.

Baselines for CDM projects are unlikely to be the same as JI projects. In a renewable energy project, say, a wind-farm generating electricity in southern Spain, the baseline for CO_2 emissions reduction in terms of tonnes per GWh_e could be the average of carbon dioxide emissions from all electricity generating sources in the country – these would include nuclear (52%), thermal (35%) and hydro (13%).

The baseline for a similar wind project in India could be the average of thermal (80%), hydro (19%) and nuclear (1%) – which would be higher in terms of tonnes of CO_2 per GWh_e. A CDM investor in India would hence achieve higher emission reductions than in Spain, at comparative lower project costs.

Specific baselines

A specific baseline reference would be appropriate for direct conversion or replacement of an emissions source. For example, a combined cycle gas turbine plant may replace an old coal-fired plant in Bosnia and Herzegovina – emissions reductions, in theory, would the difference between the new and old plants.

Specific baselines may be adopted for renewable energy projects

A small hydro or wind project in, say, Chile, may replace or supplement an ageing thermal power plant – the baseline in this scenario could be the quantitative emissions per electrical output from the thermal plant itself and not the average of all electricity sources in the country. This specific baseline may, however be applied where there is an existing thermal emissions source in the vicinity of the project.

7.1.3 Conceptual Reality of Multiple Baselines in Project Location

Two or more baselines may form reference datums for measuring emissions reductions in the project site. For example, a site may consist of a city, a conventional coal-fired power plant and a landfill. If x and y tonnes of greenhouse gas emissions were released during the twelve months of 2000 from the power plant and landfill respectively, and if municipal solid waste (MSW) replaced energy production from the power plant and obviated the need of the landfill over the next twelve months to 2001, then emissions reduction by the MSW Energy-from-Waste plant (power or CHP) would be the difference between (x + y) emissions in 2000 and emissions from the EfW plant in 2001.

Example of project with two conceptual baselines

Year 2000

CO_2 emissions from coal-fired power plant (100 MW_e) = 500,000 metric tonnes
MSW to landfill = 1,500,000 metric tonnes (20% carbon content)
CH_4 emissions from landfill = 150,000 metric tonnes
CO_2 emissions from landfill = 300,000 metric tonnes

Total emissions
= 800,000 t CO_2 + 150,000 t CH_4 (3,150,000 t CO_2 equiv) = 3,950,000 t CO_2 equiv.

Year 2001

Total emissions from Energy-from-Waste (EfW) plant = 1,100,000 t CO_2 (20% carbon content × 1,500,000 metric tonnes MSW × 3.67 tonnes CO_2/t carbon).

Total emissions reduction in year 2001 compared to year 2000 = 2,850,000 t CO_2 equiv.

A possible third baseline

If municipal raw sewage was anaerobically digested and used as biogas for the EfW plant or a separate power/CHP plant, then greenhouse gas emissions from untreated sewage would represent the third baseline for emissions reduction.

Profitability implications for Municipal Solid Waste and EfW projects with methane capture

Initial calculations by the author suggest remarkably superior returns on investments in CDM projects which involve methane capture, due to the very high prices commanded by this gas due to its global warming potential compared to carbon dioxide. Internal rates of return of between 25% and 50% can be achieved in such projects – hence host governments are more likely than not to seek a larger claim of proceeds from projects involving capture of methane, nitrous oxide, sulphur hexafluoride, HFCs and PFCs.

If the host government and local stakeholders consider the market price of carbon credits to be low or unattractive during project commissioning – i.e. "when the switch is turned on" to reduce GHG emissions – they may well demand *document entitlements* of actual tonnages of GHG reductions on agreed delivery dates, so that they are able to sell these assets at higher market prices in the future. It is hence important for stakeholders to decide whether or not they wish to receive assets in cash terms or document entitlements of GHG reduction credits as the price of these credits can not be predicted at the onset of the project.

7.1.4
Case for Inadmissibility of Multiple Counting

Annex I countries have two reference years – 1990 as the base year, and 2010 as the nominal commitment year (average of 2008–2012, the first commitment period, to simplify parameters for the case), to reduce the three dominant greenhouse gases targeted by the Kyoto Protocol.

Within the political boundaries of Annex I countries, local companies will have to reduce GHG emissions by that amount specified by national environmental authorities, the balance of which could be achieved via the flexible mechanisms of CDM, JI and Emissions Trading. Annex I firms are expected to meet their domestic obligations by absolute caps, emissions reduction improvements per unit output (this may result in higher total volumes of emissions if the company constructed additional plants) or via other methods Annex I governments may decide to honour their Kyoto Protocol commitments. Individual country methods may not have to be approved by UNFCCC, as long as "emission pluses and minuses" result in overall compliance with national targets. However, the issue of "emissions subsidy" to large firms, as exemplified by negotiated agreements in the UK, could conflict with equal treatment to SMEs. Excessive price subsidy on renewable energy projects could also distort competition in terms of financial flows amongst Annex I countries.

The measurement of emissions reductions in Annex I and non-Annex I countries should be based on the difference of emissions from the source operating at base load over a certain year (say, 1990 for old plants or 2001 for new plants) and reduced emissions from the same source if retrofitted, or perhaps a new source replacing the old source, under the same base load condition. Any improvements in the retrofit or new technology which enabled emissions reduction in the first place, would qualify for additional emission reductions – these additional reductions are expected to be marginal unless the plant is replaced with another technology.

Double, triple or multiple counting of emissions reductions in Annex I countries would simply defeat the primary objective of the Protocol. For example, if emissions sources in the United Kingdom were to be represented by one gas turbine plant, emitting, say, 100 million tonnes of carbon dioxide equivalent in 1990, conversion to a CCGT plant in 1991 may result in emissions reduced to say, 600 million tonnes of carbon dioxide equivalent. If this reduction is counted every year for 20 years, to end of year 2010, the United Kingdom might claim in total 12,000 million tonnes of carbon dioxide equivalent reduction in 2010, against say, a sustained emissions level of 600 million tonnes of carbon dioxide equivalent in the same year. The multiple counting approach would enable all Annex I countries to rely only on a few large-scale plant improvements to deliver the required commitment, while actual emissions levels could far exceed national targets, accelerating global warming.

The same negative effects apply for non-Annex I countries, which do not have targets nor reference years for emissions reductions. If Annex I firms invested in non-Annex I countries, with a view of using "multiple counting" to reduce emissions in their Annex I country of origin, they would be able to take advantage on emissions reductions of magnitudes ten to twenty times the levels they would achieve with "once only counting" at home. Moreover, European and US companies - technology and financial institutions - would gladly invest in countries like India and China and realise large profits on emissions trading.

7.1.5 GHG Baseline for the International Transportation Sector

In its existing framework, the Kyoto Protocol does not cover greenhouse gas emissions from international transportation carriers via air, sea, land and subterranean passages. Extension of the Protocol to reduce emissions emanating from cross-border traffic would assist the "on land" strategies of Annex I governments to meet their obligations. Parties of the Conference could agree a commencement baseline of, say, 2002, to kick-start the initiative.

The international transportation sector should make a concerted effort to reduce the overall impact of global warming from their emissions, or else the projected growth of this industry would lead to unchecked GHG emissions in the millennium. Carriers incorporated in Annex I countries would be affected under the Kyoto Protocol, and obligated governments should welcome this private sector undertaking which will assist them to meet their targets.

The most rational baseline that could be considered by the Conference of the Parties is GHG emissions in metric tonnes or part of a metric tonne per unit distance travelled, e.g. kilometre - in preference to absolute caps. An emissions/unit distance baseline for the global transport industry should prevent a repeat of the "hot air" controversy and encourage real technology measures to be implemented, as transport companies frequently expand or reduce carrier activity in response to changing market conditions and economic cycles.

Measurements of GHG emissions, say, in metric tonnes of CO_2 per kilometre or hundreds of kilometres, could be made on a batch or continuous basis depending on instrumentation installed in the transport carrier, once it leaves the political boundaries of an Annex I country. Domestic emissions inventories are already being compiled by environmental departments of a number of Annex I countries so that internal emissions reduction targets can be set.

Allowances could be allocated for emissions of Annex I transport carriers, and any surplus reductions in tonnage terms could be traded per Article 17 of the Kyoto Protocol.

Role of UN and non-governmental organisations

Coordination of the proposed emissions reduction initiative for the global transport industry could be spearheaded by the International Civil Aviation Organisation, International Maritime Organisation, Air Transport Association of America and Lloyds Register of Shipping.

7.2 Project Eligibility Criteria – CDM Projects

Three key criteria govern the certifiability of CDM projects – "environmental additionality", "financial additionality" and the "sustainability" requirement. The factors that determine these criteria, however, are difficult for external investors to assess as they are project and host country specific. A CDM project should meet the following criteria:

1. Emissions additionality.
2. Financial additionality.
3. Economic, social and environmental sustainability.

7.2.1 Emissions Additionality

"Emissions additionality" is a misnomer often quoted in a number of working papers on the Kyoto Protocol – it should be rephrased "emissions savings additionality" or "emissions non-additionality" of a project. A CDM project should not increase but reduce emission levels from a current emitter or "reference" emissions entity. However the "reference" entity may not be the actual technology in a plant or factory which the project is trying to reduce emissions, but a state-of-the-art technology which is more effective in cutting emissions.

For example, an investor plans to install a combined cycle gas turbine plant in China, to replace an existing conventional coal-fired plant. His emissions reduction baseline would be the coal-fired plant, but if several open-cycle gas turbine plants are already operating in the project area, UNFCCC might insist that the baseline should be the average of emissions from neighbouring open-cycle gas turbine plants. If so, the price of emission trading contracts per tonne of GHG abated would be more than that price based on the coal-fired plant.

The criteria is hence site and site-economy dependent, and would be highly debatable in terms of a threshold value for establishing whether a "superior" technology would be regarded as the baseline for measuring emission reductions; i.e. in the absence of CDM projects.

The author has made recommendations to UNFCCC that the source emission should be treated as the baseline if a particular superior technology does not constitute more than 50% of prevalent technologies in the site or region at the time of project investment.

7.2.2 Financial Additionality (Investment Additionality Context)

"Financial additionality" requires investment in projects which do not appear financially attractive – initially as a stand-alone project – in terms of financial returns.

Although the term "financial additionality" does not appear in the text of article 12, it has been advocated by a number of non-governmental organisations because it provides additional benefits to the host economy. The author supports the concept of financial additionality.

For example, a municipal solid waste (MSW) incineration project, which is more capital intensive than an open-cycle power plant (hence the MSW project's internal rate of return would be lower compared to commercial projects in the absence of CDM), may prove an astute investment if could claim methane reduction credits (preventing methane emissions), on the basis of diverting waste refuse from landfills.

The financial threshold could be evaluated in terms of net present values of projects; those projects which exhibit small or negative net present values at low discount rates - such values calculated without the benefits of CDM credits - would meet the criteria.

Countries like India and China are likely to campaign for the financial additionality clause as it reduces fiscal expenditure, hence domestic taxation, for the supply of energy and power, industrial commodities and market consumables.

If not agreed in COP VI or future meetings of COP, then financial additionality could be imposed as an investment requirement in CDM projects by individual non-Annex I countries - nothing in the Protocol prevents a developing country from enacting its own set of investment criteria to receive additional revenue if it so wishes.

UNFCCC would treat financial additionality as a positive contribution in its evaluation of CDM projects, hence investor companies would be encouraged to make "financially additive" investments to reduce the risk of non-certification of projects. As demonstrated in the next section, financial additionality can provide extra utility and material resources to the poorest communities through international relief agencies - particularly to non-energy based sectors such as water, sanitation and medical aid projects.

Mandatory or voluntary contributions that are financially additional to the humanitarian sector of the host economy should qualify as "acts of charity" by Annex I governments - hence such contributions could be deducted from imputed tax liabilities of investor firms.

Annex I countries which decide to waiver the financial additionality requirement - if this is not written into article 12 - will divert financial flows away from non-energy projects and further impoverish those communities regardless of their locations with respect to energy projects. Developing countries have to decide if they wish to compete for CDM projects on the basis of zero, partial, voluntary or mandatory financial additionality. Governments which prefer short-term energy benefits usually pay a heavier price on neglecting long-term social poverty.

7.2.3
"Economic-Social-Environmental" Sustainability

The project should not increase environmental pollution nor decrease the living and social standards of the community or region in which the project is implemented. Environmental impact assessments would be required, as well as on-the-ground surveys on economic and social impacts on the local population.

Baseline for Joint Implementation (JI) projects

The Joint Implementation programme should benefit from AIJ projects which have provided valuable data and working associations between "East-West" and "North-South" countries. AIJ projects did not receive a warm welcome from India, which favoured the CDM.

Article 6 of the Protocol specifies emissions savings "additionality" but does not stress financial nor sustainability additionality – hence JI and CDM projects face the arbitrary criteria in terms of reference baselines.

7.3 Determination of the Discount Rate

The discount rate at which CDM/JI financiers expect financial returns should reflect both business and sovereign risks. Other major risks are based on how the terms and conditions of the Kyoto Protocol are interpreted – whether the project emissions baseline is acceptable to UNFCCC in meeting the emissions additionality criteria, sustainability requirements – and market factors that will determine the price of emission credits. Lowering the discount rate will mean lower prices of the main stream product (e.g. electricity, steel) and vice-versa.

The UNFCCC should take into consideration all the financial risks which CDM/JI project sponsors undertake before it sets the "qualifying" discount rate for these projects. Project sponsors should identify all risks involved in a potential CDM/JI project and refer them to UNFCCC for their evaluation and determination of the appropriate discount rate.

At end 1999, the average cost of capital (debt and equity) in ASEAN and South American countries were estimated at 20%–25%, China and India about 20%, Eastern Europe 15%, while for Russia, CIS and African countries, discount rates of above 25% have been quoted. These rates exclude risks of non-ratification of the Kyoto Protocol or non-certification of projects failing to meet financial/emissions additionality and sustainability criteria, price risk of the mainstream product, and price volatility of emissions credits.

Financial additionality – lower discount rates for main product, competitive prices

The Kyoto Protocol's financial additionality criteria may require CDM and JI power project sponsors to generate or supply electricity at prices below average market levels in the host country or host country location (on the basis of negative net present value of the project calculated at a low discount rate, possibly specified by UNFCCC). Charts in the next chapter on cost-benefit analysis of carbon reduction technologies demonstrate how this "sacrificial" loss of originally expected IRR of electricity revenue can produce higher net IRRs with sales of electricity and carbon credits.

Impact on IRR of global generators

There are four significant impacts on different types of generators in the electricity market, provided total risks of CDM/JI projects can be managed by financial instruments and do not exceed commercial projects by a large margin.

1. Existing generators will have to reduce electricity prices to avoid loss of market share in a country or country location to new joint-venture generators which are willing to accept lower electricity revenues because of the financial additionality criteria. Prices would be cut by existing generators to a level which new market entrants would not be able to recover the "sacrificial loss of IRR" by emission reductions sales alone. However, under favourable conditions and baselines, new entrants may realise higher revenues than existing generators in landfill and landfill/Waste-to-Energy projects.
2. New generators using the Kyoto Protocol framework will be able to increase market share at lower electricity prices at higher revenues and profits with the combined sales of electricity and emissions reductions.
3. Customers will enter new power purchase agreements with new generators with Kyoto Protocol-assisted profit enhancement strategies.
4. All generators hence must prepare plans for strategic positioning in the market where they currently or aim to operate.

More competitive prices of energy, commodities and manufactured goods

Average prices of coal, gas, raw commodities and consumer products are also expected to decrease if companies utilise the benefits of the Kyoto Protocol – provided the prices of primary energy do not escalate as a result of production controls by OPEC.

Lower electricity and energy prices will promote global industries

If CDM and JI projects attract more capital from the financial markets in preference to conventional power projects because of higher net returns, the average cost of electricity in developed and developing countries will fall. Reduced greenhouse gas emissions will assist slowing down the global warming effect, and lower electricity and process heating prices will stimulate global economic growth.

7.4 Cost or Price Revisions Pro-Rata Rate of Change of Market or Subsidised Prices

For CDM and JI projects, financial additionality needs to be re-interpreted if the market or subsidised price increases after purchase contracts have been signed. Investors stand to incur large losses compared to a higher market rate, even with proceeds from the sales of emission credits. For example, an increase of US 0.1 cent in the market due to an increase in the fuel price represents a significant financial risk to project investors if they are unable to recover higher fuel costs at fixed power prices.

In a free market economy, financial additionality requires the project company to price its electricity below the market price, but in a centrally planned or market economy in which the government provides a subsidy because the commercial cost is beyond the consumer's means, the project company may not be able to meet its investor's rate of return if it were to supply the utility below the subsidised price. In a CDM/JI project, the benefits from emission credits sales would have to be substantial to offset the loss in revenue at costs of electricity below the subsidised price.

Review of financial additionality under changing market conditions

The financial additionality criteria may have to be redefined for an environment under which the subsidised price of a mainstream product changes in response to demand, supply or other associated factors after a product purchase agreement (power purchase agreement in the case of electricity) has been concluded between parties. The term "subsidised price" has to be defined before applying cost or price revisions to the project's mainstream product due to changes in economic conditions. This can be represented as:

Affordable consumer cost (ACC) + sovereign subsidy (SS) = subsidised price (SP).

Case study: power sector

Two examples of how subsidised pricing may operate are shown in the following scenarios:

Wind energy plant, e.g. location in north-west of China
ACC = US 2 c/kWh$_e$, SS = US 2 c/kWh$_e$, SP = US 4 c/kWh$_e$

Combined cycle plant, e.g. location in east of India
ACC = US 3 c/kWh$_e$, SS = US 1 c/kWh$_e$, SP = US 4 c/kWh$_e$

It is assumed that the subsidised price equals that price at which the commercial rate is calculated in the absence of Kyoto Protocol projects (US 4 c/kWh$_e$ in both cases).

To meet the financial additionality criteria, the cost or price of electricity must be below the price at which the domestic or industrial consumer pays for it. If the project company sells electricity at US 3 c/kWh$_E$ in the China or India location, the Chinese consumer will not be able to afford this price in the absence of a subsidy of at least US 1 c/kWh$_e$. The Indian government would welcome this price as subsidies are no longer required. *However*, selling electricity at a price below the commercial rate of return to achieve a negative net present value (financial additionality criteria) incurs a financial risk which may not be offset by future project sales of emission credits. The risk-return profile analysis of CDM/JI projects is fully discussed at the end of Chapter 8.

The UN's executive board on the Clean Development Mechanism and Joint Implementation initiatives may hence have to incorporate the element of flexible price adjustments to the reference product price on the criteria of financial additionality, so that investors can reduce the degree of financial risk to their investments. The price adjustment should be made at the same rate of price escalation, and prices should be revised at regular (say, three monthly intervals) after project contracts have been concluded. Price revisions under the financial additionality criteria should be included in CDM/JI project documentation – any escalation in the project reference price is hence absorbed in pro-rata price increment, but any falls in the reference price may have to be hedged through financial instruments by project investors. The types of derivatives for hedging price risks are described in Chapter 9.

Market price fluctuations represent risks to investors – remedies

In a deregulated market or centrally planned economy – where prices may be supported by subsidies – any increases or decreases in local prices will affect the rate of return to CDM/JI project investors. In the latter case, any decrease in local prices raises financial risk to all project stakeholders unless price risk can be hedged by financial instruments.

In summary, investors may not consider CDM/JI projects in which the mainstream product price is subsidised by the government – unless the same degree of subsidy is offered so that the project price is close to that at which the commercial rate is calculated. Price risk to investors is consequently reduced, and future sales of emissions credits could then recover the difference between the project price and the commercial price. Although the removal of price subsidies has been advocated, there exists regional locations in a number of countries in which inhabitants can not meet the market cost of utilities, commodities or essentials.

7.5 Equitable Treatment

The issue of "equity" (not associated in its conventional use to describe ordinary shares nor emissions per capita) to the host country in terms of financial proceeds of projects and transactions undertaken under the Clean Development Mechanism, Joint Implementation and Emissions Trading has taken a high centre-stage profile in current negotiations on the Kyoto Protocol. India, the second largest market for carbon dioxide reductions, has constantly voiced its insistence on more equitable treatment of the share of proceeds from project investments.

This section deals with a number of relevant factors which impact on "equitable treatment" of proceeds. Such proceeds could be revenue or profit, depending at which stage proceeds are determined in the investment cycle by stakeholders. In the text of the Kyoto Protocol, stakeholders include winners as well as losers of projects stimulated by the Protocol, and countries which experience severe economic hardship due to adverse climate change.

7.5.1 Allocation of Part Proceeds to Charities and Humanitarian aid – CDM Projects

Although the Protocol stipulates the sharing of proceeds of CDM investments with the host developing country, the author is advocating part of these proceeds – which arise principally from the traded value of greenhouse gas emission contracts and (potentially) as secondary contributions in the form of surplus production of goods/commodities/utilities in the project – to be apportioned to charities so that equitable distribution to the poorest communities can be realised. The author has written to UNICEF, UNHCR and OCHA for their assistance to coordinate this effort, and to UNFCCC to include this equity allocation to charities in the Protocol. CARE, CAFOD, Islamic Relief, Red Crescent, International Tree Foundation, International Children's Trust and Just World Partners for South Pacific Islands are amongst the NGOs which have expressed support for this initiative. OXFAM International plans to hold a Board Meeting end of 2000 to establish their position on climate change issues.

In most densely populated urban communities in China, India and semi-centrally planned economies, the state or provincial price of electricity, raw materials and essential commodities are normally fixed. At end 1999, the average price at which the State Electricity Boards will purchase power at generators' busbars is about US 6 cents/kWh$_e$; in China, provincial grid authorities pay US 2–3 cents/kWh$_e$ to the plant. A power project could qualify for financial additionality if an additional quantity of electricity, say, x GWh$_e$, was supplied to the state transmission and distribution authority without any charge – this would reduce the sales price of electricity, effectively the expected revenue to the project, and result in a negative NPV of project cash flows and a lower project IRR. The "free electricity" could be sold to current customers or to the national grid, and the revenue could be donated to charities and NGOs; or supplied to poverty stricken regions which are not able to afford the state or regional price of electricity. The additional electricity could be used for desalination plants, sterilisation and detoxification of water supplies, or tertiary treatment of wastewater so that clean water can be made available. In combined heat and power projects, the additional amount of district heating from plants could also be used to heat up uninsulated dwellings during the winter. The period of supply of these extra resources may be limited to one year, if project entities were only allowed a one-off sales of greenhouse gas credits to improve on the overall IRR.

Humanitarian Aid Project 1 – Utility Sector

An example of equity distribution from a potential CDM project is provided next.

Project – A combined cycle power plant in India, using natural gas, year 2001.

Plant data:	US$ 700/kW$_e$ (capital cost); power output of 600 GWh$_e$ a year
Unit power cost:	US 6 cents/kWh$_e$ = state purchase price = revenue to project
Extra power prod.:	650 GWh$_e$ a year
	600 GWh$_e$ sold at 6 c/kWh$_e$ or less to state board.
	50 GWh$_e$, at reduced price or no charge to humanitarian agencies.

Effective sales price per unit of electricity, (CDM financial additionality)
$$= \frac{600\,\text{GWh}_e \times 6\,\text{c/kWh}_e}{650\,\text{GWh}_e} = 5.5\,\text{c/kWh}_e$$

Equitable distribution of primary or secondary resources of investment projects through the increased production of these resources to meet the CDM financial aditionality criteria can also be applied to other industries, e.g. food, paper, pharmaceuticals, construction materials. Assuming US$ y is the unit price of a product, e.g. food item, printed paper or medicine; and that the project consists of a greenhouse gas emissions plant, e.g. power or CHP, and a manufacturing process; the project's average cost of capital will require the product to be sold at a minimum unit price of US$ y (y could be a fraction of a dollar). This minimum price may have to be the average market or lowest price in the project location to meet the financial additionality criteria. The production of extra units of the food item, writing paper or medicine will increase the project's average cost of capital, resulting in a unit price greater than US$ y, as additional financing is required. However, if the extra units of goods are supplied without cost to the poorest communities in the economy through charities and NGOs, then the project meets the CDM's financial additionality criteria as its products are nominally sold at less than US$ y, based on the increased total number of units produced.

Humanitarian Aid Project 2 – Food Sector

To illustrate a potential CDM project which includes a power source and manufacturing plant, the previous example is employed.

Project – A combined cycle natural gas plant supplying electricity to a food factory making food and nutrients to infants, year 2001. All electricity produced is consumed by the factory. The project will require steam for sterilisation of ingredients, other ancillary equipment and services – the cost of these operations are not included in the analysis.

Unit power cost:	US 6 cents/kWh_e
Total power cost/year:	US 6 cents/$kWh_e \times 600\,GWh_e$ = US$ 36 m
Food production cost:	Average of US 3 cents/item
Total food prod. cost/year:	US$ 40 m. (about 1,300 million items)
Market or state-regulated price of infant foods:	Average of US 3 cents/item

Project produces 100 million additional food items, at extra cost of US 0.2 cents per item; this includes electricity and food manufacturing costs = average of US 3.2 cents/item.

If the company sells 1,300 million items at US 3 cents/item to the market, and donates 100 million items to humanitarian aid agencies, its overall revenue falls from US 3 cents/item to US 2.8 cents/item. This effective sales price meets CDM's financial additionality criteria.

Other project options

Alternatively, the project entity may sell the original output of products (without incurring the extra costs of increased production) at less than US$ y to meet financial additionality – part of the sales could be made to charities or NGOs which may have budgets to assist the most disadvantaged sections of the community.

7.5.2
The LDC/IRISK Climate Change Fund

The Least Developed Countries/Islands-at-Risk Climate Change Fund is an initiative proposed by Euro-Asia Pacific Corporation Ltd – a UK company specialising in energy technology, environmental and sustainable development – to allocate part proceeds of Clean Development Mechanism, Joint Implementation and Emissions Trading projects of the Kyoto Protocol to those countries which experience abject poverty and are disadvantaged from relatively few hydrocarbon resources (that release greenhouse gas emissions under thermal treatment). Such countries are hence heavily handicapped in terms of attracting foreign investments aimed at reducing GHG emissions under the Kyoto Protocol framework.

The 48 LDCs identified by UNCTAD are listed in the table at the end of this section in terms of GNP per capita and fossil fuel reserves. Countries which have the lowest GNP per capita and little or nil hydrocarbon reserves should merit the highest priority in receiving financial benefits from GHG projects including emissions trading. In addition, two non-LDCs – Marshall Islands and Micronesia – which have been identified by Euro-Asia Pac as amongst the most vulnerable islands-at-risk from sea level rise – should also qualify as beneficiaries.

Funding via the Kyoto Protocol

Pending agreement by Conference of the Parties (participating governments of the Protocol), part of monetary proceeds direct from the trading of each and every greenhouse gas contract between sellers and buyers (1% has been recommended by Euro-Asia Pac to UNEP) could co-finance the Fund. Trades may have to be certified on the basis of emissions and financial additionality, and environmental sustainability.

Funding via non Kyoto-Protocol initiatives

Contributions through local events and corporate marketing – e.g. emergency fund-raising collections, annual festive celebrations and shows, handycraft sales, promotion of corporate products – would assist capitalisation of the LDC/IRISK Fund, if the Protocol is not ratified. It is true to state that the North-South economic gap is widening at an unprecedented pace, regardless of the Protocol, hence such a fund would greatly increase resources to international relief agencies working especially in the African continent.

Proposed charitable status and governance of the Fund

To promote global capitalisation of the fund, it is proposed that the LDC/IRISK Climate Change Fund should operate under a non-profit, charitable organisation status – a board of trustees and an administration team would be appointed. Nominal costs incurred by trustees including trustee governments would be met by the Fund.

As a result of letters sent to industrialised and developing countries, and economies-in-transition since January 1999, with regard to Euro-Asia Pac Corp's proposals of foreign investments via the Clean Development Mechanism and Joint

Implementation initiatives of the Kyoto Protocol; voluntary national emissions permits trading schemes; an international environmental trading exchange; promotion of emissions trading conferences; the following countries have responded with interest in the proposals:

1. The government of Belgium.
2. The government of Belarus.
3. The government of Brazil.
4. The government of the Czech Republic.
5. The government of Estonia.
6. The government of India.
7. The government of Indonesia.
8. The government of Germany.
9. The government of Kazakhstan.
10. The government of Kenya.
11. The government of the Kyrgyz Republic.
12. The government of Latvia.
13. The government of Malaysia.
14. The government of Nepal.
15. The government of the Netherlands.
16. The government of Mozambique.
17. The government of Mauritania.
18. The government of the Philippines.
19. The government of the Russian Federation.
20. The government of Singapore.
21. The government of Sri Lanka.
22. The government of Switzerland.
23. The government of Thailand.
24. CARICOM, the Caribbean Community and Common Market.

These countries, upon agreement, could be appointed as government trustees for the Fund; further additions to the panel from other interested governments in the future. Professor Feng Jun of People's University of China in Beijing has expressed interest in foreign investments in the energy and environmental sectors of the PRC.

Charities and non-profit organisations

Since January 2000, communications with the following charities were made with regard to Euro-Asia Pac Corp's initiatives on increasing humanitarian aid to those developing countries that will receive less investments as a result of the existing text in Kyoto Protocol:

ACTIONAID, Action Against Hunger, Afghan Aid, AFRICA NOW, Aga Khan Foundation, CAFOD, CARE, Concern Worldwide, FARM AFRICA, International Children's Trust, International Red Cross and Red Crescent Movements, International Tree Foundation, Islamic Relief, Medical Relief Emergency Relief International, Medecins Sans Frontieres, OXFAM, Save the Children, SKILLSHARE AFRICA, SOS SAHEL, Tree Aid, WaterAid and World Wildlife Fund. This list of charities is repeated at the end of the table "LDCs and Islands-at-Risk Statistics".

As there may be a conflict of interests between the roles of charities as fund trustees and potential end recipients of fund allocations, charity bodies with general policy making or pyramidal funding objectives which themselves are not direct end-users of funds – such bodies known as "foundations" – could represent trustees and advisors of the fund. If such foundations are not able to provide professional advice, then end-user charities could be considered.

Limitations of the Fund

Due to the unpredictability of climate conditions, the *internal* allocation of the proposed 1% allocation to LDCs and most vulnerable islands-at-risk should be adjusted annually in terms of increased weighting to those economies afflicted by climatic force majuere. It is may also include provision of aid to other developing countries (e.g. India, Honduras, Kenya) which experience frequent damage from tropical storms and persistent droughts.

Non-applicability of emissions and financial additionality on LDC/IRISK projects

As will be demonstrated later, emissions additionality and financial additionality increases the cost of investments in CDM projects. However, UNFCCC may decide to make an exception on both additionality rules on CDM projects on the poorest of nations. These countries have received less than 2% of global financial flows of foreign direct investments in historical terms, and this percentage is forecast to further contract in the millennium. It is against these disparaging statistics that the author is advocating non-applicability of additionality for LDCs/IRISK countries.

Estimates of proceeds to the Fund

If global trade of contracts met 10% and 50% of Annex I country targets, respectively, the annual turnover is estimated at US$ 2.2 billion (10^9) and US$ 11 billion for a first single tranche of traded contracts at US$ 10/t CO_2 – revenues for the fund (1% turnover) would be US$ 22 million and US$ 110 million, respectively. These proceeds could be multiplied by the frequency of subsequent trading of each and every contract during the same year.

Assuming 10% of Annex I country obligations were met by CDM, and emissions contracts were traded at, say, an average of 10 contracts per day for 300 days in a year from 2001 to 2012, the value of contracts traded at US$ 10/t CO_2 in 2001 would be US$ 6,600 billion or US$ 6.6 trillion, amounting to US$ 79.2 trillion for 2001–2012. 1% of GHG contracts traded in 2001 would contribute to US$ 66 billion to the LDC/IRISK fund. These estimates exemplify the huge potential of emissions trading in assisting the poorest economics.

Potential projects that could be financed by the Fund

To maximise the utility of the fund, a separate feasibility study is being performed by Euro-Asia Pac Corp for the financing of:

1. One agricultural tractor manufacturing plant based on possible joint-cooperation between the countries of Mauritania, Guinea, Mali, Senegal, Gabon and Nigeria in West Africa, and another plant in the East/Southern African region based on possible joint-cooperation between the countries of Tanzania, Zambia, Mozambique, Madagascar, Malawi and Kenya.
2. The supply or manufacturing of plants for the desalination of sea water and land brackish waters by evaporative or reverse osmosis techniques, so that clean water can be provided to the most needy in the local communities.
3. One pharmaceutical/vaccine manufacturing plant in a West African country and another plant in the East/Southern African region – to produce medicines and drugs to combat HIV and AIDS, acute respiratory (especially pneumonia in infants and children) and diarrheal diseases, sexually transmitted diseases, tuberculosis, malaria, diptheria, measles, parasitic diseases and infections. Nigeria, Kenya, and South Africa currently have the resources to produce pharmaceuticals but not in sufficient quantities to meet very severe shortages.

For the Asian region, Euro-Asia Pac Corp has met with key representatives or received official correspondence from Bangladesh, Cambodia, Laos and Nepal. An Asian LDC panel could be formed by these countries to identify areas which are vulnerable to climate change in the Asia-Pacific region, including islands-at-risk in the Indian Ocean and the South Pacific.

The financial proceeds (1% of the value of all emissions contracts traded annually) could be divided equally amongst the 50 beneficiary countries by a democratically elected body for immediate humanitarian needs, energy and environmental sectors, climate change adaptation and a proportion, 0.14%, as recommended by the author (see section "individual country's distribution strategy of 1% contract", at end of the next table) for individual or joint-country projects to manufacture agricultural machinery/parts, pharmaceuticals and desalination plants. The 1% contribution may be negotiated to a higher percentage at the annual meetings of the Conference of the Parties.

The emergence of an "Adaptation Fund"

At the 12th and 13th Meeting of the UNFCCC Subsidiary Bodies in Bonn and Lyons in 2000, an "Adaptation Fund" was proposed to finance countries that are vulnerable to the effects of climate change. The geographical coverage of this proposed fund extends beyond the political boundaries of the nations categorised as Least-Developed-Countries by UNCTAD and small islands-at-risk from sea-level rise. Kenya, India, China, Philippines, Honduras, Venezuela and other countries which have experienced flash floods and prolonged droughts in recent years will also qualify for the Adaptation Fund. The next table illustrates the economic and energy resources of LDCs.

7.5.3
Comparison of LDC Fossil-Fuel Resources

Table 7.1. Economic and fossil-fuel resources of least-developed-countries

Country	GNP/capita US $	Hydrocarbon resources	Hydropower resources	MW_e 1998
Afghanistan	270	Gas/coal (ncp).	Yes.	494
Angola	260	Oil/gas (cp).	No.	617
Bangladesh	360	Gas/oil (cp).	Small.	3,300
Benin	380	Oil/gas (s) (cp).	Yes.	15
Bhutan	430	None.	Yes.	361
Burkina Faso*	250	None.	Yes.	78
Burundi*	140	None.	Yes.	43
Cambodia*	300	None.	Yes.	35
Cape Verde	1,090	None.	No.	7
Central African R.*	320	None.	Yes.	43
Chad*	230	Oil (exp)(ncp).	No.	29
Comoros*	400	None.	Yes.	5
Dem. Congo (Zaire)	110	Oil/gas (cp).	Yes.	3,200
Djibouti	750	Gas (exp) (ncp).	No.	85
Eq. Guinea	1,060	Gas/oil (cp).	Yes.	5
Eritrea*	230	None.	No.	Not available.
Ethiopia*	110	Gas/oil (ncp).	Yes.	464
Gambia*	340	None.	No.	29
Guinea	550	None.	No.	176
Guinea-Bissau*	230	None.	Yes.	11
Haiti*	380	None.	No.	153
Kiribati*/irisk	910	None.	No.	2
Laos*	400	None.	Yes.	256
Lesotho	680	None.	Yes.	5 (est)
Liberia*	330	None.	No.	332
Madagascar*	250	None.	Yes.	220
Malawi*	210	Some coal.	Yes.	185
Maldives*/irisk	1,180	None.	No.	14
Mali*	260	None.	Yes.	87
Mauritania	440	None.	Yes.	105
Mozambique*	140	Coal, gas (ncp).	Yes.	2,400
Myanmar	1,500	Oil/gas (cp).	Yes.	1,300
Nepal*	220	None.	Yes.	292
Niger*	200	Some coal.	No.	63
Rwanda*	210	Oil (exp) (ncp).	Yes.	34
Samoa	1,140	None.	No.	19
S Tomé/Principe*	290	None.	No.	6
Sierra Leone*	160	None.	No.	126
Solomon Islands	870	None.	No.	12
Somalia*	100	Oil/gas (exp) (ncp).	No.	70
Sudan*	290	Oil/gas (ncp).	Yes.	500
Tanzania	210	Oil/gas (exp) (cp).	Yes.	543
Togo*	340	Oil/gas (ncp).	Yes.	34
Tuvalu*/irisk	330	None.	No.	Not available.
Uganda*	330	Oil (exp).	Yes.	162

Table 7.1. *Continued*

Country	GNP/capita US $	Hydrocarbon resources	Hydropower resources	MW_e 1998
Vanuatu	1,340	None.	No.	11
Yemen	270	Oil/gas (cp).	No.	810
Zambia*	370	None.	Yes.	2,400
Non-LDC Islands-at-Risk from sea-level rise				
Marshall Is./irisk*	1,610	None.	None.	Not available.
Micronesia/irisk*	1,920	None.	None.	Not available.

7.5.4 Percentage Contribution to LDC/IRISK Countries

Distribution strategy of 1% of contract value:

Choice between "contract value" and "Certified Emissions Reduction (CER) units"

Terminologies and their differences:

The final monetary values of contributions, as calculated on "contract value" basis and "CER unit" basis may vary significantly for LDC governments and international relief agencies, hence to the ultimate beneficiaries.

Case examples to illustrate monetary differences:

- "Contract value" means the value at which Certified Emissions Reduction (CER) units, the actual commodity (greenhouse gas emissions reduction) units, are sold – this is equal to the number of CERs multiplied by the market price at which the seller is willing to sell CERs.

100 metric tonnes of carbon dioxide equivalents × US$ 20/t CO_2 equivalents
= US$ 2,000.

- Who sells the CERs?

As the plant or project reduces GHG emissions by 100 tonnes of carbon dioxide equivalent, the recipient of CERs is likely to be the joint-venture company or consortium, as the CDM requires the participation of at least one Annex I technology company.

Stakeholder A may want to sell as CERs as soon as they are realised because he may need cash to meet their own needs, while stakeholder B is prepared to wait for more favourable market prices. If the market price increases to US$ 25/t CO_2 equivalent in the following week or month, stakeholder B's revenue would be US$ 2,500.

- "CER" unit, or entitlement, is hence, the preferred choice for stakeholders, as they can hold on to these commodity assets when they are realised by the project until the market offers a more favourable price – of course, the downside could be that prices may fall.

The LDC/IRISK Climate Change Fund would be a small stakeholder of CER and ERU (Emissions Reduction Units) assets under the CDM, Joint Implementation and Emissions Trading initiatives of the Kyoto Protocol – the decision to sell these assets should be made both by the trustees and an advisory team which is able to forecast CER and ERU market prices (there may be price divergence between these markets) with reasonable accuracy or based on rational assumptions of the market and associated factors. For these reasons, the term "*value of contracts or traded contracts*" – with reference to the LDC/IRISK Fund – is that value at which the trustees are prepared to sell CERs or ERUs with consideration of advice from an advisory team, to provide maximum benefit to LDC/IRISK end beneficiaries.

The proposed 0.1% of the value of traded contracts which are UNFCCC certified or pending certification, could be distributed within recipient countries for the following objectives:

1. Proposal: 70% (effectively 0.7% of contract value)
   Humanitarian aid – e.g. water treatment and sterilisation, sanitation, medical care and training of local nurses, medicines. Financing local production of drugs or vaccines to lower the cost of imported products, improving domestic crop yields by manufacturing of more agricultural machinery, treatment of sewage to avoid cross-contamination in the food chain, desalination plant construction in countries where raw materials are available for making components, and providing medical skills and training to the local workforce to assist foreign doctors, in the following recommended ratios – 80% (effectively 0.56% of contract value) to the country's immediate humanitarian needs, of which 60% is directed to international relief charities and organisations (effectively 0.33% of contract value) and the balance 40% to the government (effectively 0.23% of contract value, for projects such as upgrading clinics and hospitals, improving sewage works, purchasing air transport carriers to gain access to remote regions during emergencies); and 20% (effectively 0.14% of contract value) to individual or joint-government projects for construction of manufacturing plants to produce medicines, pharmaceuticals, desalinated water and agricultural machinery.
2. Proposal: 20% (effectively 0.2% contract value), of which:
   Energy strategy – investing in "natural renewables" energy projects (solar, wind, wave, geothermal, hydro), say, 50% (effectively 0.1% contract value). Combustion/gasification of biomass/regrowth biomass, combustible renewable wastes (Crw); Energy from Waste "EfW" – munipical solid municipal raw sewage and industrial wastes, gasification of waste refinery oils/fossil fuels; waste heat recovery; improved electrical and electronic controls, insulation and automation, 50% (effectively 0.1% contract value).
   – The 0.2% of contract value could be allocated 50% to the foreign sector (effectively 0.1% of contract value) and 50% (effectively 0.1% of contract value) to the local enterprise/host government sector as capital to finance the projects.
   – Such projects may qualify for emissions credits if the Kyoto Protocol is ratified.
3. Proposal: 10% (effectively 0.1% contract value)
   Environmental pollution control, conservation and sustainability projects – e.g. treatment of industrial effluent wastes, recovery and disposal of landfill

leachate, afforestration and land reclamation, conservation of wildlife fauna and flora.
- The 0.1% of contract value could be allocated to the host government for purchasing or manufacturing environmental pollution control equipment and components, and implementing regional conservation and sustainability programmes.
4. Other projects beneficial to the host location or economy. For islands-at-risk, priorities of strengthening or constructing sea-wall defences, and planning evacuation/resettlement of island inhabitants may take precedence to humanitarian and energy objectives.

All allocations from the 1% contract value should be approved by an independently elected body, whose responsibility also includes the supervision of all projects, regular monitoring of performance, and evaluation of project results to UNFCCC.

Analysis of proceeds allocation

The distribution strategy of 1% of emissions contract value – adopting the author's allocation priorities – amount to about 0.57% of contract value to the host government and about 0.43% to the foreign sector, which includes international relief charities and technology companies. The proportional balance of contract proceeds in favour of the governments of LDCs and islands-at-risk should enable their wide acceptance of this international community effort.

Direct investments in LDC/IRISK countries: additional revenue to LDC/IRISK Fund

If an Annex I firm invests in energy-related projects in say, Mauritania, and if it achieves GHG reductions, then it may earn an increased return to his investment with sales of CER units under the Kyoto Protocol framework. The LDC/IRISK Fund is a stakeholder in 1% of CER units created within Tanzania, and hence would benefit from this investment, in addition to 1% of CERs and ERUs created in other host countries external to Mauritania.

7.5.5
LDCs and Islands-at-Risk Statistics

Priority sectors are identified in this section for funding by the proposed 1% of GHG contract trading under CDM and JI initiatives, and contributions from the international community under business-as-usual conditions.

Operational framework of the LDC/IRISK Climate Change Equity Fund

Under the auspices of the Kyoto Protocol, investors could benefit from the proposed 1% mandatory contribution of emissions transactions, from the year 2000 to the first commitment period 2008–2012. Each contribution would be recorded and an account of investment would be made for each project financed by the investor. The next table illustrates the vulnerability characteristics of LDCs, areas of operation by international relief agencies, and potential energy developments.

Table 7.2. Energy, humanitarian and environmental statistics of least-developed countries and islands at risk

Country	Energy Strategy	Humanitarian sector/ Locations/agencies	Environmental/other
Afghanistan	Renewables. EfW.	High infant mortality, medicare.	Industrial pollution.
		Badakshan, Badghiz, Ghazni, Hazarajat, Herat, Kandahar, Oruzgan (MSF).	
		Ghanj, Sharh-e-Buzerg (Islamic Relief).	
		Dehdorooz, Denshahr, Ghachan, Ghenew,Katik, Kharmany, Kushpak, Morkan, Neechim, Petow,Turbat, Yakshira; Badakshan, Baharak, Ghor and Nuristan regions (Afghan Aid).	
		Abazon, Chah Ab, Khoja Khair, Langer, Rustaq (Concern Worldwide).	
		Action Against Hunger.	
		Save the Children.	
		Access to safe water.	12% population.
		Average daily kilocalories/person	1,696 (UK: 3,216).
		Priority diseases/medical concerns	Respiratory, digestive.
Angola	Renewables. Wave. EfW.	High infant mortality, medicare.	Drought, landmines.
		Bibala, Caala, Camacuio, Camacupa, Chitembo, Ekunha, Huambo, Luanda, N'Dalatando (MSF).	
		Bie Province (CARE).	
		Bie, Huambo, Luanda, Malange (Concern Worldwide).	
		Action Against Hunger.	
		Save the Children.	
		Access to safe water.	32% population.
		Average daily kilocalories/person	1,756.
		Ratio of doctor to population	1/14,300 (UK:1/667).
		Priority diseases/medical concerns	Malaria, diarrhoea, respiratory.
Bangladesh	Wind. Wave. Crw gasif. Renew. EfW.	Medicare.	Floods, cyclones.
		Cox's Bazaar, Dhaka, Shamlapur, Sonarpara (MSF).	
		Derai, Hoshendi, Norshingdi, Pairabondh, Rangpur, Ranipukur – (Islamic Relief).	
		Tree planting projects (ITF).	
		Health and rural development projects	
		Aga Khan Foundation.	
		Gowainghat, Itna, Khaliajuri, Saidpur, – (Concern Worldwide).	

Table 7.2. *Continued*

Country	Energy Strategy	Humanitarian sector/ Locations/agencies	Environmental/other
Bangladesh		Save the Children.	
		CARE. WaterAid.	
		Priority diseases/medical concerns	Diarrhoea, parasitic.
Benin	Renewables. EfW.	Medicare. HIV/AIDS.	Desertification.
		Lalo (MSF).	
		Ratio of doctor to population	1/10,000.
		% of cropland irrigated	0. (UK:2).
		Priority diseases/medical concerns	Diarrhoea, malaria, AIDS.
Bhutan	Renewables. Wind. EfW.	High infant mortality.	Deforestration.
		Average daily kilocalories/person	Data not available.
		Priority diseases/medical concerns	Tuberculosis, malaria, diarrhoea.
Burkina Faso	Renewables. EfW.	Medicare.	Desertification.
		Ouagadougou, Soum (MSF).	
		Barsalogho Tree Project, Loroum Tree Project, Signoughin Village (Tree Aid).	
		Save the Children.	
		WaterAid.	
		Ratio of doctor to population	1/57,300.
		Priority diseases/medical concerns	Malaria, diarrhoea.
		% of cropland irrigated	1.
Burundi	Renewables. EfW.	Medicare. Lack of clean water.	Deforestration.
		Bujumbura, Cankuzo, Karusi, Kayanza, Kinyinya, Makamba, Matongo, Murore, Ruyigi (MSF).	
		Bururi, Cibitoke (Concern Worldwide).	
		Action Against Hunger. CARE.	
		Access to safe water.	32% population.
		Average daily kilocalories/person	1,756.
		% of cropland irrigated	1.
		Priority diseases/medical concerns	Parasitic, infections.
Cambodia	Renewables. EfW.	Medicare. HIV/AIDS.	Deforestration.
		Anlong Veng, Phnom Penh (MSF).	
		Banteay Meachey, Chhnang, Pursat, Siem Reap, Speu (Concern Worldwide).	
		Action Against Hunger.	
		Save the Children. CARE.	
		Access to safe water	36% population.
		Average daily kilocalories/person	1,805.
		Priority diseases/medical concerns	Circulatory, AIDS.
Cape Verde	Wind. Wave. Renew. EfW.	Medicare.	Drought.
		Access to safe water.	Data not available.
		Priority diseases/medical concerns	Circulatory, tuberculosis.

Table 7.2. *Continued*

Country	Energy Strategy	Humanitarian sector/ Locations/agencies	Environmental/other
Central African R.	Renewables. EfW.	Medicare.	Deforestration.
		Bangui (MSF).	
		Access to safe water	18% population.
		Average daily kilocalories/person	1,959.
		% of cropland irrigated	0.
		Priority diseases/medical concerns	Parasitic, infections.
Chad	Renewables. Geothermal.	Medicare.	Desertification.
		East Logone, Koumra, Mayo Kebbi (MSF).	
		Action Against Hunger. CARE.	
		Access to safe water.	24%, 38% pop. (2 info. sources).
		Average daily kilocalories/person	1,840.
		% of cropland irrigated	0.
		Ratio of doctor to population	1/16,667.
		Priority diseases/medical concerns	Diarrhoea, parasitic.
Comoros	Wind. Wave. Geothermal. Renew. EfW.	Medicare.	Land conservation.
		CARE.	
		Access to safe water.	Data not available.
		Ratio of doctor to population	1/10,000.
		Priority diseases/medical concerns	Malaria, infections.
Dem. R. Congo (Zaire)	Crw gasif. Geothermal. Wave. EfW. Renewables.	Medicare. HIV/AIDS.	Deforestration.
		Bas Congo, Equateur, Kinshasha, Kimpese, Kisangani, Lubumbashi, Nsona-Mpangu (MSF).	
		Maniema, South Kivu (MERLIN).	
		North Kivu, Uvira (Concern Worldwide).	
		World Wildlife Fund.	
		Save the Children.	
		Access to safe water.	27%, 42% pop.(2 info. sources).
		% of cropland irrigated	0.
		Ratio of doctor to population	1/10,000.
		Priority diseases/medical concerns	Malaria, respiratory, AIDS.
Djibouti	Renewables. Geothermal.	Medicare. HIV/AIDS.	Waste accumulation.
		Access to safe water.	Data not available.
		Priority diseases/medical concerns	Circulatory, respiratory, AIDS.
Eq. Guinea	Renewables. EfW.	Medicare.	Deforestration.
		Bata (MSF).	
		Access to safe water.	Data not available.
		Average daily kilocalories/person	Data not available.
		% of cropland irrigated	0.
		Priority diseases/medical concerns	Diarrhoea, respiratory, malaria.

Table 7.2. *Continued*

Country	Energy Strategy	Humanitarian sector/ Locations/agencies	Environmental/other
Eritrea	Geothermal. Renewables. EfW.	Medicare.	Deforestation.
		Riverine Forest Project in Gash Barka, West Lowlands (SOS SAHEL).	
		WaterAid.	
		Access to safe water.	Data not available.
		Average daily kilocalories/person	1,608.
		Priority diseases/medical concerns	Malaria.
Ethiopia	Crw gasif. Geothermal. EfW. Renew.	Medicare.	Deforestation.
		Adigrat, Jijiga, Mekelle, Salamago, Tigray, Wag-Hamra, Woldya (MSF).	
		Tofik (ACTIONAID).	
		Meket Food Project, Koisha Soil Fertility Project and Borana Forest Project (SOS SAHEL).	
		Dairy Goat Project, Conservation Project (Bonga, Chilimo), Rural Dewvelopment Project (Tigray), and Capacity-building Project at Konso (FARM AFRICA).	
		Addis Ababa, Adwa, Gondar, Hitosa, Robe-Maliyu, Tereta (WaterAid).	
		Debre Birhan Tree Nursery Project, Shoa Robit Women's Reforestration Project, Tree Aid.	
		Tree planting projects (ITF).	
		Addis Ababa, Damot Weyde, Kalu Woreda, – (Concern Worldwide).	
		Action Against Hunger.	
		Save the Children. CARE.	
		Access to safe water.	25% population.
		Average daily kilocalories/person	1,661.
		% of cropland irrigated	1.
		Ratio of doctor to population	1/25,000.
		Priority diseases/medical concerns	Diarrhoea, respiratory, malaria, tuberculosis, infections.
Gambia	Wind. Wave. Geothermal. Renew. EfW.	Medicare.	Land conservation.
		Microfinance for Women (AFRICA NOW).	
		% of cropland irrigated	1.
		Priority diseases/medical concerns	Tuberculosis, malaria.

Table 7.2. *Continued*

Country	Energy Strategy	Humanitarian sector/ Locations/agencies	Environmental/other
Guinea	Wind. EfW. Geothermal. Renewables.	5th highest infant mortal. 120/1000. Average life span is 47 years. Forecariah, Guinee Forestiere, Labe, Mamou (MSF). Action Against Hunger. Priority diseases/medical concerns	Deforestration. Malaria, tuberculosis, diarrhoea.
Guinea-Bissau	Wind. EfW. Renewables.	3rd highest infant mortality 130/1000. Average life span is 45 years. Bissau (MSF). Priority diseases/medical concerns	Drought. Locust plagues. Malaria, diarrhoea, parasitic.
Haiti	Crw gas. EfW. Wind. Wave. Geothermal. Renewables.	Medicare. Bois Neuf, Cite Simon, Jeremie, Port-au-Prince, Saint Mare, Sapotille (MSF). Cite Eternel, Gonave Island, St Martin, – (Concern Worldwide). Action Against Hunger. CARE. Access to safe water. Average daily kilocalories/person Ratio of doctor to population Priority diseases/medical concerns	Deforestration. 28% population. 1,721. 1/10,000. Malaria, tuberculosis, parasitic.
Kiribati	Wind. EfW. Wave. Geotherm. Renewables.	Nutritional shortages. UK Foundation for South Pacific, – (Just World Partners). Priority diseases/medical concerns	Coral reef destruction. Drought (El Niño effect). Sea level rise. Circulatory, diabetes.
Laos	Renewables. EfW.	Medicare. Sanitation. Bokeo, Champassak, Dakchung, Kalum, Sekong (MSF). Bokeo, Khammoune, Savannakhet, Thakhek, – (Concern Worldwide). Action Against Hunger. Save the Children. CARE. Priority diseases/medical concerns	Deforestration. Diarrhoea, malaria, parasitic.
Lesotho	Renewables. EfW.	Medicare. CARE. SKILLSHARE AFRICA. % of cropland irrigated Ratio of doctor to population Priority diseases/medical concerns	Deforestration. Drought. Within El Niño outreach. 1. 1/20,000. Tuberculosis, parasitic.

Table 7.2. *Continued*

Country	Energy Strategy	Humanitarian sector/ Locations/agencies	Environmental/other
Liberia	Renewables. EfW.	High infant mortality. Average life span is 47 years. Bong, Kolahun, Margibi, Monrovia, Tubmanberg (MSF). Nimba, Tappeta (MERLIN). Bomi County (Concern Worldwide). Action Against Hunger. Save the Children.	Deforestration.
		Average daily kilocalories/person	1,728.
		% of cropland irrigated	1.
		Priority diseases/medical concerns	Circulatory, diarrhoea, malaria.
Madagascar	Wind. Wave. Renew. EfW.	Malaria, bubonic plague, HIV/ AIDS.	Deforestration.
		Ambovombe, Betioky, Morondava (MSF). Tree planting projects (ITF). World Wildlife Fund. CARE. WaterAid.	El Niño zone.
		Access to safe water.	29% population.
		Ratio of doctor to population	1/10,000.
		Priority diseases/medical concerns	Malaria, digestive.
Malawi	EfW. Coal gas. Renewables.	2nd highest infant mortal. 135/ 1000. Average life span is 39 years. Blantyre, Chiradzulu, Mwanza, Thyolo (MSF). CARE. WaterAid.	Drought. Deforestration.
		Access to safe water.	37% population.
		Average daily kilocalories/person	1,911.
		Ratio of doctor to population	1/50,000.
		Priority diseases/medical concerns	Respiratory, parasitic.
Maldives	EfW (sewage, MSW, coconut husk/shell).	Medicare.	Average height of islands is only 5 feet. Sea-level rise.
	Renewables.	Priority diseases/medical concerns	Tuberculosis, parasitic.
Mali	Renewables. Geotherm. EfW.	High infant mortality. Ansongo, Bamako, Kayes, Koulikoro, Yanfolilia-Selingue (MSF). Ghourma Rharous, Segou, Timbuktu, – (Islamic Relief).	Drought. Deforestration.

Table 7.2. *Continued*

Country	Energy Strategy	Humanitarian sector/ Locations/agencies	Environmental/other
Mali		Bankass Food Project and Tominian Environmental Project (SOS SAHEL). Joliba Trust Project; Reforestration and Forestration projects at Banamba, Falembougou, Mena, Saboussire, Samory, Segou and Tominian (Tree Aid).	
		Action Against Hunger.	
		Save the Children.	
		CARE. WaterAid.	
		Ratio of doctor to population	1/10,000.
		Priority diseases/medical concerns	Malaria, respiratory, diarrhoea.
Marshall Is.	Renewables. EfW.	Nutritional shortages.	Sea level rise.
		Priority diseases/medical concerns	Circulatory, diarrhoea.
		UK Foundation for South Pacific – (Just World Partners).	
Mauritania	Wave. Wind. Geotherm. Renew. EfW.	Medicare.	Desertification.
		Tagant, Toujounine (MSF).	
		Access to safe water.	<40% pop (est).
		Ratio of doctor to population	1/10,000.
		Priority diseases/medical concerns	Diarrhoea, tuberculosis.
Micronesia	Renewables. EfW.	Adequate healthcare.	Drought. Sea level rise.
		Priority diseases/medical concerns	Circulatory, digestive.
		UK Foundation for South Pacific – (Just World Partners).	
Mozambique	Wind. Geotherm. Renew. EfW. Coal gasif.	Medicare. HIV/AIDS.	El Niño victim, 2000.
		Mandimba and N'gauma (Niassa), Muturara in Tete (MSF).	
		Buzi, Machenga (Action Against Hunger).	
		Chimoio, Gondola, Majune, Nampula, – (Concern Worldwide).	
		SKILLSHARE AFRICA.	
		Save the Children. ACTIONAID. WaterAid.	
		Access to safe water.	33% population.
		Average daily kilocalories/person	1,685.
		Ratio of doctor to population	1/50,000.
		Priority diseases/medical concerns	Tuberculosis, digestive, AIDS.
Myanmar	Crw gasif. Geothermal. Renew. EfW.	Medicare. HIV/AIDS.	Deforestation.
		Action Against Hunger.	
		Save the Children. CARE.	
		Ratio of doctor to population	1/10,000.
		Priority diseases/medical concerns	Malaria, circulatory, AIDS.

Table 7.2. *Continued*

Country	Energy Strategy	Humanitarian sector/ Locations/agencies	Environmental/other
Nepal	Crw gasif. Wind. EfW. Renewables.	Medicare. Save the Children. Care. WaterAid. Britain-Nepal Medical Trust. World Wildlife Fund.	Deforestation.
		Average daily kilocalories/person	2,126.
		Priority diseases/medical concerns	Respiratory, diarrhoea.
Niger	Coal gasif. Geothermal. Renew. EfW.	High infant mortality.	Drought. Deforestation.
		CARE. Takieta Forest Reserve, Bermo Pastoral Project and Zinder Plantation Project (SOS SAHEL). World Wildlife Fund.	
		Ratio of doctor to population	1/33,333.
		Priority diseases/medical concerns	Malaria, tuberculosis.
Rwanda	Geothermal. Renew. EfW	4th highest infant mortality 124/1000.	Deforestation.
		Average life span is 41 years. Bushenge, Butare, Cyangugu, Janja, Mataba, Nyagatare, Ruhengeri (MSF). Butare, Cyangugu, Gitarama, Kibilizi, Ruhengeri (Concern Worldwide). Save the Children. CARE.	
		Access to safe water.	Data not available.
		Average daily kilocalories/person	1,861.
		% of cropland irrigated	0.
		Ratio of doctor to population	1/40,600; but adequate network.
		Priority diseases/medical concerns	Malaria, diarrhoea.
Samoa	Renewables. EfW.	Adequate healthcare.	Deforestation.
		Priority diseases/medical concerns	Circulatory.
Sao Tomé & Principe	Renewables. EfW.	Adequate healthcare.	Deforestation.
		Access to safe water.	Data not available.
		Priority diseases/medical concerns	Malaria, diarrhoea, respiratory.
Sierra Leone	Renewables. EfW.	*Highest infant mortality 170/1000.*	Derelict farmlands.
		Average lifespan is 37 years.	Land mines.
		Bo, Bonthe, Freetown, Kambia, Moyamba, Portloko, Pujehun (MSF). Kenema (MERLIN).	

Table 7.2. *Continued*

Country	Energy Strategy	Humanitarian sector/ Locations/agencies	Environmental/other
Sierra Leone		Freetown & Waterloo, Northern Province – (Concern Worldwide).	
		Action Against Hunger.	
		Save the Children. CARE.	
		Access to safe water.	34% population.
		Average daily kilocalories/person	1,888.
		Ratio of doctor to population	1/14,300.
		Priority diseases/medical concerns	Malaria, infections.
Solomon Islands	Renewables.	Adequate healthcare.	Deforestration.
	EfW.	Access to safe water.	Data not available.
		Average daily kilocalories/person	Data not available.
		% of cropland irrigated	0.
		Priority diseases/medical concerns	Adequate control.
Somalia	Wind.	Medicare. Nutritional shortages.	Drought.
	Geotherm.	Average lifespan is 47 years.	
	Renew.	Aden Yabal, Gaalkayo, Giohar Mahaday, Kismayo (MSF).	
	EfW.	Afgoye (Islamic Relief).	
		Lower Shabelle (Concern Worldwide).	
		Action Against Hunger.	
		Save the Children. CARE.	
		Access to safe water.	31% population.
		Average daily kilocalories/person	1,533 (*lowest in LDC group*).
		Ratio of doctor to population	1/14,300.
		Priority diseases/medical concerns	Diarrhoea, parasitic.
Sudan	Crw gasif.	Medicare.	Drought.
	Geothermal.	Average lifespan is 28 years.	Deforestration.
	Renew.	Bahr El Ghazal, Kala Azar, Kordofan, Jonglei region, Lakes region, Leer, Meiram (MSF).	
	EfW.	Tambura, South Sudan (MERLIN/ CARE).	Sleeping sickness.
		El Obeid, Khartoum, Um Dam and Muzdalifa in North Kordofan, – (Islamic Relief).	
		Wau (CARE, CAFOD).	
		N. Kordofan (CARE).	
		Khor Arba'at Rehabilitation Project in Port Sudan; Al Ain, Ed Dammer and El Ain Forestry Projects (SOS SAHEL).	
		El Ain Forestry Project (Tree Aid).	
		Aweil West, Gogrial, Nuba, Yei, Yiroi, – (Concern Worldwide)	
		Action Against Hunger.	
		Ratio of doctor to population	1/10,000.
		Priority diseases/medical concerns	Parasitic, infections.

Table 7.2. *Continued*

Country	Energy Strategy	Humanitarian sector/ Locations/agencies	Environmental/other
Tanzania	Geothermal. Crw gasif. Renew. EfW.	Medicare. HIV/AIDS.	Deforestration.
		Kasulu, Kigoma, Lindi, Mtawara, Nshamba (MSF).	
		Dairy and edible oil processing, - (AFRICA NOW).	
		Dairy Goat Programme, Babati Agro Project, Nou Forestry Project, - (FARM AFRICA).	
		Dodoma, Kalola, Kiteto Mongorama, - (WaterAid).	
		Babati Escarpment Reforestration and Project, Karagwe Tree Planting Project, - (Tree Aid).	
		Health and rural development projects, - (Aga Khan Foundation).	
		Iringa, Ismani, Karago, Kibondo, Kigoma, Malangani, Masasi, Mtwara, - (Concern Worldwide).	
		Action Against Hunger.	
		Save the Children. CARE.	
		Ratio of doctor to population	1/25,000.
		Priority diseases/medical concerns	Diarrhoea, malaria, AIDS.
Togo	Renewables. EfW.	Medicare.	Desertification.
		CARE.	
		% of cropland irrigated	0.
		Ratio of doctor to population	1/10,000.
		Priority diseases/medical concerns	Malaria, diarrhoea, parasitic.
Tuvalu	Renewables. EfW.	Medicare.	Sea-level rise.
		Priority diseases/medical concerns	Malaria, diarrhoea.
		UK Foundation for South Pacific - (Just World Partners).	
Uganda	Renewables. Geotherm. EfW.	Medicare. HIV/AIDS.	Land conservation.
		Adjumani, Bundibugyo, Omugo, Moroto, - (MSF). Dairy Goat Project (FARM AFRICA). Kahangi, Kitimba, Nyamihoti, Rwakabengo, - (WaterAid).	
		Kampala, Opportunity International.	
		Tree planting projects (ITF).	
		Health and rural development projects, - (Aga Khan Foundation).	

Table 7.2. *Continued*

Country	Energy Strategy	Humanitarian sector/ Locations/agencies	Environmental/other
Uganda		Kampala, Mbarara, Mpigi, Rakai, - (Concern Worldwide). Action Against Hunger. Savethe Children. CARE.	
		Access to safe water.	34%, 38% pop. (2 info. sources).
		% of cropland irrigated	0.
		Ratio of doctor to population	1/25,000.
		Priority diseases/medical concerns	Malaria, respiratory, AIDS.
Vanuatu	Renewables. Wind. EfW.	Lack of clean water.	Land conservation.
		Ratio of doctor to population	1/14,100, but adequate network.
		Priority diseases/medical concerns	Circulatory, cancers, malaria.
Yemen	Wind. Geotherm. Renew. EfW.	Adequate healthcare. Aden, Az Hazm, Dhi Sufal (MSF). Al Hodeidah (Islamic Relief). CARE.	Wildlife conservation.
		Ratio of doctor to population	1/10,000, but adequate network.
		Priority diseases/medical concerns	Diarrhoea, tuberculosis, malaria.
Zambia	Crw gasif. Renewables.	Medicare. HIV/AIDS. Kaputa, Mwange (MSF). Lusaka (CARE). Luansobe Primary School, -(International Children's Trust). Chula, Monze (WaterAid). Kitwe (Opportunity International).	Wildlife conservation.
		Access to safe water.	27% population.
		Average daily kilocalories/person	1,954.
		% of cropland irrigated	1.
		Ratio of doctor to population	1/10,000.
		Priority diseases/medical concerns	Respiratory, malaria, AIDS.

Information sources

International relief charities working in the humanitarian sector, NGOs and multilateral agencies.

Percentage of population's access to safe water: Economic and Social Surveys of Africa 1995–96, United Nations Economic Commission for Africa; UNDP; WaterAid; World Resources Institute 1998–1999.

General statistics: Survey of Energy Resources 1998, World Energy Council; WHO; UNDP; World Bank Atlas.

ACTIONAID

Table 7.2. *Continued*

Action Against Hunger
Afghan Aid
AFRICA NOW
Aga Khan Foundation
Britain-Nepal Medical Trust
CAFOD
CARE
FARM AFRICA
International Children's Trust
Islamic Relief
ITF International Tree Foundation
MERLIN Medical Emergency Relief International
MSF Medecins Sans Frontiers
Opportunity International
OXFAM
SKILLSHARE AFRICA
SOS SAHEL
STC Save the Children
Tree Aid
UK Foundation for South Pacific – Just World Partners (renamed)
VSO Volunteer Service Organisation
WaterAid
World Wildlife Fund

IRCRC, OXFAM, VSO – The International Red Cross and Red Crescent Movements, OXFAM and VSO are present in most of the countries above.

If the Kyoto Protocol is ratified

Corporates or private individuals could receive beneficial tax advantages by deducting the 1% mandatory contribution from the amount of tax owed to the tax authorities in the Annex I country's financial year. Support by Annex I countries for this scheme would be vital for the fund's operation, as CDM projects require the financial participation of Annex I corporates.

If the Kyoto Protocol is not ratified

The Fund would simply end up as a theoretical initiative of global cooperation if the Kyoto Protocol is not ratified by Conference of the Parties. As a hedge against non-ratification of the Protocol, the author proposes other means of channelling financial flows into the fifty countries identified.

A double tax advantage to corporate or individual contributions to the Fund – providers of capital in excess of US$ 1,000 (or a higher minimum amount to be decided by investor and recipient countries to adjust for administrative costs):

1. A first tax advantage – by deducting contributed amounts from the investor's imputed tax due to the appropriate taxation authority in the financial year.

2. A second tax advantage – contributions invested in the Fund in humanitarian and energy projects in the ratio 70% to 30% (to promote investments in water, medical and infrastructure – building sectors) respectively, may provide profits to investors. Any profits resulting from projects – some of the humanitarian and energy projects may be financed by soft loans and equity to encourage enterprise productivity and entrepreneurship thinking (hence dispel notion of "easy money") and introduce the concept of market dynamics into the host economy – could be shared between the project recipient and investor. The additional tax advantage could be realised by the investor if zero tax is applied on dividends and interest, subject to double tax agreements between the tax authorities of the investor's country and the project host economy for the purpose of this fund.

Role of non-Annex I capital and tax implications

Some of the investors in projects described in the scenarios described above will originate from non-Annex I countries (e.g. local investment funds, private investors, project bonds) but their capital participation is unlikely to match the amounts funded by Annex I investors. Potential tax advantages for non-Annex I investors in Kyoto Protocol or non-Protocol projects – deductions of contributions from investors' imputed tax; zero or reduced tax rates on profits on dividends and interest – are corporate benefits non-Annex I governments could consider.

7.5.6 Benefits Comparison Between Host Country, Annex I Investors and Traders

Can CDM host country increase benefits beyond "maximum threshold" CER receipts?

The amount of carbon dioxide equivalent units that can be reduced per unit capital investment in a CDM project is limited in terms of modern technology used – according to the author's calculations based on a number of potential projects. This means that for every dollar invested in an energy project in a non-Annex I country, the amount of carbon dioxide equivalents that can be reduced against an existing baseline – essentially, emission credits that can be traded for cash – can not exceed a theoretical maximum threshold.

For example, the amount of carbon dioxide equivalent units of emissions reduction that can be achieved in fossil-fuel plants in Asia, e.g. CCGT, varies between 0.05–0.07 metric tonnes of CO_2 equivalent per US dollar invested (capital investment in plant). For wind farm plants, the figure is slightly lower, 0.04–0.06 metric tonnes of carbon dioxide equivalent per dollar. Using the "once-only counting" reduction approach (as favoured against multiple counting of emissions reductions, explained earlier in this chapter), investors hence can only receive a finite quantity of emissions reductions or CERs, certified emissions reductions.

The technical limitation of emissions reduction credits from CDM projects to host countries reduces the upward potential for host developing countries and possibly, restricts the growth of the LDC/IRISK Climate Change Fund on a first tranche of mandatory contributions.

In macroeconomic terms for a country like India – which is likely to be a "one off" trader of emissions credits, and not an OTC or exchange trader – its benefit may be limited to about US$ 1 to 2 billion from the CDM market until the end of the first commitment period, 2012. The following case example shows how this figure is estimated:

CDM Case Study: Maximum threshold of CER entitlements for India

Assumptions: global quantity of about 2,200 million tonnes of CO_2 equivalents to be reduced. 1 CER = 1 CO_2 equivalent.
Percentage captive market by India, of 50% = 1,100 million tonnes
Percentage CER entitlement from projects by government, of 10% = 110 million tonnes
Value of CER entitlements, at US$ 10 and US$ 20/tonne = US$ 1.1 and 2.2 billion.

Comparison of cost savings by Annex I corporate investor, NWE vs Asia

The cost savings of an Annex I investor, say, from North-West Europe (average baseline of about 400 metric tonnes of carbon dioxide equivalent/GWh_e in the power generation sector – estimated by the author for the year 2000 based on 100, 600 and 440 t CO_2/GWh_e for France, Germany and UK, respectively – would be the difference between the average cost of carbon dioxide reduction (CO_2 used as reference gas for power sector) in North-West Europe using 400 t CO_2/GWh_e as baseline, and the average cost in the Asian region using conventional coal plant emissions of 964 t CO_2/GWh_e (normal value adopted by several scientific institutions). A factor of 2, instead of 2.4, could be assumed for comparing CO_2 reduction costs between NWE and Asia, excluding China where the cost differential would be much higher.

CDM case study on India: Cost savings by NWE power project investors in India, and commissions to emissions traders on Indian project-originated CERs, would far outweigh a maximum benefits threshold to the Indian government – is there a cost opportunity India can impose on NWE investors and emissions traders to minimise "North-South exploitation"?

NWE benefits vs India benefits – case study on power sector:

Assumption: North-West Europe (France, Germany and UK) has to reduce 550 million tonnes of CO_2 emissions to meet Kyoto Protocol targets.

Average cost of investment faced by NWE investors within NWE (fossil-fuel and renewable energy technologies; based on author's calculations and assumptions of 70/30, respectively), of US$ 100/t CO_2.

Average cost of investment in India (fossil-fuel and renewable energy, 70/30 ratio) of about US$ 25/t CO_2.

Cost savings to NWE investors by investing in India = US$ 75/t CO_2.

Total cost savings = 550 million × 75 = US$ 41,250 million.

Benefit to India if 10% of CERs is negotiated as share of project CERs = 55 million t CO_2.

Value of CER entitlements to India, at US$ 10/t and US$ 20/t = US$ 550 million and US$ 1,100 million.
Ratio of NWE investors' cost savings to CER benefit to India
at CO_2 market price of US$ 10/t = $ 41,250 m/$ 550 m = 75 times.
at CO_2 market price of US$ 20/t = $ 41,250 m/$ 1,100 m = 37.5 times.

The results suggest that NWE investors can attain their Kyoto Protocol targets and conduct, perhaps, profitable business in the power generation sector in India at a price advantage of some 30 to 70 times compared to costs at home. This could invoke further implications of "North-South" inequity by India, if the author's calculations and comparisons were correct.

In such a scenario, a possible "balancing" act to address equitable treatment for India could be 50% sharing of NWE's cost savings of US$ 41,250 million in the ratio of, say,

- 30% of NWE's cost savings to the Indian government – of which 10% could be directed at the poverty and humanitarian sectors via international relief agencies working in India;
- 10% of NWE's cost savings to assist the development of renewable energy sector and to enable SMEs to compete for foreign investments with large enterprise projects;
- 10% of NWE's cost savings to the LDC/IRISK Climate Change Fund/Adaptation Fund, for its worldwide beneficiaries including India.

Practicality of 50% sharing scheme? Possibly very remote, as the scheme would require NWE investors to part with some US$ 20 billion in addition to their project investments.

Emissions traders' benefits vs. India's benefits

In the section on international emissions trading in Chapter 9, the author provides a range of forecast revenues to global emissions traders in terms of commissions. It is possible to predict the gains to Annex I countries with the following calculations.

Using the previous NWE investors case example,

550 million tonnes of CO_2 emissions represents about 25% of the global target of emissions reductions of 2000–2500 million tonnes.

At a market price of US$ 10/t CO_2, commissions at 2% per trade to emissions traders 10 times a day for 300 days/year would amount to

550 million × US$ 10 × 0.02 × 10 × 300 = US$ 330,000 million annually (see * below)

This total commissions figure for emissions traders, only for one year's trading, is 8 times the NWE investor group's one-off cost savings of US$ 41,250 million!

Assuming (for a simulated three-country NWE corporate investment in India, which does not reflect global CDM financial flows):

1. *This trading volume is achieved in 2005, commissions over the period over next 5 years, from 2006 to 2010 (notional average year in the first commitment period,

2008–2012), would multiply fivefold based this trading frequency to US$ 1,650 billion.
2. Half of this trading volume is achieved in 2002, and this reduced trading frequency is maintained from 2003–2010, commissions would multiply ninefold to US$ 1,485 billion.

Ratio of commissions to emissions traders on secondary trading of CERs originating from India, to CER benefit to India, on the basis of

1. 2006–2010 period, US$ 1,650 billion/US$ 550 million (at US$ 10/t CO_2) = 3,000 times.
2. 2003–2010 period, US$ 1,485 billion/US$ 550 million (at US$ 10/t CO_2) = 2,700 times.

If emissions traders had to allocate 1% of the value of every contract traded to the Indian government, on CERs originating from India, this would not affect their trading commissions as they would simply extract 1% from the value of the traded contract in addition to their commission. The task would require additional work for traders but should not entail any marked increase in commissions in a mature industry, if mandatory contributions formed part of the Emissions Trading initiative under the Protocol.

The market price of carbon credits is crucial for project sponsors and investors – the technical cost of carbon dioxide reduction in "North" countries is expected to exceed the average market price in global markets – however, if surplus carbon credits are generated from the project (e.g. Waste-to-Energy and landfill projects), then these projects would be favoured. Potentially higher IRR in low cost countries, compared to Annex I countries, would offset differences in technical abatement costs and market prices of carbon.

Practical solutions to equitable sharing of proceeds with non-Annex I countries?

The preceding case examples clearly demonstrates the inequity of benefits between Annex I and non-Annex I countries in terms of magnitudes of cost advantage of some 20 to 70 times to an Annex I (North-West European) investor grouping in the power sector in India, and commissions benefits of 2,500 to 3,000 times to emissions traders from trading CERs that are "Made-in-India" – one possible compromise the Conference of the Parties may consider is the mandatory contribution of 1% or more of the value of every CER contract traded to India, the country of CER origin.

Mandatory contributions of contract values to India – how they stack up with NWE investor group's cost savings in the power sector case study on India, and commissions from trading these power-sector CERs originating from India:

1. If 1% of every CER contract – originating from India – is traded at US$ 10/t CO_2, on the basis of US$ 165,000 million revenue per year [refer to case conditon (1) in the preceding section)] from 2006–2010 = US$ 825,000 million = mandatory contribution to India.
2. 1% of every CER contract – originating from India – is traded at US$ 10/t CO_2, on the basis of US$ 82,500 million (basis of halved trading frequency) revenue per

year [refer to case conditon (2) in the preceding section] from 2003–2010 = US$ 742,500 million = mandatory contribution to India.

This exceeds the NWE investor group's one-off cost savings of US$ 41,250 million by about 20 times, but trails emissions traders' commissions of US$ 1,485–1,650 billion by half.

Annex I investor countries, however, may demand equitable treatment if emissions traders' commissions and non-Annex I countries' revenues (based on 1% contribition of the value of every contract traded) outweigh their one-off cost savings – they may hence insist on the same percentage contribution of 1% mandatory contribution as non-Annex I countries.

Emissions traders will be the biggest winners of the Kyoto Protocol, unless . . .

The author's analysis, based on a set of case studies on the relative benefits to CDM host countries, suggests that emissions traders will realise the highest financial gains from the Kyoto Protocol, and not CDM countries nor Annex I investors. The same results could be derived for higher GNP and lower GNP Annex I countries.

The Conference of the Parties may consider a mandatory 1% contribution of contract value – 1% each to non-Annex I countries and Annex I countries, and perhaps, leave market forces to determine the commissions rate to emissions traders. Non-Annex I countries and lower GNP Annex I countries would argue that without their lower costs of labour and materials, emissions trading would not occur – hence the proverbial "goose that lays the golden egg" – the eggs being CERs and (lower GNP host country originated) ERUs – takes precedent.

COP may hence decide mandatory contributions on a sliding scale to host economies which offer low costs for GHG reduction, i.e. CDM countries, followed by low GNP Annex I and medium GNP Annex I countries, before considering high GNP Annex I investors. Possible percentages could be 1% for CDM countries, 0.5% and 0.4% for low and medium GNP Annex I countries, and 0.3% for high GNP Annex I investors. Non-Annex I negotiators may ask for higher percentages – based on a multiple of three, the percentage contributions may escalate to 3%, 1.5%, 1.2% and 0.9%, respectively.

To encourage ratification of the Protocol by non-Annex I countries, the author recommends a mandatory percentage contribution from every CER contract traded to the CER country of origin – this surcharge will dilute the high ratios of benefits to "North" investors compared to "South" countries. The percentage for this contribution, and any other contributions to low GNP Annex I host countries and Annex I investors – could be decided by Conference of the Parties. A one percent contribution is sought for the LDC/IRISK Climate Change or precursor Adaptation Fund.

A go-stop-go strategy on setting percentages for equitable share of proceeds

If non-Annex I countries oppose any sharing of CER contract traded values with Annex I countries, then one possible solution COP could consider is application of an experimental 1% percentage contribution each to the CER country of origin and the LDC/IRISK Climate Change Fund for a trial period of 2 January–30 June 2001, or any subsequent six-month period – after which percentages could be renegotiated for all parties.

7.5.7 Equitable Balancing to Stimulate Investments in SMEs

Case A – Relaxing the financial additionality requirement

The first tranches of foreign investments in CDM and JI projects are likely to be directed at the largest enterprises in the host country. Small and Medium Size enterprises (SMEs) hence suffer from a distinct disadvantage in attracting foreign investments because their average costs of production are much higher than those of larger enterprises.

A possible solution to address this equity concern of "favouring the largest companies" may be the waivering of the financial additionality requirement for foreign investments in SMEs. Project investments in SMEs could be considered as additional in developing countries if, in the absence of CDM (business-as-usual scenario), no foreign investor is willing to put any money in these companies. The criteria of employee numbers or asset values of SMEs vs. big enterprises may not represent a practical basis of applying a "financial additionality block exemption" as some SMEs may be highly profitable organisations given their flexibility and operational structures. However, classifying SMEs and large enterprises on the basis on their average commercial production costs may offer an equitable solution to address the imbalance of investment attractiveness between the two groupings. An example of how a "financial additionality block exemption" could be applied to a SME is provided next.

Country: A developing country, 1999 fiscal year

Average costs of electricity of large company-owned generator and SME generator, calculated at 10% discount rate, are 3 c/kWh$_e$ and 4 c/kWh$_e$, respectively. The capacity of the SME generator is less than that of the large company generator.

If the "business-as-usual" state or market price of electricity was, say, 4.5 c/kWh$_e$, then the IRR for the large generator would be much higher than the SME generator. With the CDM, the IRR for the large generator would still surpass the SME generator's if GHG reductions could be sold by both companies – this provides no incentive for investing in the SME.

If UNFCCC specifies, say, a minimum 1000 basis point or 10% gap requirement between the electricity cost of the large and SME generators, then in order for the SME to qualify for "financial additionality block exemption", the equivalent cost of SME generated electricity would have to be at least (100% + 10%) $\times$ 3 c/kWh$_e$ or 3.3 c/kWh$_e$. The SME generator meets this requirement as its average cost of electricity is 0.7 c/kWh$_e$ above this benchmark.

Implications for foreign investors

If the financial additionality requirement was waivered for eligible SME companies, then the enterprise need not sell its product at a level below the state or market, i.e. it is permitted to earn revenue at the fixed or commercial price for the particular product. The IRR for such a SME with GHG sales would hence be more attractive for foreign investors – moreover, not all investors are in the super league to spend large amounts of capital on large projects.

Case B – Emissions and financial additionality should not apply to SMEs

A strong case for relaxing financial and emissions additionality for Small and Medium Size Enterprise projects exist, when expected returns on CDM-SME and CDM-LE investments are compared. The following case examples demonstrate the issue of equity or rather, equitability, of investment flows based on the Clean Development Mechanism of the Kyoto Protocol.

Case examples, 1999 cost estimates

Assumption: A region in a non-Annex I country, where the majority of power stati ons are single gas-fired turbine plants. This example would produce the same results if the majority of power plants in the same region were conventional coal-fired plants, as shown in case B3.

Basis of CO_2 emissions estimates: 644 t CO_2/GWh$_e$, 100 MW$_e$ GT plant, 33% efficiency; 349 t CO_2/GWh$_e$, 100 MW$_e$ CCGT plant, 54% efficiency.

In developing countries (non-Annex I countries): For 250 MW$_e$ and 10 MW$_e$ GT plants, plant efficiencies of 37% and 30% respectively, emissions of 574 t CO_2/GWh$_e$ and 708 t CO_2/GWh$_e$ respectively. For 250 MW$_e$ and 10 MW$_e$ CCGT plants, plant efficiencies of 54% and 40% respectively, emissions of 349 t CO_2/GWh$_e$ and 471 t CO_2/GWh$_e$. 68.5% load factor assumed.

Case B1: 1 large thermal power station, 500 MW$_e$
Ownership: one single or multiple (e.g. joint-venture) owners
Prime mover: gas turbines
GT power generation configuration: 2 × 250 MW$_e$
GT approx. cost: US$ 180/kW$_e$
GT plant cost: about twice equipment cost = US$ 360/kW$_e$
GT power station efficiency: 37%
CCGT packaged system: US$ 440/kW$_e$, efficiency 54%
CCGT plant cost:
About 1.5 times (add construction) CCGT packaged system cost = US$ 660/kW$_e$
Upgrade of GT power station to CCGT: Additional US$ (660 – 360) = US$ 300/kW$_e$
Additional investment: US$ 300 × 500,000 = US$ 150 million
CO_2 reduction/GWh$_e$ from GT to CCGT = (574 – 349) = 225 t CO_2
CO_2 reduction/year = 3000 GWh$_e$ × 225 t/GWh$_e$ = 675,000 t
Undiscounted capital cost of CO_2 reduction = US$ 222/t

In the above example, the cost advantage of non-Annex I countries has not been taken into account. If the cost of capital investment in a non-Annex I country is deflated by 50% due to more competitive costs of sub-manufacturing, materials, construction and labour costs;

Undiscounted capital cost of CO_2 reduction = US$ 75 million/675,000 t = US$ 111/t.

Case B2: Portfolio of 50 "small and medium size" power stations, 10 MW$_e$ each.
Ownership: separate ownerships of each plant – each plant may be singular or multiple owned
Prime mover: gas turbines

GT power generation configuration: $50 \times 10\,MW_e$
GT approx. cost: US$ 300/kW_e
GT plant cost: about twice equipment cost = US$ 600/kW_e
GT power station efficiency: est. 30% in developing country (32% in NWE)
CCGT packaged system: US$ 900/kW_e, efficiency 40%
CCGT plant cost:
About 1.5 times (add construction) CCGT packaged system cost = US$ 1,350/kW_e
Upgrade of GT power stations to CCGT: Additional US$ (1,350 – 600) = US$ 750/kW_e
Additional investment = US$ 750 × 500,000 = US$ 375 million
CO_2 reduction/GWh_e from GT to CCGT = (708 – 471) = 237 t
CO_2 reduction/year = 3,000 GWh_e × 237 t/GWh_e = 711,000 t
Undiscounted capital cost of CO_2 reduction = US$ 527/t

If cost competiveness of host country (e.g. capital cost deflation by 50%) is considered;

Undiscounted capital cost of CO_2 reduction = US$ 187 million/711,000 t = US$ 263/t.

If GT efficiency of singular 10 MW_e plants was less than 30%, say, 25%, and if emissions of CO_2/GWh_e were lower than 708 t/GWh_e - about 850 t/GWh_e, pro-rata 25% efficient plant;

Undiscounted capital cost of CO_2 reduction = US$ 187 m/[(850–471t) × 3000]
= US$ 165/t.

Case B3: Baseline datum shift from gas turbine to conventional coal-fired plant for SMEs.

If the baseline for emissions reductions for the SME gas turbine plants was changed from gas turbines (majority of power plants in the project region) to, say, conventional coal-fired plants as may be the case in the project host country;

Undiscounted capital cost of CO_2 reduction = US$ 187 m/[(1100–471 t) × 3000]
= US$ 99/t.

(CO_2 emissions from a 10 MW_e conventional coal, steam cycle plant would be much higher than a more efficient 250 MW_e plant operating with the same steam cycle – estimate of 1,100 t CO_2/GWh_e compared to normally quoted values of 964 t CO_2/GWh_e in energy literature).

If financial additionality is not required of SME power plants, then the reduction of the sales price of electricity of the large power plant compared to its commercial rate of return (in the absence of CDM), and the subsequent, potentially higher internal rate of return on the large power plant could be the same as the IRR on the SME power plants.

Analysis of results

Based on simulated findings of the case study above for power projects in an non-Annex I developing country, the return on CDM investments in Small and Medium Size Enterprises may be comparable or better than investments in large enterprises – if emissions and financial additionality were relaxed for SMEs. The author's results should enable many SMEs in all sectors of industry (e.g. power, paper, chemicals, cement, food manufacturing, steel, auto) to actively participate and not refrain from CDM projects – hence dispelling the notion that large enterprises are the only entities to benefit from inward investments.

The Conference of the Parties meeting at COP VI in the Hague in November 2000 should vote for non-applicability of emissions and financial additionality to Small and Medium Size Enterprises so that CDM can function on a level playing ground.

An alternative solution for the host government – "SME surcharge" on LE projects

A host government may be able to create equally attractive investment conditions for SMEs and large enterprises if a small tax – "a SME equality surcharge" – was levied on, say, up to 3% of the total capital value of investments in large enterprises, but zero surcharge on SME projects. The optimum level of the surcharge may have to be worked out carefully to enable SMEs to compete effectively with LEs.

If a SME power project was 2% more expensive than a LE project in the same sector, say, US\$ 510/kW$_e$ compared to US\$ 500/kW$_e$, then a SME surcharge of US\$ 10/kW$_e$ could be levied on investors of the LE project (additional to the value of plant investment capital) but not on the SME project. The financial returns from each project may differ due to operating parameters, efficiency and energy consumption.

Equitable balancing to promote foreign investments in "Natural Renewables" sector

At end 1999, the average commercial price of solar photovoltaics (solar modules plus balance of system) in North-West Europe was about US\$ 4 per peak Watt, or US\$ 4,000 per kW peak. This is roughly eight times the cost of (100 MW$_e$ rated) CCGT plants, four times that of CHP plants, and 1.3 times that of municipal solid waste plants – the multipliers would be reduced in real practice if solar PV plants reached outputs of 100 MW$_e$!

The "natural renewables" energy resources – solar, wave, geothermal and to a lesser extent, wind (as wind turbine production costs have fallen steeply in recent years) – shall need some financial support to enable them to compete with fossil-fuel based technologies, but only in circumstances which encourage competition or assist poverty-stricken locations:

1. An environment in which fossil-fuel plants command a large share of the market, e.g. in an urban or commercial environment.
2. Remote and non-grid connected zones of the economy where the average income per head is too low to afford commercial prices.

Financing "Natural Renewables" projects

A number of options exist for governments which intend to promote foreign investments in tried-and-tested "natural renewables" – solar, wind, wave and geothermal energy:

1. If renewable energy resources are available and exploitable, the host government could apply a "renewable energy-equality surcharge", say, of up to 3% to 5% on top of the value of plant investment capital in large enterprise projects. The surcharge could be managed in a fund to provide financing to those locations where the cost of electricity from natural renewables are not economically viable or where the community experience abject poverty.
2. An initiative launched by Euro-Asia Pac Corp – the LDC/IRISK Climate Change Fund, described earlier in this section – aims to raise capital from Kyoto Protocol traded emissions contracts and business-as-usual (non-Kyoto Protocol) project finance. Of the proposed 1% mandatory contribution from all traded contracts, 0.7% is directed to humanitarian sectors, and 0.1% to the "natural renewables" energy sectors.

Application of SME-equality and renewable energy-equality surcharges

Applying an appropriate surcharge or a combination of ratios of these surcharges may offer a level-playing field for SMEs and large enterprises, and fresh investment opportunities in inaccessible parts of the country. Regardless of the Kyoto Protocol, LDCs and islands-at-risk require capital to develop secure, long-term supplies energy and access to humanitarian aid.

Summary of "Equitable Treatment" issues

The Conference of the Parties must provide necessary inclusions within the Kyoto Protocol, in the form of new sub-clauses or the creation of new clauses:

1. To ensure equitable financial flows to developing countries whose economies are likely to deteriorate further as a result of the existing Protocol framework – as countries with abundant fossil fuel reserves and high utilisation of these reserves are the prime targets;
2. To enable Small and Medium Size Enterprises (SMEs) to compete with large enterprises on a level footing in attracting foreign investments – as financial returns on large scale enterprise investments are higher than SME investments;

so that LDCs, and SMEs in developing countries and LDCs, can accept a global agreement which will provide equitable treatment to all entities to attain its objectives.

In the presence or absence of the Kyoto Protocol, financial contributions to humanitarian sectors in LDCs could be made direct to international relief organisations and charities, or through proposed funds such as the LDC/IRISK Climate Change Fund.

A series of charts is provided in the next pages to illustrate "equitable treatment" issues between North and South countries.

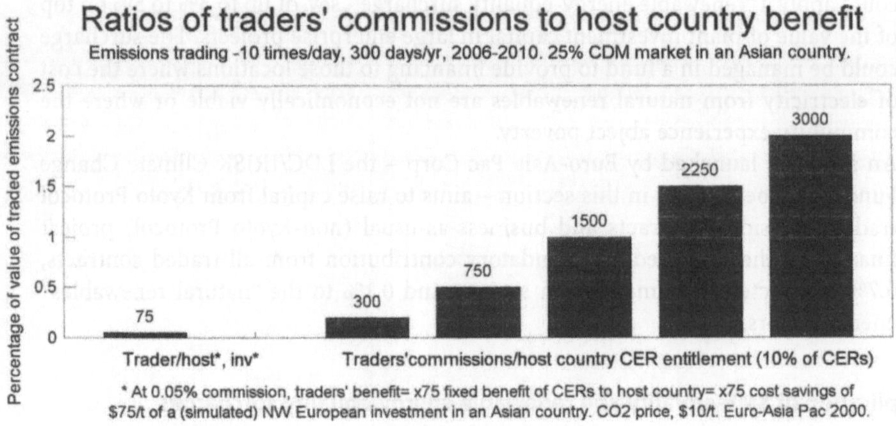

Fig. 7.1. Comparison of ratios of traders' commissions to CDM host country benefit

In a simulated case study of a North-West European (France, Germany and UK) corporate group's investments in India assuming (i) 25% of Annex I reduction obligations of 2,200 million tonnes of CO_2 equivalent, i.e. 550 million tonnes of CO_2 equivalent, were met by the CDM initiative in India, via Certified Emissions Reduction units, (ii) average CO_2 equivalent price of US$ 10/t, (iii) cost savings of US$ 75/t based on estimated technology cost of US$ 100/t in NW Europe and US$ 25/t in India, and (iv) secondary trading of India-originated CERs ten times/day, 300 days/annum, 2006–2010 period; gross commissions to traders will outperform by 75 and 3000 times CER entitlements to India, based on 0.05% and 2% gross commissions, respectively. Average net commissions of 0.1% to 0.15% of contract values would still be able to generate high benefits to traders – of magnitudes more than 50 times the fixed CER benefits to the host country, India. These findings are shown in the figure

If the cost savings were US$ 50/t [(US$ 100 – 50)/t] of CO_2 equivalent, based on the difference of estimated technology cost between the NW European investor consortium and India, the first bar in the chart would only change – to a ratio of 50 times, instead of 75 times

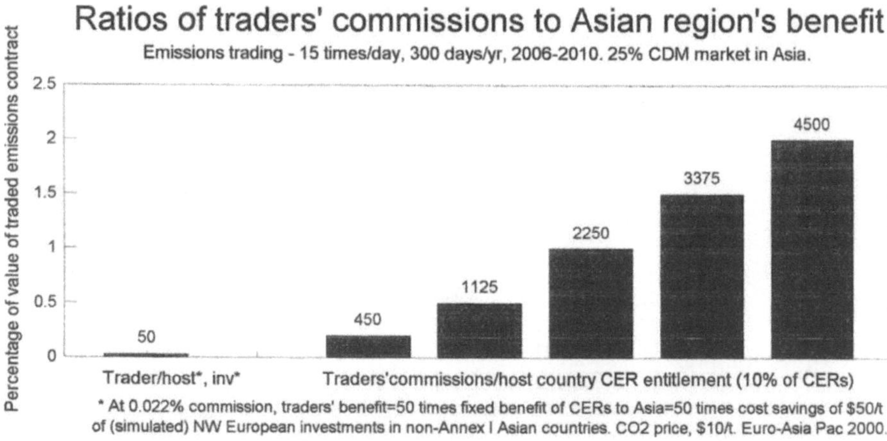

Fig. 7.2. Comparison of ratios of traders' commissions to host Asian region's benefit

The inequity of benefits distribution is demonstrated in a region-wide investment programme, based on the previous case simulation of the North-West European investment in India with same underlying conditions and assumptions. Traders' total commissions dwarfs the fixed benefits to host nations in the Asian region – Central, South/South-East Asia and East Asia including China. As some Asian countries have higher GNPs/capita than India or China, technology cost savings of US$ 50/t of CO_2 equivalent or less to a NW European investor consortium would be more appropriate than US$ 75/t CO_2 equivalent. The chart illustrates the rationale for a mandatory percentage contribution of all CER contracts originating from the host Asian countries to be channelled back to the CER project community or government

The itinerary for the chart would be:

In a simulated case study of a North-West European (France, Germany and UK) corporate group's investments in Asia, assuming (i) 25% of Annex I reduction obligations of 2,200 million tonnes of CO_2 equivalent, i.e. 550 million tonnes of CO_2 equivalent, were met by the CDM initiative in Asia, via Certified Emissions Reduction units, (ii) average CO_2 equivalent price of US$ 10/t, (iii) technical cost savings in CO_2 abatement (not cost savings in market price of CO_2, which is predicted to remain fairly constant within a fluctuation band – these technical cost savings are accompanied by sales of primary products of projects in Asia) of US$ 50/t based on estimated technology cost of US$ 100/t in NW Europe and an average of US$ 50/t in Asia, and (iv) secondary trading of Asia-originated CERs fifteen times/day, 300 days/annum, 2006–2010 period; gross commissions to traders will outperform by 50 and 3000 times CER entitlements to Asia, based on 0.02% and 1.35% gross commissions, respectively. Average net commissions of 0.1% to 0.15% of contract values would still be able to generate high benefits to traders – of magnitudes more than 50 times the fixed CER benefits to Asia

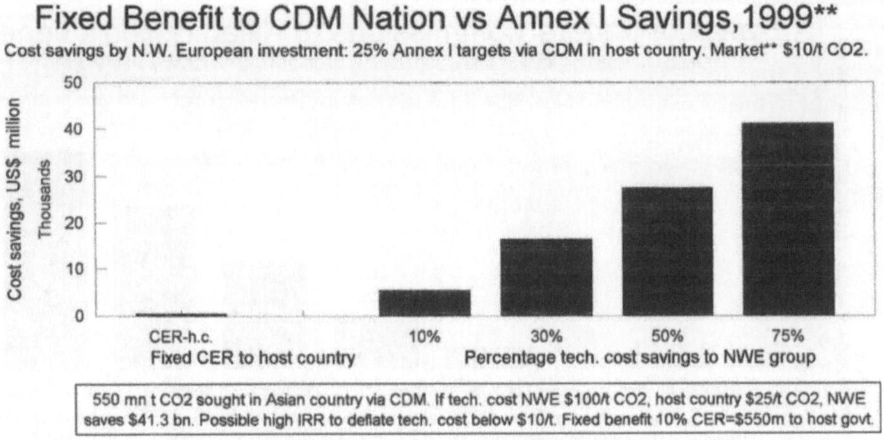

Fig. 7.3. Fixed benefit to CDM nation versus Annex I savings, 1999

The cumulative financial benefits to carbon contract traders could amount to about US$ 40 billion – as shown in the figure – on the basis of 75% technical cost savings, assuming 25% of Annex I obligated reductions were achieved in a low cost region such as Asia. Although the market price of carbon may be lower than the technical cost of carbon abatement, a huge surplus of carbon credits from landfill or combined cycle plants would easily balance out the difference and improve on the project's internal rate of return

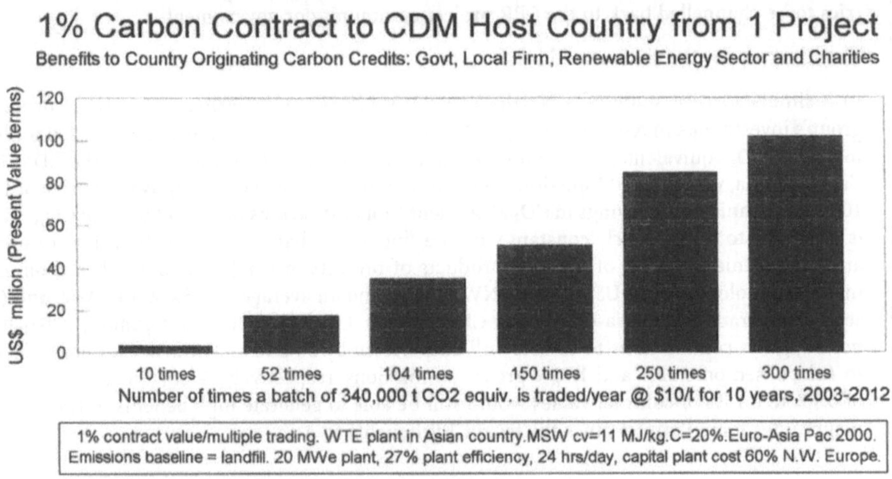

Fig. 7.4. 1% carbon contract to CDM host country from 1 project

In quantitative terms, if non-Annex I countries were to ask for a mandatory contribution of 1% of the settled value of all carbon contracts originating from its soil – either by approval of Conference of the Parties or by a special clause inserted in the "seed" carbon contract – then revenues of some US$ 20 to 80 million (at 10% discount) could accrue to the host country based on a case waste-to-energy project. The reward to the host government is shown in the figure

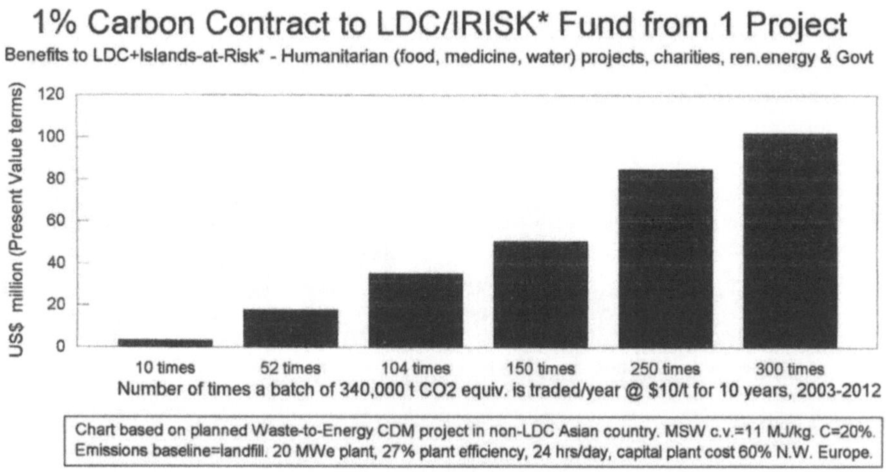

Fig. 7.5. 1% carbon contract to LDC/IRISK fund from 1 project

The benefits to an adaptation fund to vulnerable countries, such as a proposed LDC/IRISK Climate Change Fund, from a 1% mandatory contribution of a "seed" carbon contract (basis of project conditions stated in the figure) could replicate much needed capital aid to these endangered economies. This amount of benefit, however, shall have to be divided between beneficiary countries, as it arises from a single project. A 10% discount is used for present values.

The benefits to the host economy can be substantial to Third World economies assuming Annex 1 countries jointly or unilaterally agree on mandatory contributions from carbon contracts – acts of charity which North countries should perform to alleviate global poverty.

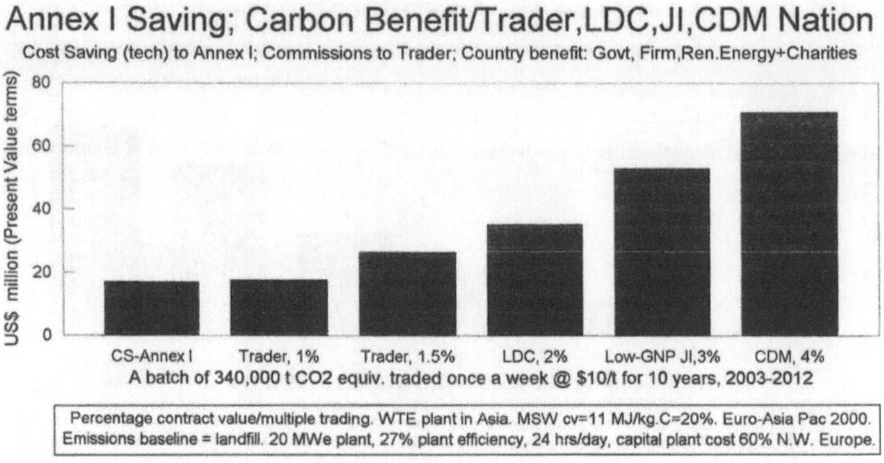

Fig. 7.6. Annex I saving; carbon benefit/trader, LDC, JI, CDM nation

The relative benefits of the Kyoto Protocol's flexible mechanisms assumimg a set of conditions enabling project proceeds to be allocated from a Waste-to-Energy plant in Asia are shown. The benefits are: technical cost savings of US$ 17 million to an Annex I investor (based on savings of US$ 50/t of CO_2 equivalent compared to domestic action); revenues to traders based on 1% and 1.5% commissions; 2% carbon contract value contribution to an adaptation fund for Least-Developed-Countries, 3% to low GNP Annex I countries, and 4% to non-Annex I countries. The percentages for the last two entities may have to be negotiated in COP 6 or COP 7 – there are two possible solutions: either these percentages are agreed by all Parties of the Protocol for each and every secondary trading of the "seed carbon contract" originating from the project host country, or failing such an agreement, that percentage is written into the contract for buyers and sellers to consider as a transaction cost. There are two political incentives for including these mandatory contributions – the current text of the Protocol does not exclude the improvisation of emission trading price techniques to encourage participation of "South" countries, which would welcome such revenues; secondly, the impact of these contributions on the price of "South" carbon credits compared to "North" technical or market price cost savings would only amount to a small dent. A 10% discount is used for present values.

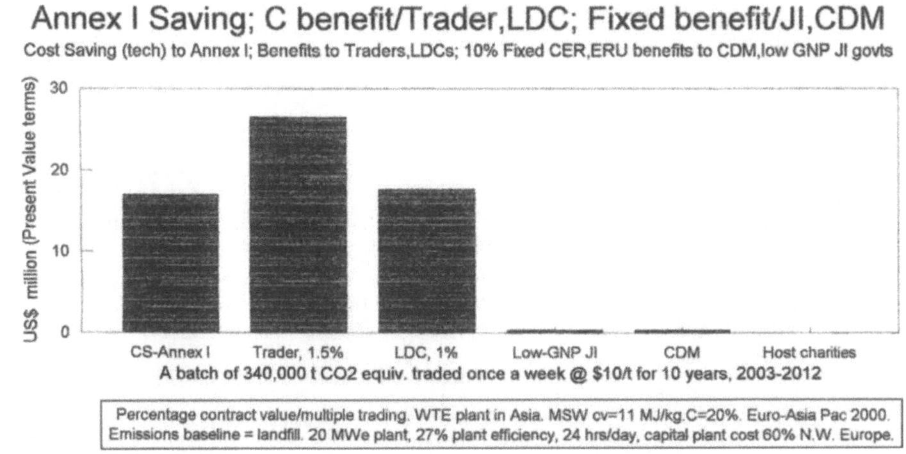

Fig. 7.7. Annex I saving; carbon benefit/trader; fixed benefit/JI, CDM

If CDM and low GNP/capita JI countries do not demand mandatory percentage contributions of the traded value of each and every carbon contract, then they will not receive additional revenue from the "seed" carbon contracts originating from energy saving projects undertaken on their soils. If traders recieved 1.5% gross commissions, and LDCs or the Adaptation Fund received 1% of the transacted value of every contract in the secondary market, then these host countries receive insignificant benefits compared to traders and countries vulnerable to climate change. CDM and low GNP/capita JI countries should hence insist on mandatory percentage contributions of carbon contracts as a standard requirement of the Kyoto Protocol framework or impose these percentage surcharges unilaterally on all carbon contracts originating from projects conducted in their sovereign territory. Local or international relief charities operating in these countries should also receive benefits to ensure distribution to the regions affected by climate change and the poorest communities. A 10% discount is used for present values.

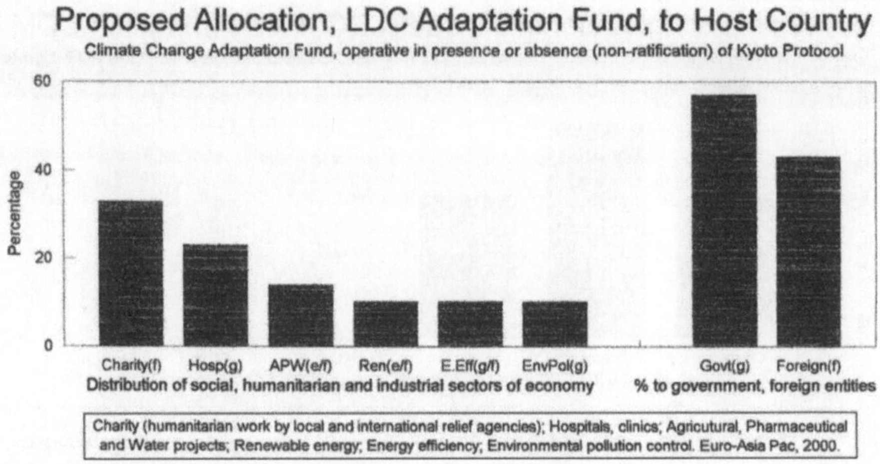

Fig. 7.8. Proposed allocation, LDC/Adaptation Fund, to host country

A proposed scheme of allocation from the proceeds of secondary trading in the emissions market, CDM and JI projects to the LDC & Islands-at-Risk Climate Change Fund/Adaptation Fund is shown in the figure. The government's share is about 60%, and participating foreign share 40%, which should be agreeable to the host country. The humanitarian sector should receive the highest proportion, followed by socio-economic and environmental programmes.

8
CO_2 Reduction Technologies and Financial Analysis

This chapter describes modern energy efficiency technologies; the recycling industry – which elongates the life-span of raw materials; energy saving techniques for the process industries; biogas energy; electricity storage pilot systems; and presents in the form of a comprehensive table, a comparative cost-benefit analysis of carbon dioxide reduction technologies. The table is also a useful guide for project financiers in terms of assessing the amount of carbon credits that can be acquired from individual technologies. The last part of the chapter examines the rate of return of simulated projects in "North" and "South" countries.

8.1
Thermal Energy and Renewable Energy Technologies

Power and heat generation industry

Although combined cycle gas turbine plants and co-generation provide the highest reductions in carbon dioxide emissions, these technologies rely on availability of natural gas which most developing countries do not possess.

Atmospheric/pressurised fluidised bed combustion, and integrated gasification combined cycle technologies – which operate at 42%–45% thermal efficiencies – offer practical solutions to cutting both carbon and sulphur emissions from conventional coal-fired power plants. Where coal is not a major energy resource, particularly in LDCs, gasification of biomass to generate electricity in combined cycle plants would improve energy efficiency, reduce carbon dioxide emissions and imports of diesel.

"Carbon sink" creation and biomass technologies

Carbon dioxide emissions reduction or avoidance with minimal leakage can be achieved with high cost technologies, such as solar photovoltaic and solar thermal systems. To lower the cost of acquiring carbon credits, a number of Annex I countries – namely the USA, Japan, Netherlands and Australia – have opted for carbon sequestration and forestry projects.

"A country should conserve its forests anyway" – is a popular rhetoric of NGOs advocating higher cost technical solutions to reduce greenhouse gas emissions, compared to soft options which result in non-guaranteed removal of GHGs. The hard options involve more expensive equipment and machinery which permanently reduce GHG emissions (e.g. a biomass thermal conversion plant) compared to protecting an area of forest for say, ten years, after which it may be razed to the ground for cash crop cultivation.

Short rotation coppice (e.g. willow, hazel and wood) is considered a carbon-neutral fuel-regrowth carbon sequestration process because of the short-life crop cycle, typically three year rotation cycle with annual harvests. However, the distinction between SRC and forestry projects may pale if proportions of forestry products - normally used for paper and packaging were to be used as fuel in waste incineration plants. The issue of "sinks" is highly debatable in COP negotiations - all "man-made" sinks in practice cannot be regarded as permanent.

The possible exclusion of forestry projects for CDM and JI on such rationale is a topical issue which COP must resolve. Investors and financiers of potential CDM projects should be advised on the accreditability of forestry projects before money is parted.

8.2 Cost of Carbon Dioxide Reduction

Published data on CO_2 control costs vs. author's calculations - "overnight" costs

Marginal abatement costs of AIJ projects are currently the only available data for comparison - it is frequently termed the "overnight" cost by industry, which includes a small percentage in contingency cost, technically at zero percent discount rate. At a discount rate of 8% to 10% over the project life, the present value of control costs (annuitised or leveraged cost) is about 4 to 5 times less than the marginal cost. The "present value" control costs are less than overnight costs because the cost of financing is distributed over a period of years.

Certain overnight and discounted costs calculated by the author are some two to three times higher than average AIJ project costs, particularly in fuel-switching technologies. The author suggests that such cost differences may well be reflected in future JI or CDM projects if AIJ projects have been selectively "cherry-picked" to incur the least costs.

The "discounted" cost of emissions abatement varies according to the method of calculation and discount rate used. The most frequently employed technique is dividing the present value of emission reductions by the present value of costs involved.

Table 8.1. Comparison of AIJ project costs with author's calculated costs

Data source	Technology/ methodology	Quoted AIJ costs. Average undiscounted unless indicated in brackets. US$/t CO_2	Author's calc. costs. Average undiscounted unless indicated in brackets. US$/t CO_2 (see notes*)	Comments
1.	Carbon sequestration	2–17 0.3–7	– –	Annex B countries. Tropical countries.
2.	Carbon sequestration	6 (4.4%) 4 (0%)	– –	
2.	Fuel-switching/ high tech. including renewables	44 (4.4%) 29 (0%)	100–150 (10%)	Developed country cost. Author's estimate – see row 3.
2.	Fuel switching/ high tech. including renewables	108	100–150 (10%)	Developed country cost. 108 is average based on various sources.
3.	Carbon sequestration	4.3	–	JI project costs, third column.
	Renew. energy	71.4	132	Author's estimate: developed country cost. Average value of geothermal, wind and wave at year 2000+ costs. Solar excludeddue to low radiation levels in most "North" countries.
	Energy eff.	40.3	–	
	Fugitive gas capture	49.2	–	
	Fuel-switching	204	154	
	Biomass Heat & Power	n/a	560	
	Combined Heat & Power	n/a	190	
	IGCC	n/a	660	Basis of more efficient IGCC systems.
4.	All technologies	2.5–2.7	15–20 (10%)	Global average cost quoted from ref. 4. Author's estimate for developed countries is based on technology shifts to open cycle, combined cycle and co-generation plants, but excludes carbon sequestration projects.

Table 8.1. *Continued*

Data source	Technology/ methodology	Quoted AIJ costs. Average undiscounted unless indicated in brackets. US$/t CO_2	Author's calc. costs. Average undiscounted unless indicated in brackets. US$/t CO_2 (see notes*)	Comments
4.	Carbon abatement	16.3	–	OECD.
		13.6	–	Developing nations (ex-PRC).
		1.1	–	Developing nations. 12% reduction, 2020.

Notes

* As calculated for Table 8.1 and from results of case examples in Table 8.6, at the end of this chapter.

Marginal costs are assumed if no discount rates are applied.

Energy efficiency projects reduce the consumption of fuel, electricity or heating in buildings, households and industrial establishments; it also includes projects in process plants where energy is saved, e.g. cement kiln, iron and steel works.

Fuel-switching is a generic term used to describe a change from an existing fuel to a less carbonaceous fuel or renewable energy resource. This broad definiton hence covers "high technology" – e.g. construction of an open cycle gas turbine, combined cycle gas turbine plant or a solar electric facility.

Brunello's data [data source (3)] on specific technologies is based on 78 Joint Implementation projects with widely divergent costs of carbon dioxide abatement – it should hence be used with discretion until more significant numbers of projects are analysed.

References

1. K. Munlongoy *et al*, Are Joint Implementation and the Clean Development Mechanism Opportunities for Sustainable Forest Management Sustainable through Carbon Sequestration Projects? Paper to International Academy of the Environment, 1998.
2. M.A. Ridley, Lowering the Cost of Emission Reductions: JI in FCCC, Kluwer, 1998. S. Fankhauser's data is quoted by Ridley.
3. A. Brunello, Cost-Effectiveness of Joint Implementation Projects for Securing Reductions of Greenhouse Gases, MSc thesis, University College, London, 1998.
4. ECON Centre for Economic Analysis, Oslo, 1997.

General reference: D. Pearce *et al*, The Clean Development Mechanism – Benefits of CDM for Developing Countries, The Centre for Social and Economic Research on the Global Environment (CSERGE), London, 1998.

Declining costs of electricity generation from renewable energy technologies

According to T.B. Johansson *et al*'s Renewable Energy: Sources for Fuels and Electricity, EarthScan/Island Press, 1993 – probably the most authoritative work on renewable energy – future capital costs of renewable energy technologies are expected to compete with fossil fuelled electricity in 2050 as gas and oil resources become increasingly scarce.

The cost of electricity from solar and wind energy in 2000–2002 will fall by 20%–25% and 60%–70%, respectively, from 1990 levels due to improvements in technology. Although cost projections are based on large-scale modules - magnitudes higher than current facilities - the data provides indicative electricity prices which could compete with thermal power plants as oil and gas prices start to escalate between 2005–2010.

The commercial costs of electricity production from geothermal, hydro, wave and tidal sources are very much site-dependent; tidal plants are the most capital intesive. Small hydro, wave and geothermal offer good opportunites, but prospecting for geothermal reservoirs bear the same risks as for oil exploration.

The profile of declining electricity costs from solar and wind is provided in the next table.

Waste-burn and wave power costs in Asia could become commercially competitive

In early 2000, the author's estimates of average capital costs of waste-burn and wave power technologies in South-East Asia were 4–6 c/kWh$_e$ and 3 c/kWh$_e$, respectively; corresponding costs in China, 3–4 c/kWh$_e$ and 2 c/kWh$_e$ (wave power based on LIMPET technology on 24 hour load factor). These estimates assume relatively generous discounts on turbine-generator equipment manufactured in Europe which are imported into these Asian countries, low costs of construction and O&M overheads.

Table 8.2. Falling costs of power production from solar and wind technologies

Technology	Year	Cost elect. US cents/ kWh$_e$ (6% disc. rate)	Cost data, conditions Capital \$/kW$_e$	Capacity factor	Comments
Solar – parabolic trough	1990–92	14.2	3,000	0.23	80 MW$_e$ to year 2000.
	1995/00	13.5	2,700	0.22	200 MW$_e$ post-2000.
	2000/05	9.8	2,200	0.24	
Solar – central receiver	1995	16.0	3,300	0.32	Cost in 2005 subsidised.
	2005	8.0	2,600	0.35	
	2005/10	5.5	2,600	0.59	
Wind	1990	7.2	1,100	0.2	6% discount rate. Hub height raised from 25 m in 1990 to 50 m in 2020. Includes insurance and land cost.
	1995	5.0	1,000	0.28	
	2000	4.3	950	0.3	
	2010	3.6	850	0.33	
	2020	3.1	800	0.34	

8.3 Role of Recycling Plants on Greenhouse Gas Emissions

Investments in construction of recycling plants for biodegradable domestic solid or mixed waste would result in reduction of carbon dioxide and methane emissions from landfills - such reductions could be claimed by investors as emission credits under the Clean Development Mechanism or Joint Implementation initiatives.

Average energy savings through recycling the main constituents of municipal solid waste in North-West Europe in 1999 are estimated in Table 8.3. From the table, the recycling of aluminium and steel would represent the most attractive CDM/JI projects based on emission reductions of carbon dioxide. These technologies would provide

Table 8.3. Energy savings via the recycling industry

Material	Energy usage, production from raw material	Energy savings, GJ/t compared to prod. from raw material	Emission red. t CO_2/t material recycled	Estimated cost for recycling (av. UK, 1999) US \$/t	Implied cost of CO_2 red. US \$/t
Aluminium	120	55	10.0	24.8	2.48
Glass	30	3	0.25	12.3	49.2
Paper	25	see notes	0.66 see notes	15.2	22.7
Plastic (PET)	150	30	3.0	56.1	18.7
Steel	40	15	1.5	12.2	8.13

Notes

Average of US\$ 1.6 = £ 1 used for currency conversion.

For glass, the author estimates energy savings of about 3 GJ/ton vs. on-line production, after consultations with key UK reprocessors – a much lower figure than quoted by the European Commission.

PET = Polyethylene terepthalate, a polyester plastic used for liquid beverage containers.

Energy savings from recycling paper vs. commercial paper production from virgin fibre are marginal, based on author's discussions with paper trade associations and recycling plants in the UK. Almost the same amount of energy is required to recycle paper as to produce commercial paper, resulting in 0.7–0.8 tonne of carbon dioxide emitted per tonne of paper produced (depending on the type of fuel used).

In paper production, a unit weight of timber with moisture needs to be converted to bone-dry pulp, to which starch and fillers are added. Thus, a total of about 1.2 tonnes of commercial paper results from 1 tonne of timber, i.e. 1 tonne of paper is produced from about 0.8 tonne of timber. If this 0.8 tonne of timber is not cut down, then about 0.66 tonne of carbon dioxide – based on 0.83 tonne of carbon dioxide released by burning 1 tonne of wood – is saved within its physical structure as part of a tree. In greenhouse gas reduction terms, the recycling plant is saving trees from their "business-as-usual" fate of decomposition into carbon dioxide and methane in a landfill or combustion in a biomass power/CHP furnace – such activity, in the Kyoto Protocol context, would be equivalent to GHG reductions from landfill sites below 1990 levels if the plant was operating in an Annex I country.

Sources: DGXI (Directorate General XI, a division of E. Commission) Working Document. Energy usage of manufactured product – *Industrial Energy Conversion*, C.M. Gottschalk, Wiley, 1996. Estimated cost of recycling obtained from OM Environmental Exchange, London.

higher rates of financial return than combined cycle gas turbines and cogeneration. To qualify for emission credits, investors should ensure that the project meets both financial and emissions additionality criteria. In North-Western Europe, market prices of commercial grade aluminium and steel at end 1999 were multiples of their recycling costs. If UNFCCC considered the "original project" as baseline, then additionality on the recycling plant's capital cost and emission levels may be required – but this would prove financially less advantageous to investors.

The recycling of PET plastic bottles and paper may also appear potentially good projects if investors were entitled to carbon dioxide and methane credits (for preventing these gas emissions) from landfills – the amount of credits could be calculated according to average compositions of these materials in domestic waste and reductions in emission levels compared to nearby landfill sites. As plastic and paper can not be recycled more than four or five times because of their physical properties, carbon dioxide and methane credits could be calculated pro-rata the period these materials do not meet their fates at landfills.

8.4 Energy Saving Strategies for Process Industries: Potential Emissions Reductions

CDM/JI investments can be also made in non-power or thermal producing plants which are sources of greenhouse gas emissions – these are primary process industries with "chimneys" from which gases are emitted: iron and steel works, paper mills, cement kilns, chemical and petrochemical plants, refineries, metallic ore smelters,

Table 8.4. Energy-saving strategies in north-west Europe, 1999

Industry /product	Energy audit Prod. MJ/kg	Pot. sav.%	Energy saving strategy
Aluminium	120	5–10	Waste heat recovery.
Brick	2	15–20	Waste heat recovery.
Cement	9	10–30	Conversion from wet to dry process. Waste heat recovery, process controls.
Ceramic ware	20	15–20	Waste heat recovery.
Chemicals	150	15–20	Invest in cogeneration plant for steam and power requirements.
Coconut/ vegetable oil	3	15–30	Waste heat recovery; improvement in evaporation.
Copper	45	5–10	Waste heat recovery.
Fertiliser	8	20–25	Waste heat recovery/reformer; recovery of hydrogen and carbon monoxide.
Glass	30	10–15	Replace old and obsolete burners to improve combustion efficiency.

Table 8.4. *Continued*

Industry /product	Energy audit Prod. MJ/kg	Pot. sav.%	Energy saving strategy
Glass			Improve insulation of furnace chamber and all ductings. Supply optimal excess air for combustion. Check boiler efficiency. Waste heat recovery to preheat combustion air and/or feedstock.
Oil & gas exploration /production	5	5–10	Reduce flaring; replace diesel fuel with gas for generators.
Oil refinery /petrochemicals	2	15–25	Improve sealing to prevent oxidation of intermediates. Reduce hydrocarbon losses to atmosphere in wastewater treatment plant. Incineration of industrial and chlorinated effluent sludges. Recycling of materials, particularly rubber scraps. Waste heat recovery.
Lead	40	5–10	Waste heat recovery.
Mining	15	5–15	More effective drilling techniques and explosives.
Paper	25	10–15	Invest in cogeneration plant for steam and power requirements. Recycle/re-use: bark from debarking, woody residues from pulping process.
Plastic (PET)	150	15–25	Invest in cogeneration plant for steam and power requirements.
Rubber	150	15–25	Reuse old tyres and scrap rubber.
Steel (iron /steel plant)	40	15–20	Improve coke oven and blast furnace efficiencies. Convert existing Open Hearth to Basic Oxygen Furnaces (savings of 2 to 3 GJ per tonne of casted steel product). Invest in combined cycle plant for electricity requirement.
Sugar	20	15–30	Waste heat recovery. Invest in cogeneration plant for steam and power requirements.
Textiles	25	15–20	Waste heat recovery, improvement of water and liquor systems. Invest in cogeneration plant for steam and power requirements.

Notes

GHG	greenhouse gas
Oil/gas exploration	energy used to locate 1 kg of crude oil (per column 2)
Oil refinery	energy used to produce 1 kg of petroleum (per column 2)

vehicle manufacturing plants, textiles factories, heavy/light machinery production lines and transportation vehicles.

A survey of primary and secondary process industries is provided in Table 8.4, which shows the average energy consumption per unit weight of the product, and energy efficiency measures which can result in greenhouse gas emissions reduction. Improvements in fuel efficiency for transportation vehicles is discussed separately.

Transportation – improvement in fuel economy of cars and trucks

The average mileage per gallon of fuel for cars and trucks in North America and Western Europe at end 1999 was 23 and 12, respectively, but in "South" countries, the corresponding figures were about 10 and 5. Hence much improvement is needed in these countries in the design and use of lightweight materials for vehicle production, and development of electric and gas fuels should result in less carbon dioxide and methane emissions to the atmosphere.

Greenhouse gas reductions – North-West Europe process industry experience

The European Union is considered as the world leader in addressing environmental issues posed by global warming and creating long-term solutions to these issues. It has proposed to member states that 50% of national emissions targets under the EU "bubble" agreement should be met by domestic action. Some of the largest companies in Germany, France, UK, Netherlands, Denmark and Sweden, in conjunction with local governments, have implemented a series of measures to reduce energy consumption. The next table describes the strategies employed by European firms in meeting their greenhouse gas reduction obligations.

Table 8.5. Energy-saving strategies of European firms in the process industries

Industry/source	Specific corporate action/results – key achievements
Brewing /drinks	Replacement of higher CO_2 content beers/drinks with more nitrogenated/smoothflow products.
	CHP schemes to reduce energy demand on electricity and process heating.
Buildings	Double skin insulated walls, automatic control of gas space heating.
	Active and passive solar energy – reinforced glass roofs and windows, reduction of lighting requirements during daytime with transparent roof designs.
Construction	Replanting trees; land reclamation; use of recycled and waste materials – pulverised fuel ash and blast-furnace slag in pre-cast blocks and ready-mixed concrete. Use of demolition waste, asphalt for road planings, by-products silt and clay (land restoration).
	Erection of totally enclosed netting system over landfills to discourage scavengers.

Table 8.5. *Continued*

Industry/source	Specific corporate action/results – key achievements
Construction	Conversion to Ultra Low Sulphur Diesel Fuel (ULSF) for road vehicles.
	Promote use of train and river barges for transporting materials.
	Replacing old lime kilns; using natural gas for firing; waste heat recovery.
	Generating electricity with methane "captured" from landfill sites.
	Conversion from diesel to natural gas fuel for aggregate-hauling rolling stock.
	Producing usable aggregates from waste Cement Kiln Dust (CKD).
Food and plantations	Avoidance of planting on steep hillsides to allow uninterrupted growth of dense vegetation.
	Allowance of natural forest growth on both sides of river banks.
	"Terrace" planting where appropriate; leguminous cover crops to enhance nitrogen-fixation.
	Natural rotting of old plants to enrich soil nutrients.
	Use of traditional animals where possible to avoid heavy machinery which would damage soil.
Glass	High performance glass in double glazing of houses and buildings to reduce heat lossess.
	Structural/solar cell-embedded glass for walls/roofs to reduce electricity requirements.
	Quarryland – remediation where possible, landfill gas generation site operation.
	Increased use of LPG and immersed electric heaters in glass production.
	Glass recycling preserves natural resources and reduces energy consumption in furnaces.
Gas transm/ distribution	Setting 20% reduction target from 1992 methane leakage levels by 2000 for one country.
	Replacing gas-activated valves and gas-powered starters on standby generators with closed hydraulic systems. Remediation and reclamation of wasteland.
	Creating storage facilities from depleted gas fields and desalted underground cavities.
Iron/steel	Adoption of environmental management standard ISO 14001 for 50% of operations by 2000.
	Industrial scrap steel and cans from domestic waste are melted in in electric arc furnaces.
	Mining site remediation, clean-up and regeneration.
	Product improvements to reduce energy losses – new steel alloy rail castings, car body design using ultra-light steel auto body (ULSAB), and improved steels for electric motors.
Mineral/ smelting	Reduction of perfluorocarbons (PFCs) – mainly CF_4 and C_2F_6 – by about 70% below 1990 levels in 1997. Reduction per tonne of carbon dioxide equivalent emissions in aluminium production from 2.7 tonnes in 1990 to 0.63 tonne in 1997.
Oil explor./ oil refining/ chemicals	Reduction of drilling mud and cuttings; use of water-based drilling muds.
	Decrease in CO_2 from diesel oil combustion (to power drills). Solid waste disposal.

Table 8.5. *Continued*

Industry/source	Specific corporate action/results – key achievements
Oil explor./ oil refining/ chemicals	Reduced gas flaring by installation of automatic pressure sensors.
	Inject oily drill cuttings from borehole into separate subsea cavity to contain emissions; likewise for CO_2 stripped from a primary well into separately excavated well.
	Use of wind machines and wave turbines, and solar energy resources.
	Diversifying into renewable energies including biomass for land-based refineries.
	CHP schemes to improve plant efficiency in chemicals production.
Paper/pulp	Paper from virgin fibre: reduction of energy used to 17 MJ/kg of paper, net weight basis; reduction to 0.65 tonne of carbon dioxide/tonne of paper.
	Increase of percentage of solid waste recycled to 90%.
	Increase recycled fibre content of commercial paper to 10% (80% wood pulp, 10% filler).
	Re-using water thrice in the paper-making process.
	Up to 100% re-use of effluent plant sludge to agricultural land spread.
	Waste heat recovery from shrink film packaging unit. Self-generation of electricity.
Supermarkets	Use of CNG (compressed natural gas, which also includes Mobil's trademarked CNG Cleanerburn Natural Gas) and LPG in vehicle fleets. Shifting transportation to rail.
	Increase of vehicle distance covered to 3.1 km per litre of road fuel.
	Building new stores over derelict buildings; re-use of bricks, roof slates and materials.
	Waste heat recovery from refigeration (highest energy usage) for hot water/space heating.
	CHP schemes to provide total energy requirements – refrigeration, heating and lighting.
	Ammonia refrigerant for warehouses and compressed nitrogen for trailers.
	Developed the world's first solar-powered refrigeraterd lorry.
	Adoption of the Integrated Crop Management (ICM) and Forest Stewardship Council's (FSC) standards to protect crop and timber origins.
Textile/ dept stores	Reduction of waste disposal costs in 1998 by up to 50% compared to 1991 as a result of packaging minimisation and materials recycling.
	Supplying energy-from-waste incinerators with non-recyclable materials.
	Replacing mechanical compressors with liquid nitrogen refrigerant.
Vehicles (road)	Speed limitation to 90 km per hour resulting in 6.5% improvement in exhaust emissions.
	Upgrading capacity from 32 tonnes to 38 tonnes and increased trailer lengths, resulting in reduced mileage/vehicle movement.
	Replacing old with new engines, achieving 10% increase in vehicle km per litre.

8.5 Traditional Biomass Energy Technology

Methane from anaerobic digestion

High tech steam cycle and gasifier/STEG (steam and gas) cycle plants require relatively high capital investment costs compared to traditional biogas plants using anaerobic digestion of organic waste. The average estimated cost in developing countries is about US$ 1000/kW_e for microbial anaerobic digestion of farm/animal manure/livestock waste; domestic sewage; industrial organic wastes/sludge which include waste from bottling/brewing/soft drinks/dairy/distillery/food processing/meat & fish/pharmaceutical/pulp & paper plants.

The capital cost falls if traditional biogas is used for combined cycle electricity or CHP, but this would require very high volumes of biogas production. The average cost of electricity/thermal heating can be reduced if additional revenue is obtained from sales of liquid fertiliser and product fibre nutrient, and landfill tax or tariffs imposed on waste disposers.

A reduced average capital cost of US$ 500/kW_e may be attainable for biomass projects if foreign manufactured steam and gas turbines were the only power equipment (gasifiers and digesters) imported from established manufacturers in Europe/USA, and purchased at large discounts; or if power turbines, boiler equipment and plant ancillaries were locally produced and/or assembled by domestic/Western joint-venture firms in developing countries. In the latter case, industrial turbines could be made at a fraction, perhaps half, of Western f.o.b. prices. Very low labour costs in certain developing countries, e.g. China, could be capable of achieving substantially reduced plant capital costs.

Coal bed methane extraction

The capital cost is estimated to be the same or less than natural gas sourcing from a nearby gas or associated oil/gas field, and the cost of power generation could match CCGT plants – US$ 500/kW_e for plants less than 50 MW_e.

Capital plant costs are based on 50% of prices of power generation equipment made in North-West Europe (main power items are gas and steam turbines) – this 50% reduction may be achieved by a combination of 40%–45% capital cost of NWE f.o.b. (free-on-board) prices and 5%–10% of sub-manufacturing and/or assembly of key power equipment and accessories. If 50% of NWE power equipment prices are not attainable, then project capital costs should be based on 60%–70% of NWE f.o.b prices – the corresponding average costs of electricity, costs of financing, and costs of greenhouse gas emission reductions would increase pro-rata.

Methane emission reduction from landfills

About 38,500 tonnes of methane per year is saved from emissions from landfill sites and domestic/industrial sewage effluents (each source produces 200–400 m^3 of biogas per tonne of each respective waste), based on author's estimates in the UK and methane density of 0.7 $kg\,m^{-1}$. The methane content of biogas from these sources vary between 50%–60%, resulting in the prevention of nearly 40,000 tonnes of the greenhouse gas per year if it was used as fuel in a 101 MW_e, 54% efficient, CCGT power plant at 68.5% load factor. This reduction of methane, if traded at, say, the same price as CO_2 reductions at US$ 25/t, could generate an additional revenue of US$ 1 million to the project.

If the methane, in the form of biogas produced from similar landfills/industrial effluents was used to fuel a CCGT plant of the same operational capacity in coal abundant CDM countries as described previously, the reduction in CO_2 emissions compared to a conventional coal plant (assuming the coal plant represented the emissions baseline or plant to be replaced) would be 615 tonnes CO_2/GWh_e. Multiplied by a market or negotiated price of CO_2 emissions reductions, an additional revenue to project investors could be realised.

8.6 Electricity Storage from Renewable Energy Sources

If electricity can be stored, then power continually generated during periods when there is little or no demand for the utility – wave, wind and geothermal energy are perennial sources – can supplement normal and peak supplies during the daytime.

Indicative costs in Europe are US$ 150–300/kWh_e, depending on frequency charging and discharging frequency, cost of current converters, and replacement of fuel cells/electrolytes. The lifespan of fuel cells and conducting media vary between 500–2,000 charge and recharge cycles; lead/acid accumulators for automotive and road transport lasts between 4 to 5 years, depending on usage conditions. National Power, a UK firm, is developing an experimental regenerative fuel-cell plant, Regenesys, with a working life of 15 years. Estimates of capital cost of less than US$ 150 per kWh_e of storage capacity has been quoted by the firm – this may fall to, perhaps, US$ 100/kWh_e with progressive R&D and sales of pilot plants.

Based on National Power's power storage cost data, the average cost of electricity for wave and wind energy applications are estimated by the author in the following two case studies. Assumptions:

1. In the most ideal wave and wind sites – offshore wave and offshore wind plants – energy is available 24 hours a day. Power production from the wave device "LIMPET" assumes a 24 hour day energy supply, and the UK wind farm, 35% wind load factor.
2. Power is stored in a rechargeable electrolyte system, and discharged to business, industrial and domestic consumers according to demand on the following day. Electricity demand for consumers is substantially reduced over the weekend, but electricity is continually stored over this period at incident wave and wind load factors, ready for discharge on the following Monday morning or next working day.

Storage of electricity from wave energy

Basis of wave energy device LIMPET, 0.5 MW_e, North Scotland

In 1 year, the LIMPET wave device delivers about 2 GWh_e of electricity, based on 24 hours per day generation of electricity.

However, in the night time, when electricity demand is substantially less than the day time, electricity produced from wave energy, from say, 8 pm to 6 am (no. of hours of electricity generation and storage = 10 hours) could be used in 24 hour desalination plants, horticultural plants, or sold at a cheaper rate than the day rate to consumers who may use it for heating/pre-heating hot-water or industrial boilers for the next day.

If there is low demand for electricity generated from wave energy at night, then electricity storage facilities – costs of which may not commercially viable at the time of completing the manuscript – could be considered.

Based on data supplied by Wavegen, a UK company, its LIMPET model generates 1 to 1.2 GWh_e per annum during night time. If not used, this power could be wasted, hence storage systems – depending on costs – may represent a means to conserve this energy for future use.

Costing:

Assuming about half of the 2 GWh_e annual output produced at night by the LIMPET wave energy device is not used (because of lack of demand),

- About 1 GWh_e could be potentially stored in the Regenesys plant.
- Lifetime of the Regenesys plant is 15 years (Regenesys project office, NatPower/ETSU, Harwell) – half the lifespan of the LIMPET wave device (30 years). Hence 2 Regenesys units × (1 GWh_e/365 or 2.74 MWh_e) are required to support 1 LIMPET device.

The additional cost for installing the Regenesys plant would be

$$\text{US\$ } 100 \text{ (average)} / \text{kWh}_e \times 2 \text{ Regenesys units}$$
$$\times\ 1 \text{ GWh}_e / 365 \text{ (storage capacity per day)}$$
$$= \text{US\$ } 200 \times 10^6/365 = \text{ US\$ } 548{,}000$$

Annuitised cost method approximation, 10% discount factor:

$$\frac{\dfrac{0.1\times \$\ 1.1\text{m}+0.1\times \$\ 548{,}000}{\left[1-(1.1)^{-30}\right]=0.9427}}{2\ \text{GWh}_e}+1.65\text{c/kWh}_e$$
$$=8.74\ \text{c/kWh}_e+1.65\ \text{c/kWh}_e=10.39\ \text{c/kWh}_e$$

(compared to 7 c/kWh_e without Regenesys, but this assumes consumers are willing to buy at 7 c/kWh_e at night time – probably not. In practice, the cost of electricity could be much higher than the theoretical cost of 7 c/kWh_e as less than 2 GWh_e is sold per annum, due to <100% demand for 2 GWh_e/annum on a 24 hour basis).

Note: The O&M cost of Regenesys plant is not included.

In this scenario, stored electricity from wave energy can be sold at day-time, commercial or peak prices compared to reduced tariffs if electricity were to be sold at discounted prices at night. The internal rate of return would be much higher under these circumstances.

Investment appraisal of storage potential:
If the wave power intensity is higher than 2 GWh_e per annum (based on the conditions assumed in the previous case study), and the cost of the Regenesys plant is lower than US$ 100 per kWh_e, then the average cost of electricity could be less than 10 cents/kWh_e.

Cost for developing countries – The cost of the LIMPET wave device is expected to much lower than N.W. European prices, hence LIMPET may be commercially viable in these countries with or without the Regenesys plant.

Storage of electricity from wind energy

Basis of Carno wind farm, Wales, UK: Relocated to high wind-swept site in N.W. Europe (about 140 GWh_epa, 50% load factor)

Daytime generation to consumers	= 100 GWh_e
Nightime generation and storage per night	= 40 GWh_e/365 = 0.11 GWh_e
Wind turbine life	= 25 years
Regenesys plant life	= about 15 years (Regenesys/NatPower/ETSU Harwell)

Hence 2 Regenesys units ×0.11 GWh_e capacity are required to support 1 wind turbine.

The additional cost of installing the Regenesys plant would be

$$\text{US\$ } 100 \text{ (average) per } kWh_e \times 2 \text{ Regenesys units}$$
$$\times\ 40\ GWh_e\ /\ 365 \text{ (storage capacity per day)}$$
$$= \text{US\$ } 200 \times 40 \times 10^6/365 = \text{ US\$ } 21.9\text{m}$$

Annuitised cost method approximation, 10% discount factor:

$$\frac{\dfrac{0.1\times\$\ 39\ \text{m}+0.1\times\$\ 21.9\text{m}}{0.9427}+\$\ 0.375\text{m}}{140 GWh_e}$$
$$= 4.9\ \text{c}/kWh_e \text{ (more highly wind-swept region than Wales, North-West Europe)}$$

Note: This does not include O&M cost of Regenesys plant.

Comparison with Carno wind farm, Wales, UK: 35% load factor, average cost of electricity 4.6 c/kWh_e.

Investment appraisal of storage potential:
If the capital plant cost for Regenesys plant could be reduced below US$ 100/kWh_e, and if its O&M costs were small, and wind load factors increased beyond 50% (e.g. highly wind-swept coast locations, off-shore sites), then the average cost of electricity could be lower than 4.9 cents/kWh_e, based on project conditions as described in the wind farm case study. Cost for developing countries – the capital costs for wind farms are declining in countries such as India and China, as more wind projects are commissioned. Power storage facilities like the Regenesys plant should enable wind energy to compete with CCGT plants by 2010.

8.7
Costs and Benefits of Carbon Dioxide Reduction Technologies

Several studies performed by the World Bank, research institutes and NGOs on the cost of carbon dioxide abatement are principally based on discount-rate financial models. Some of the research groups suggest that discount rates should not be used to compensate for increase of external or social costs, while others argue that – in Kyoto Protocol terms – baselines will shift lower towards the commitment period of 2008–2012.

Table 8.6. Comparison of costs, benefits and carbon credits of power and co-generation technologies

Technology	Capital cost	Fixed & Variable O&M cost	Efficiency	Load Factor	Levelised cost	CO_2 emissions, reductions Overnight cost		
	Plant capacity on which capital cost is based				Basis of 10% disc. rate, fossil fuel & renew. energy, for compar.	Total CO_2	CO_2 emissions reduction below baseline	Overnight cost /t CO_2 reduction (*basis capital cost*)
	US$/kW$_e$ /kW$_{e+t}$ CHP plant where quoted. MW$_e$, MW$_t$ $ m total.	US cents /kWh	%	%	US cents /kWh	t/ GWh	t /GWh total tonnage	US$ /t
Baseline case								
Coal (convent.) minus SOx and particle removal	1,400 100 MW$_e$ 140	0.85	35	68.5	4.42	964	baseline	baseline
Energy Technologies								
AFBC + SOx and particle removal	1,500 100 MW$_e$ 150	1.87	35	68.5	5.72	963	negl. –	n/a –

In this section, a multi-sector analysis of costs and benefits of carbon dioxide technologies is provided in an extensive table. The contributions of the value of carbon credits to projects are also estimated at different market prices. It is clear from this analysis that very large differentials between actual costs of abatement and market prices exist. The author employs a uniform emissions benchmark for measuring emissions reduction, based on carbon dioxide emisisons per unit power output of a conventional coal-fired power plant, for most cases. For Annex I OECD countries, the baseline will be lower due to the average carbon content of the different types of fossil fuels, renewable energy and nuclear plants.

CO_2 costs and benefits to investors
(Total CO_2 emissions/GWh: World Energy Council and author's figures)

Levelised cost /t CO_2 reduction *(basis capital cost)* PV cost @ 10% /PV red @ 10%		CO_2 red. cost per tonne based on AIJ project data Costs are undiscounted: average, high and low costs for projects which were conducted in (mostly) developed countries		Approx. CO_2 return on investment assuming plant operates at its specific load factor. *Note: Reduction or parts of reduction would be traded at the market price, but prices may be negotiated between seller and buyer. One-off sales or part sales over fixed duration pending UNFCCC approval. No repetition of sales on first reduction, unless further investments are made on the technology which delivered the first reduction.*					
				Return market rate	Return market rate	Return on cap. cost if	Return on cap. cost if	Return on cap. cost if	Return on cap. cost if
US$ /t	Av.	US$/t High	Low	US$ 25/t Benefit US$ m	US$ 50/t Benefit US$ m	market rate US$ 25/t %	market rate US$ 50/t %	reduction sold @ levelised cost %	reduction sold @ 1/2 level. cost %
		baseline		baseline		baseline		baseline	
	-	n/a	-	negl.	negl.	negl.	negl.	negl.	negl.

Table 8.6. *Continued*

Technology	Capital cost	Fixed & Variable O&M cost	Efficiency	Load Factor	Levelised cost	CO_2 emissions, reductions Overnight cost		
	Plant capacity on which capital cost is based				Basis of 10% disc. rate, fossil fuel & renew. energy, for compar.	Total CO_2	CO_2 emissions reduction below baseline	Overnight cost /t CO_2 reduction (*basis capital cost*)
	US\$/kW$_e$ /kW$_{e+t}$ CHP plant where quoted. MW$_e$, MW$_t$ \$ m total.	US cents /kWh	%	%	US cents /kWh	t/ GWh	t /GWh total tonnage	US\$ /t
Biomass Steam cycle Power only (non-regrowth)	2,600 5 MW$_e$ 13	1.58	28	68.5	11.7	917	47 1,402	9,629
Biomass Steam cycle Power only (re-growth)	2,600 5 MW$_e$ 13 30 GWh$_e$	1.58	28	68.5	11.7	–	964 28,920	449
Biomass Steam cycle Heat + Power (non-regrowth)	3,000 3.9 MW$_e$ + 12.3 MW$_t$ 12	3.9	80 +	68.5	11.1 (electricity only)	t/GWh$_e$ 109 GWh$_{e+t}$ calc. by author	not applicable 21,388	561
Biomass Steam cycle Heat + Power (re-growth)	3,000 3.9 MW$_e$ + 12.3 MW$_t$ 12 3,000 based on kW$_e$	3.9	80 +	68.5	11.1 (electricity only)	t/GWh$_e$ 80 GWh$_{e+t}$ quoted by source	not applicable 46,244	259
CCGT	740 101 MW$_e$ 74.7	0.62	54	68.5	2.93	349	615 372,690	200
CHP Kemsley: CHP plant	1,110 81 MW$_e$ + 200 MW$_t$	0.70	80	68.5	2.90 /kWh$_e$ (elec)	t/GWh$_e$ Equivalent heat from coal for heat and power 581,579 t CO_2 red.	not applicable	191
Coal-oil (OPEC + coal); no S or part. rem)	1,160 100 MW$_e$ 116	0.81	35	68.5	5.66	873 54,600	91	2,124

CO_2 costs and benefits to investors
(Total CO_2 emissions/GWh: World Energy Council and author's figures)

Levelised cost /t CO_2 reduction *(basis capital cost)*	CO_2 red. cost per tonne based on AIJ project data			Approx. CO_2 return on investment assuming plant operates at its specific load factor.					
				Return market rate	Return market rate	Return on cap. cost if	Return on cap. cost if	Return on cap. cost if	Return on cap. cost if
US$ /t	Av.	US$/t High	Low	US$ 25/t Benefit US$ m	US$ 50/t Benefit US$ m	market rate US$ 25/t %	market rate US$ 50/t %	reduction sold @ levelised cost %	reduction sold @ 1/2 level. cost %
2,503	19	235	5	0.035	0.07	0.27	*0.54* may not meet investor's cost of finance	27.0	13.5
121	19	235	5	0.72	1.44	6.46	*12.9* potentially viable @ $ 50/t	57.2	28.5
77	40	178	1	0.53	1.07	4.4	8.9	13.7	6.8
35	40	178	1	1.15	2.31	9.6	*19.1* potentially viable @ $ 50/t	29.7	14.8
21.2	56	159	2	*9.3*	*18.6*	*12.4* potentially viable @ $ 25/t	24.9	10.5	5.3
16.4	40	178	1	*14.5*	*29.0*	*13.0* potentially viable @ $ 25/t	26.0	8.6	4.3
225	n/a	n/a	n/a	1.36	2.73	1.2	2.3	10.6	5.3

Table 8.6. *Continued*

Technology	Capital cost	Fixed & Variable O&M cost	Efficiency	Load Factor	Levelised cost	CO_2 emissions, reductions Overnight cost		
	Plant capacity on which capital cost is based				Basis of 10% disc. rate, fossil fuel & renew. energy, for compar.	Total CO_2	CO_2 emissions reduction below	Overnight cost /t CO_2 reduction (*basis* baseline *cost*)
	US\$/kW_e /kW_{e+t} CHP plant where quoted. MW_e, MW_t \$ m total.	US cents /kWh	%	%	US cents /kWh	t/ GWh	t /GWh total tonnage	US\$ /t
Gas Turbine	250 98 MW_e 24.5	1.12	32	68.5	3.24	644	320 226,838	108
Geothermal$_e$ *Simulated plant*	1,600 100 MW_e 160 600 GWh_e/yr	1.5	10	68.5	4.3	57 5	907 44,200	132
Geothermal$_{e+t}$ *Simulated plant* Heat + Power	2,400 30 MW_e + 70 MW_t	3.0 (est)	20 (est)	68.5 6,000 hrs/yr	5.74 (est) electricity	t/GWh_e	not applicable 244,317 sep. calc. on emissions	294
IGCC (biomass) Gasifier-STEG Power only (non-regrowth)	2,800 5 MW_e 14	1.67	41	68.5	11.4	718	246 7,374	1,966
IGCC (biomass) Gasifier-STEG Power only (re-growth)	2,800 5 MW_e 14 30 GWh_e 2,800 based on kW_e	1.67	41	68.5	11.4	-	964 28,920	484
IGCC (biomass) Gasifier-STEG Heat + Power (non-regrowth)	3,350 4 MW_e + 8.3 MW_t 13.4 3,350 based on kW_e	2.02	80 +	68.5	11.3 (electricity only)	t/GWh_e 140 GWh_{e+t} calc. by author	not applicable 19,044 sep. calc. on emissions	735

CO_2 costs and benefits to investors
(Total CO_2 emissions/GWh: World Energy Council and author's figures)

Levelised cost /t CO_2 reduction *capital capital cost)*		CO_2 red. cost per tonne based on AIJ *(basis*		Approx. CO_2 return on investment assuming plant operates at its specific load factor. project data					
				Return market rate	Return market rate	Return on cap. cost if	Return on cap. cost if	Return on cap. cost if	Return on cap. cost if
US$ /t	Av.	US$/t High	Low	US$ 25/t Benefit US$ m	US$ 50/t Benefit US$ m	market rate US$ 25/t %	market rate US$ 50/t %	reduction sold @ levelised cost %	reduction sold @ 1/2 level. cost %
11.4	56	159	2	5.67	11.34	*23* potentially viable @ $ 25/t	46	10.5	5.3
31	19	235	5	13.6	27.2	8.5	*17.0* potentially viable @ $ 50/t	10.5	5.3
31	40	178	1	6.1	12.2	8.5	*16.9* potentially viable @ $ 50/t	10.5	5.3
257	19	235	5	0.18	0.37	1.3	2.6	13.5	6.7
66	19	235	5	0.72	1.44	5.1	*10.3* potentially viable @ $ 50/t	53.1	26.5
92	40	178	1	0.47	0.95	3.5	7.0	13.0	6.5

Table 8.6. *Continued*

Technology	Capital cost	Fixed & Variable O&M cost	Efficiency	Load Factor	Levelised cost	CO_2 emissions, reductions Overnight cost		
	Plant capacity on which capital cost is based				Basis of 10% disc. rate, fossil fuel & renew. energy, for compar.	Total CO_2	CO_2 emissions reduction below baseline	Overnight cost /t CO_2 reduction (*basis capital cost*)
	US\$/kW$_e$ /kW$_{e+t}$ CHP plant where quoted. MW$_e$, MW$_t$ \$ m total.	US cents /kWh	%	%	US cents /kWh	t/ GWh	t /GWh total tonnage	US\$ /t
IGCC (biomass) Gasifier-STEG Heat + Power (re-growth)	3,350 4 MW$_e$ + 8.3 MW$_t$ = 12.3 MW$_{e+t}$ 13.4 3,350 based on kW$_e$	2.02	80 +	68.5	11.3 (electricity only)	t/GWh$_e$ not applicable 61 GWh$_{e+t}$ quoted by source	38,876	344
IGCC (coal) Power only	1,573 100 MW$_e$ 157	1.22	45	68.5	5.54	783	181 108,600	1,445
IGCC (heavy fuel oil residue) Heat + Power Texaco	1,530 280 MW$_e$ + 11.6 MW$_t$ 444	0.40	46 gross 40 net	68.5	4.30	450 –470 if c.v. fuel gas 20.6 MJ/m^3 Mostly electricity, hence t/GWh$_e$ acceptable	500 approx. 760,000	584
Landfill gas Power only plant	1,133 10 MW$_e$ 11.33	3.18	26	68.5	5.18	793 emitted per GWh$_e$	171 10,260	1,104
Oil (OPEC); no S or part. rem. Gas oil basis.	1,160 100 MW$_e$ 116	0.78	38	68.5	6.49	833	131 78,600	1,476
PFBC + SOx and particle rem.	1,274 80 MW$_e$ 101	3.31 waste disp.	40	68.5	7.60	900	64	3,308

CO_2 costs and benefits to investors (Total CO_2 emissions/GWh: World Energy Council and author's figures)									
Levelised cost /t CO_2 reduction *(basis capital cost)*		CO_2 red. cost per tonne based on AIJ project data		Approx. CO_2 return on investment assuming plant operates at its specific load factor.					
				Return market rate	Return market rate	Return on cap. cost if	Return on cap. cost if	Return on cap. cost if	Return on cap. cost if
US$ /t	Av.	US$/t High	Low	US$ 25/t Benefit US$ m	US$ 50/t Benefit US$ m	market rate US$ 25/t %	market rate US$ 50/t %	reduction sold @ levelised cost %	reduction sold @ 1/2 level. cost %
45	40	178	1	0.97	1.94	7.2	*14.5* potentially viable @ $ 50/t	26.7	13.3
154	40	178	1	2.7	5.4	1.7	3.4	10.6	5.3
62	40	178	1	19.0	38.0	4.3	8.6	10.6	5.3
171	19	235	5	0.25	0.51	2.2	4.5	15.5	7.7
156	56	159	2	1.96	3.93	1.7	3.4	10.6	5.3
352	40	178	1	0.76	1.52	0.75	1.5	10.6	5.3

Table 8.6. *Continued*

Technology	Capital cost	Fixed & Variable O&M cost	Efficiency	Load Factor	Levelised cost	CO_2 emissions, reductions Overnight cost		
	Plant capacity on which capital cost is based				Basis of 10% disc. rate, fossil fuel & renew. energy, for compar.	Total CO_2	CO_2 emissions reduction below baseline	Overnight cost /t CO_2 reduction (*basis capital cost*)
	US$/kW$_e$ /kW$_{e+t}$ CHP plant where quoted. MW$_e$, MW$_t$ $ m total.	US cents /kWh	%	%	US cents /kWh	t/ GWh	t /GWh total tonnage	US$ /t
Solar PV Ford plant Bridgend Wales, UK	33,000 100 kW$_p$ 110,000 kWh$_p$ or 0.11 GWh$_p$ 3.3	3.63	12.5 solar p.v. eff.	12 load factor: UK solar peak equiv.	322	5	959 105	31,428
Solar PV (simulated plant) – North Hemis. developed country	2,300 100 kW$_p$ 110,000 kWh$_p$ or 0.11 GWh$_p$ per year 0.23	1.7	15 assume 2000–05	12 load factor: UK solar peak equiv.	23.9	5	959 105	2,190
Solar thermal (simulated plant) – North Hemis. developed country	2,000 100 kW$_t$ 110,000 kWh$_t$ or 0.11 GWh$_t$ per year 0.2	1.5	10 assume 2000–05	12 load factor: UK solar peak equiv.	19.3* c/kWh$_t$	3	Not applicable in t/GWh$_e$. Separate calc. for heat from coal in coal fired-boiler 80% efficiency	4,728 34.2 t CO_2
Small hydro	3,683 10 MW$_e$ 48 GWh$_e$ per year	0.43	33	55	4.33	7	957 45,936	802
Tidal Mersey barrage UK (under study)	2,070 700 MW$_e$ 1,450 1,400 GWh$_e$ per yr	1.04	33	23	12.0	<10 (est)	954 (e) 1,335,600 (est)	1,085

CO_2 costs and benefits to investors
(Total CO_2 emissions/GWh: World Energy Council and author's figures)

Levelised cost /t CO_2 reduction *(basis capital cost)*	CO_2 red. cost per tonne based on AIJ project data			Approx. CO_2 return on investment assuming plant operates at its specific load factor.					
				Return market rate	Return market rate	Return on cap. cost if	Return on cap. cost if	Return on cap. cost if	Return on cap. cost if
US$ /t	Av.	US$/t High	Low	US$ 25/t Benefit US$ m	US$ 50/t Benefit US$ m	market rate US$ 25/t %	market rate US$ 50/t %	reduction sold @ levelised cost %	reduction sold @ 1/2 level. cost %
3,318	19	235	5	Insignificant; market price will not justify investment – however, marketing strategy has proved very effective – car sales increased. Reduction in internal lighting costs/year.					
232	19	235	5	0.0026	0.0052	1.1	2.2	10.6	5.3
501	19	235	5	0.0010	0.0020	0.5	1.0	10.6	5.3
reduction per year/boiler eff = 42.3 t CO_2 reduction per year									
85	19	235	5	1.1	2.3	2.9	5.9	10.6	5.3
115	19	235	5	33.4	66.8	2.3	4.6	10.6	5.3

Table 8.6. *Continued*

Technology	Capital cost	Fixed & Variable O&M cost	Efficiency	Load Factor	Levelised cost	CO_2 emissions, reductions Overnight cost		
	Plant capacity on which capital cost is based				Basis of 10% disc. rate, fossil fuel & renew. energy, for compar.	Total CO_2	CO_2 emissions reduction below baseline	Overnight cost /t CO_2 reduction (*basis capital cost*)
	US$/kW$_e$ /kW$_{e+t}$ CHP plant where quoted. MW$_e$, MW$_t$ $ m total.	US cents /kWh	%	%	US cents /kWh	t/ GWh	t /GWh total tonnage	US$ /t
Waste burn MSW CHP plant Nottingham	3,200 14.9MW$_e$ 118 GWh$_e$	6.21	59	80	8.7 Nottingham (est) electricity only	175,823 pa if C in MSW 30%	10,034 t CO_2 red. per yr	13,753
	28.3 MW$_t$ 170 GWh$_t$ 28.3					124,583 pa if C in MSW 20% t/GWh$_e$ not applicable separate calc. for heat from coal to produce heat + elect.	61,274 t CO_2 red. per yr	2,252
Wave LIMPET Model WaveGen (quoted)	2,200 0.5 MW$_e$ 1.1 2 GWh$_e$ per yr	1.65	65 conversion eff. (quoted) 20 kW/m	45	7.5	<10 (est)	954 (e) 1,908 (est)	576
Wind (onshore) Carno wind farm, Powys, Wales	1,160 33.6 MW$_e$ 100 GWh$_e$ per year 39	0.37	33	35	4.60	7	957	407

CO_2 costs and benefits to investors
(Total CO_2 emissions/GWh: World Energy Council and author's figures)

Levelised cost /t CO_2 reduction *(basis capital cost)* US$ /t	CO_2 red. cost per tonne based on AIJ project data US$/t Av.	High	Low	Approx. CO_2 return on investment assuming plant operates at its specific load factor. Return market rate US$ 25/t Benefit US$ m	Return market rate US$ 50/t Benefit US$ m	Return on cap. cost if market rate US$ 25/t %	Return on cap. cost if market rate US$ 50/t %	Return on cap. cost if reduction sold @ levelised cost %	Return on cap. cost if reduction sold @ 1/2 level. cost %
504	40	178	1	0.25	0.50	0.18	0.36	3.6	7.2
83	40	178	1	1.53	3.0	1.1	2.2	3.7	7.4
						potentially viable if methane capture credits were awarded: return will depend on market price of methane reductions			
51	19	235	5	0.047	0.095	4.3	8.5	8.8	4.4
45	19	235	5	2.4	4.8	6.1	*12.3* potentially viable @ $ 50/t	11.0	5.5

Table 8.6. *Continued*

Notes to Table 8.6.

Abbreviation/acronym	Technology/technical term/definition
AFBC	Atmospheric Fluidised Bed Combustion
CCGT	Combined Cycle Gas Turbine
CHP	Combined Heat and Power
Co-gen	Co-generation – same principle as combined heat and power (CHP)
IGCC	Integrated Gasification Combined Cycle
LIMPET	Locally Installed Marine Powered Energy Transformer – a wave power technology
OPEC	Organisation of Petroleum Exporting Countries – oil is the major energy resource for power generation in OPEC countries
MSW	Municipal Solid Waste
PFBC	Pressurised Fluidised Bed Combustion
S	Sulphur
SOx	Oxides of sulphur – sulphur dioxide and sulphur trioxide
STEG	Steam and Gas Turbine
Coal (convent.)	Conventional coal-fired power plant – normally without sulphur dioxide and particulate removal equipment

Fuel costs are based on average world market prices, July 1999.

CO_2 emissions per unit electricity output are based on data sourced from *Energy Technologies for Reducing Emissions of Greenhouse Gases: Proceedings of an Experts' Seminar*, IEA/OECD, and *Environmental Emissions from Energy Technology Systems: Total Fuel Cycle*, US Department of Energy – both attributed to R.L. San Martin (1989); and independent calculations by the author.

CHP projects which use fossil fuels or biomass, and IGCC projects which convert coal and heavy oil residues into combustible gas for combined cycle plants, are categorised as energy efficiency projects. The main criteria of specific "industry" categorisation is based on energy resource origin and relative technology cost – the latter factor is employed to further differentiate projects which involve elements of renewable energy, energy efficiency and fuel switching before or during operation.

Columns 11–13

Average, high and low costs of Activities Implemented Jointly (AIJ) projects are quoted. The figures are only indicative of technology costs in (mostly) developing countries but are subject to large variations due to different capacities of thermal power plants; local prices of fuel, labour, capital equipment; and methods of calculating the cost of CO_2 reduction. AIJ project costs quoted may hence not be comparable to costs calculated by the author for North-West Europe in columns 9 and 10, but provide at best some measure of cost savings or escalations if similar projects were conducted in developing countries. A rough method of discounting AIJ project costs may be obtained by dividing the undiscounted figure by 5.

Technology notes	Data, comments/remarks
Biomass	Steam cycle: power, heat + power; IGCC gasifier-STEG cycle: power, heat + power. Non-regrowth basis: CO_2 emissions calculated acording to average carbon content of biomass (eg. baled straw, wood). Re-growth basis: Zero CO_2 emissions assuming planting on short rotation coppice cycle of an average of three years.

Table 8.6. *notes continued*

Gas turbine, CCGT North Sea, n. gas, $ 1.50/GJ (Bacton price)
Coal-oil Gas oil, diesel; $ 150/t, CIF, North-West Europe (Jul 99), $ 3/GJ, 42 GJ/t, 1 t gas oil =

3.4 t CO_2; Coal $ 30/t, marker price (July 99), $ 1.50/GJ, 23.45 GJ/t (assume), 1 t coal = 2.2 t CO_2 (60% C). Basis of calculation: 50% gas oil, 50% coal;

Average values: 187,000 t fuel/yr, $ 90/t, $ 16.8 m. fuel cost/yr, $ 2.25/GJ, 33 GJ/t, 2.8 t CO_2/t of fuel, 523,600 t CO_2/600 GWh_e.
Oil Gas oil, diesel; $ 150/t, CIF, NWE (Jul 99), $ 3/GJ, 42 GJ/t, 1 t gas oil = 3.4 t CO_2, 147,000 t fuel/yr, $ 22 m. fuel cost/yr, 499,800 t CO_2/600 GWh_e (if heavy fuel oil, higher CO_2 emissions)
CHP Kemsley Cost based on $ 90 m./81 MW_e, quoted figures; $ 1.5/GJ Bacton field price, N. Sea.
Small hydro Load factor of 55%, assume Northern hemisphere summers lack water resources for

hydro.
Landfill Landfill power plant may claim methane emission reduction credits pending UNFCCC approval.

Discount rate

The author has employed a common interest rate of 10 percent for calculating costs of electricity for both thermal and renewable energy technologies, so that direct comparisons can be made from the table between thermal and renewable projects without re-calculation. Whilst 5% to 8% has been advocated as hurdle rates for CDM and JI projects, investors would require higher returns as a result of higher sovereign, currency and GHG reduction certification risks. It would be reasonable to assume that the weighted average cost of capital of a 70 : 30 debt to equity financing package in dollar terms would require a minimum return as reflected by the author's input rate of 10%. Consultations with power project sponsors and private financiers in late 1999 suggest that this may be an extremely conservative return in dollar terms. If project financing is dominantly debt, then the appropriate discount rate could be about 5% and 8% for multilateral and commercial lenders, respectively. For combined heat and power projects, separate analysis of power and heat production is required to calculate the costs of electricity and process/district heating steam.

Costs: UK costs vs costs in developing countries
The table provides estimates of energy, equipment and project costs in the UK, which may reflect general costs in "North" countries - OECD and developed countries; average 1995–99 costs for thermal projects; and circa year 2000 + costs for biomass (except waste-burn) solar and geothermal energy projects. Costs in developing countries and LDCs could be lower by an average of one third of costs stated in the table, excluding oil and natural gas projects.

Listing sequence: Technology is listed in alphabetical order.

8.8
CO_2 Cost and Electricity Cost and Return on Capital

Developed countries vs. developing countries: carbon dioxide reduction and electricity costs

A series of charts, Figures 8.1–8.15, provides technology and cost comparisons based on the author's independent calculations for carbon dioxide abatement and electricity generation in developed and developing countries – Asia was chosen as the comparative "developing countries" region because of its huge coal reserves. "Developed countries" generally refer to Western Europe (North-Western Europe represents the more advanced economies), North America and Japan. Emission levels from conventional coal plants is used as a common baseline for emissions reduction measurements for both developing and developed countries. Separate calculations for emissions resulting from cogeneration plants were performed for sources generating power and heat. As large differences in costs of production, labour and energy resources exist within, and between the "North" and "South" regions, the results presented in the comparative charts should only be treated as general estimates with variations of 10%–20% for developed countries and up to 30% for the Asia-Pacific region.

CO_2 Costs

Figures 8.1–8.6 illustrate the CO_2 cost comparisons.
Fossil-fuel based technologies – Developed countries and Asian region (basis: capital cost)

Renewable energy technologies – Developed countries and Asian region (basis of capital cost)

NW Europe vs. Asia – Fossil fuel vs. renewable energy

The average capital cost of CO_2 reduction is lowest for waste-burn in Asia with methane credits is US$ 8/t, but this value is dependent on the organic content of MSW, compared to CCGT Asia (US$ 15/t), and CCGT NWE (US$ 32/t). The CCGT NWE figure is calculated by the formula US$ 21/t x [lower emissions baseline of 900 t/GWh$_e$ (est. average Asia) divided by 600 t/GWh$_e$ (est. average NWE)]

Electricity Costs

Figures 8.7–8.11 illustrate the electricity cost comparisons.
NW Europe vs. Asia – Fossil-fuel based technologies

As costs of labour, energy resources and supply, technology and infrastructure development vary over a wide range across Asia, comparisons with developed countries can only be made with selected economies.
NW Europe vs. Asia – Renewable energy technologies

Return on Capital

Figures 8.12–8.15 illustrate the comparative returns on capital.
Fossil-fuel technologies, NW Europe vs. Asia
Renewable energy technologies, NW Europe vs. Asia

8.9 Financial Analysis – IRR, Risk Return and Project Financing

The internal rate of return (IRR) of a carbon dioxide reduction project can be calculated based on the sum of the amount of carbon credits generated (which have intrinsic financial value as determined by the market) and the amount of revenue fetched by the primary product of the project (e.g. electricity, paper, steel, cement, steam).

If financial additionality is a pre-requisite of the project, then the risk to investors would be that amount of revenue that could be lost as a result of selling the primary product at a price which reflects a negative net present value of cash flows, assuming that these cash flows were based on the commercial discount rate for the project. For a power project, investors are unlikely to provide finance unless the price difference between the primary product price (e.g. electricity generation price at negative NPV) and the lowest market price of the primary product (lowest purchase price of electricity in location where project is to be implemented) can be fully recovered by sales of carbon credits for the same location.

Project investors hence will not finance a CDM or JI project unless the price risk between the primary product price and the lowest market price of the primary product can be fully recovered by sales of emissions credits.

IRR and Risk-Return Profiles of a CCGT Project – North-West Europe

Two charts on Internal Rate of Return (IRR) and Risk-Return of cash flows for a CCGT project in North-West Europe are provided in Figures 8.16 and 8.17. The price risk represents the premium that investors may have to forego to realise an IRR higher than the original cost of financing.

Risk-Return Profile of an Energy Project in Asia

In Asia, where most of the savings in emissions abatement costs of "North" countries are to be realised, the author has produced a risk-return profile based on a multi-project proposal to a provincial power corporation. The returns, which are sensitive to baselines and capital costs of investments, could be improved by selective project management – particularly in projects associated with high generation of methane emissions. Figure 8.18 illustrates the risk-return profile of the Asian energy project.

Project Financing

If a CDM or JI project required investors to comply with financial additionality, then the flows of investments and proceeds of carbon credits could be represented by Figure 8.19. If mandatory contributions of 1% of the value of all carbon contracts were to be directed to an adaptation fund (such as a proposed LDC/Islands-at-Risk Climate Change Fund) and a similar 1% to the host country originating the carbon contract, then the pattern of cash flows could be illustrated by Figure 8.20 – provided the host country was also a beneficiary of the fund.

It is quite likely that an adaptation fund will be approved by the Conference of the Parties in Hague in November 2000, but the case for non-LDC, non-Annex I countries benefiting from a mandatory percentage of all carbon contracts traded from projects originating from these countries is yet to be discussed.

Figures 8.1–8.20 illustrate the technical abatement costs of carbon dioxide, electricity costs, return on capital, risk-return profiles, and financing flows of greenhouse gas reduction projects on a global basis.

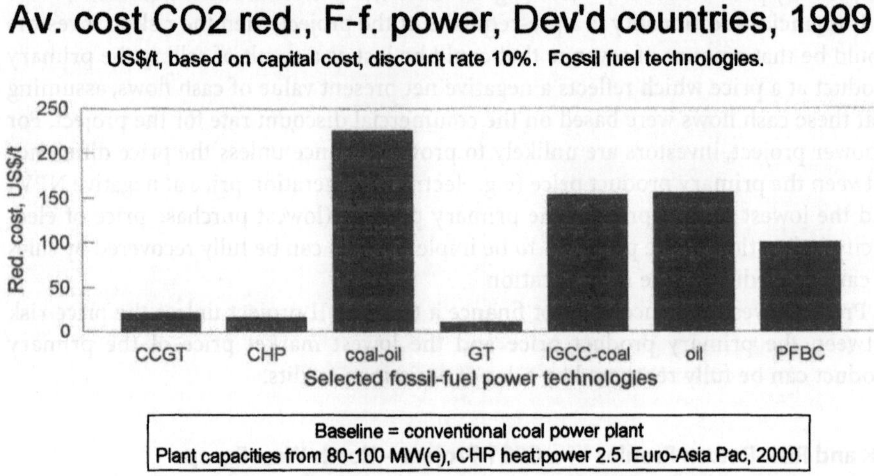

Fig. 8.1. Average cost of CO_2 reduction, fossil fuel power, developed countries, 1999. In developed countries, average costs of CCGT, CHP and gas turbine technologies in 1999 were lowest for 80–100 MW_e plants, as shown in the figure. PFBC provided least reductions

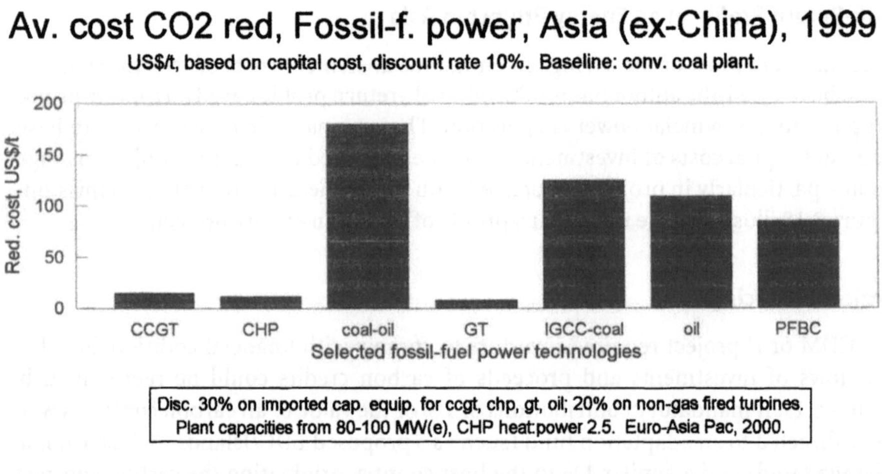

Fig. 8.2. Average cost of CO_2 reduction, fossil fuel power, Asia, 1999. In Asia, on the basis of 30% and 20% discounts on capital equipment (specifically turbines) powered by "easy-burn" and "processing/transformation required" fuels, respectively, gas technologies achieved the highest reductions per unit capital input, as shown on in the figure. Estimated costs in China are about half the values shown on the chart, but would be sensitive to gas prices

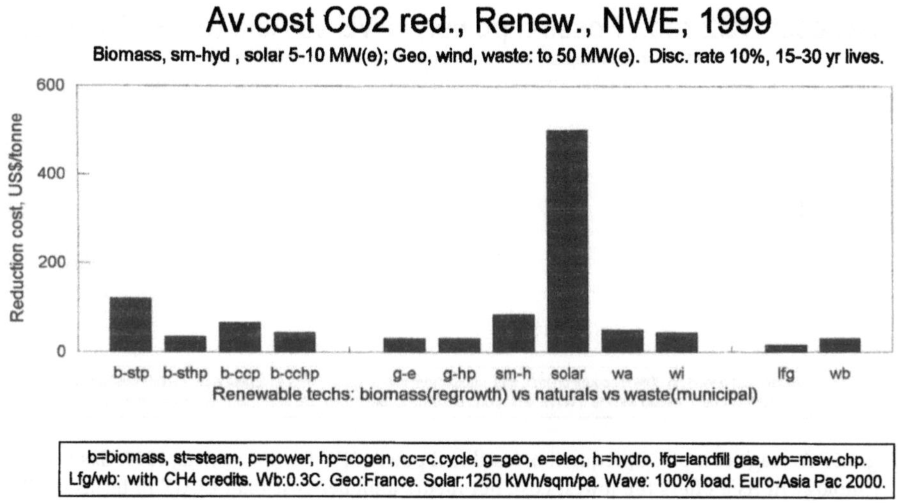

Fig. 8.3. Average cost of CO_2 of reduction, renewable energy, N.W. Europe, 1999. In North-West Europe, regrowth biomass steam cycle and IGCC-STEG heat and power schemes, geothermal (France), wave and wind energy are the most competitive technologies for carbon dioxide reduction compared to fossil-fuel thermal technologies. Wave power assumes annual full-load on the turbo device. The financial potential of waste-burn technology can be increased if methane capture can be credited to the plant owners or investors

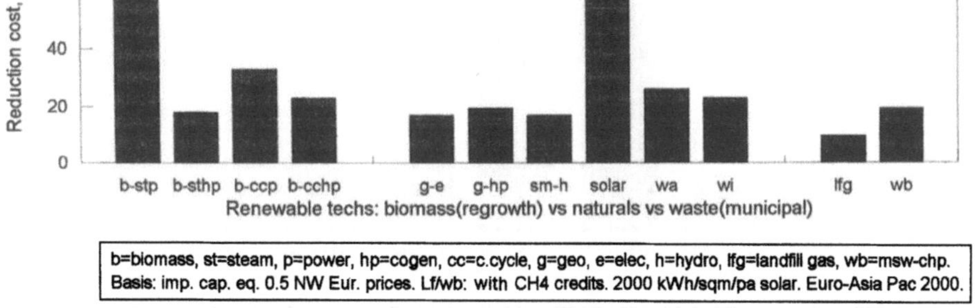

Fig. 8.4. Simulated average cost of CO_2 reduction, renewable energy, Asia, 1999. "Physical-naturally occuring" renewable sources in Asia – geothermal, small hydro, wave and wind; and regrowth biomass steam cycle and IGCC-STEG heat & power schemes offer the highest carbon dioxide reduction efficiencies, as revealed in the figure. Solar power would be competitive at annual radiation levels of 2,500–3,000 kWh_p per square metre. The benefits of methane reduction for landfill gas operations and binary gas reduction (carbon dioxide plus methane) for waste-burn plants is augmented by land protection and preservation of local flora and fauna

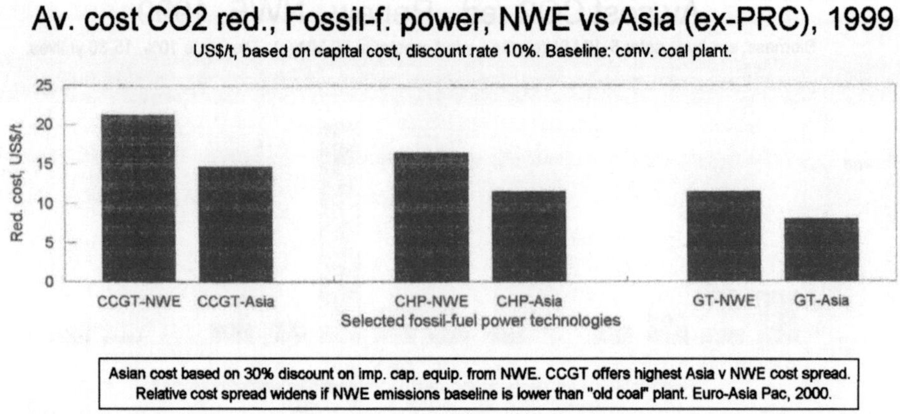

Fig. 8.5. Average cost of CO_2 reduction, fossil fuel power, NW Europe vs Asia, 1999. Gas turbine technology in Asia offers the lowest cost, but incurs a higher cost than CCGT in terms of IRR on electricity sales (GT produces less power per unit investment expenditure). CCGT plant investors can realise larger savings on CO_2 reduction technology investment in Asia than in North-West Europe

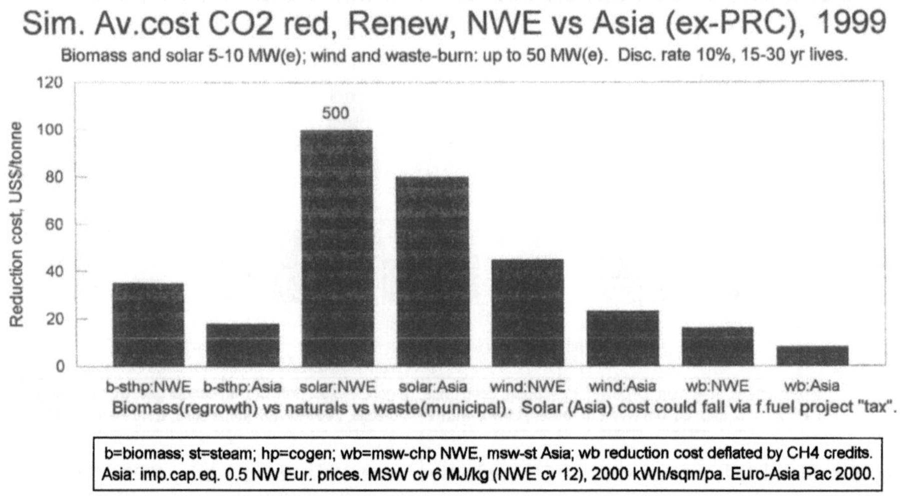

Fig. 8.6. Simulated average cost of CO_2 red., Renewable energy, NWE vs Asia, 1999. The impact of methane credits at US$ 10 per tonne of CO_2 equivalent is shown with reference to steam cycle waste-burn plants in NWE and Asia. Solar electricity costs in NWE and Asia could be subsidised by a "solar energy tax" in locations in which solar plants are uncompetitive with fossil-fuel plants (e.g. urban areas) or poverty-stricken communities in non-grid connected, remote zones

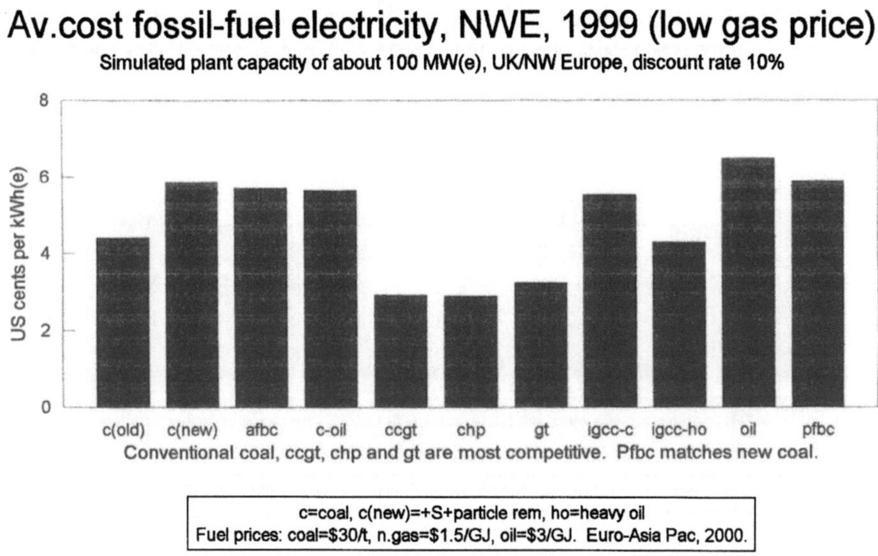

Fig. 8.7. Average cost of fossil fuel electricity, N. West Europe, low gas price, 1999. Gas-fired combined cycle and conventional coal plants, as shown, offer the lowest costs, but acid gas emissions controls may erode old coal's advantage. Heavy oil gasification technology are able to guarantee the next most competitive prices

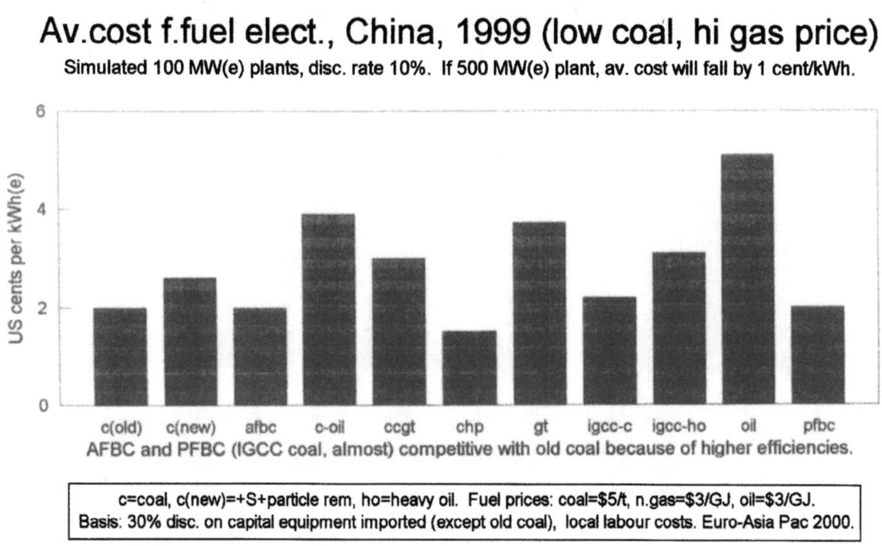

Fig. 8.8. Average cost of fossil fuel electricity, China, 1999 (low coal, high gas prices). Three "clean coal" technologies – AFBC, PFBC and IGCC compete with conventional coal plants. Gas-fired combined cycle plants are not viable unless security of gas supply and low delivery prices were guaranteed. Combined heat and power can offer low electricity prices if exported to the grid or private sector

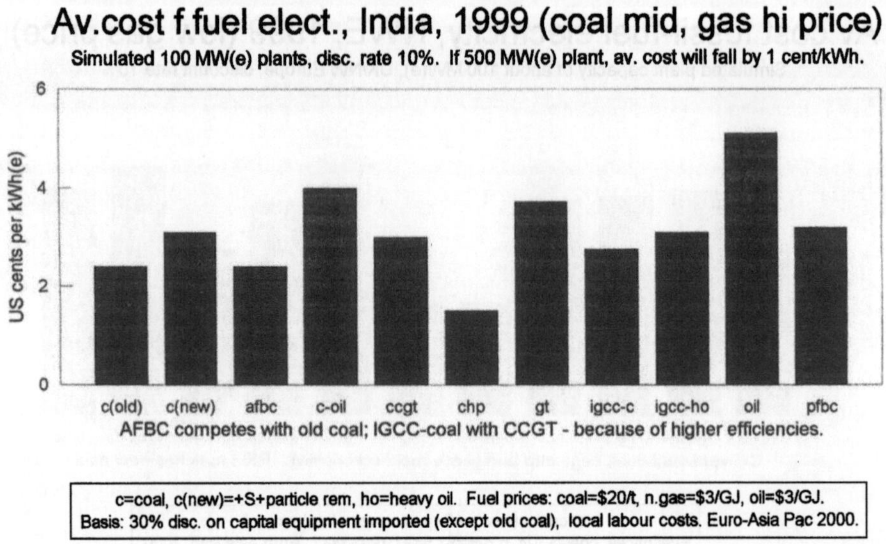

Fig. 8.9. Average cost of fossil fuel electricity, India, 1999 (mid coal, high gas prices). AFBC and conventional coal plants incur the same generation costs, and both are able to compete with CCGT, as the figure reveals. IGCC using coal or heavy oil feedstocks, and PFBC produce electricity at lower costs than single cycle gas turbines. CHP exporters offer cheap power

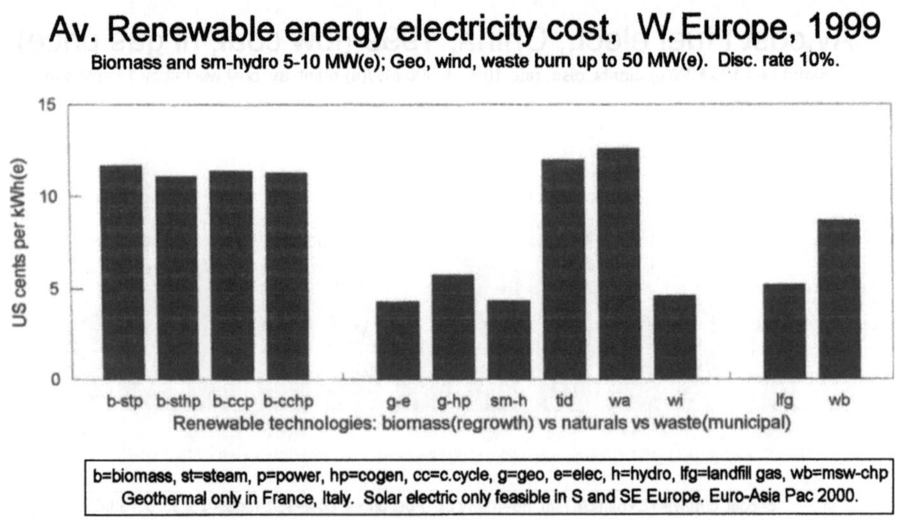

Fig. 8.10. Average renewable energy-based electricity cost, Western Europe, 1999. Of the technologies compared, small hydro, wind, landfill gas and geothermal power plants offer the lowest electricity prices. Biomass electricity generation costs should fall if larger capacity turbines are used. For CDM or JI projects, landfill projects may be able to claim methane reduction credits, and waste-burn project costs could be reduced with higher municipal waste removal charges

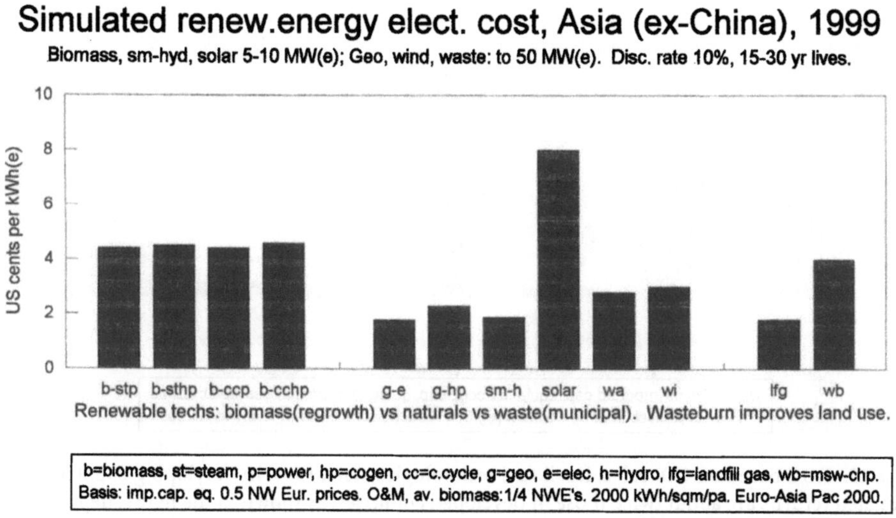

Fig. 8.11. Simulated renewable energy-based electricity cost, Asia, 1999. In locations where there is high availability and supply/intensity of biomass, geothermal, small hydro, wind, wave energy and landfill gas, the cost of electricity from these renewables as shown in the figure would be commercially viable compared to that if the locations were grid-connected. Solar power costs could be reduced by more efficient photovoltaic cells and high solar radiation levels. Municipal solid waste plants represent potential opportunities for power and heat generation. The costs for China are estimated at half of the Asian region's

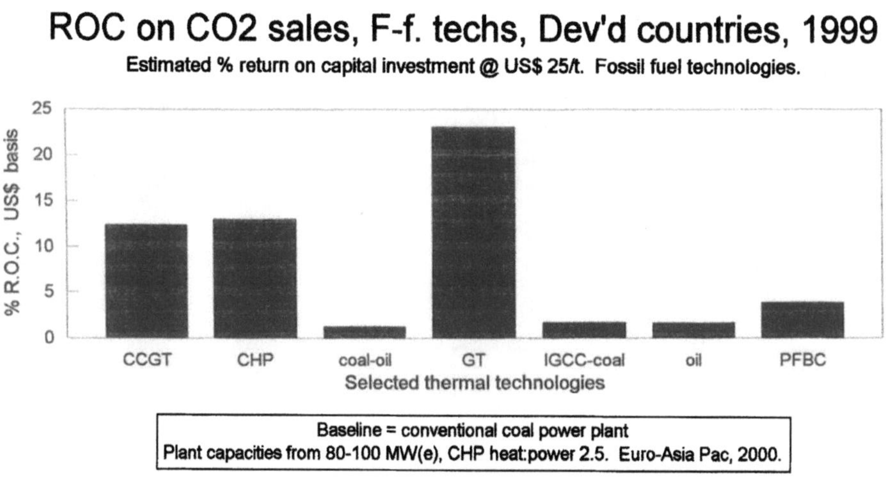

Fig. 8.12. ROC based on CO_2 sales, fossil fuel technologies, developed countries, 1999. In developed countries, single cycle gas turbine plants offer the best returns on capital, but would incur higher costs per electrical output than CCGT or CHP if gas prices escalated – as the figure reveals. The returns on "clean coal" technologies are small, but their costs of electricity would fall at low coal prices

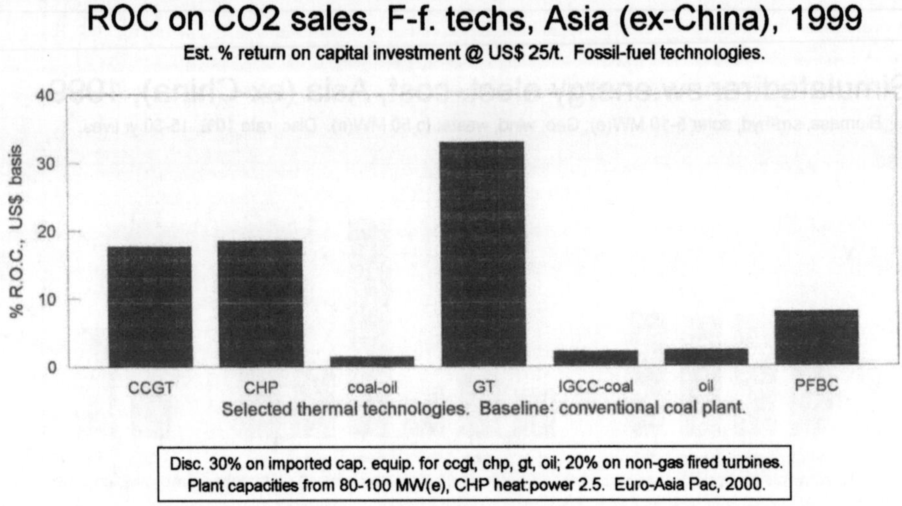

Fig. 8.13. ROC based on CO_2 sales, fossil fuel technologies, Asia, 1999. In developing countries, estimated returns are about twice those in developed countries, as shown in the figure, depending on discount rates on prices of imported capital equipment. Commercial price discounts for gas turbines would be higher than "ruggedised" turbines that operate in more severe environmental conditions

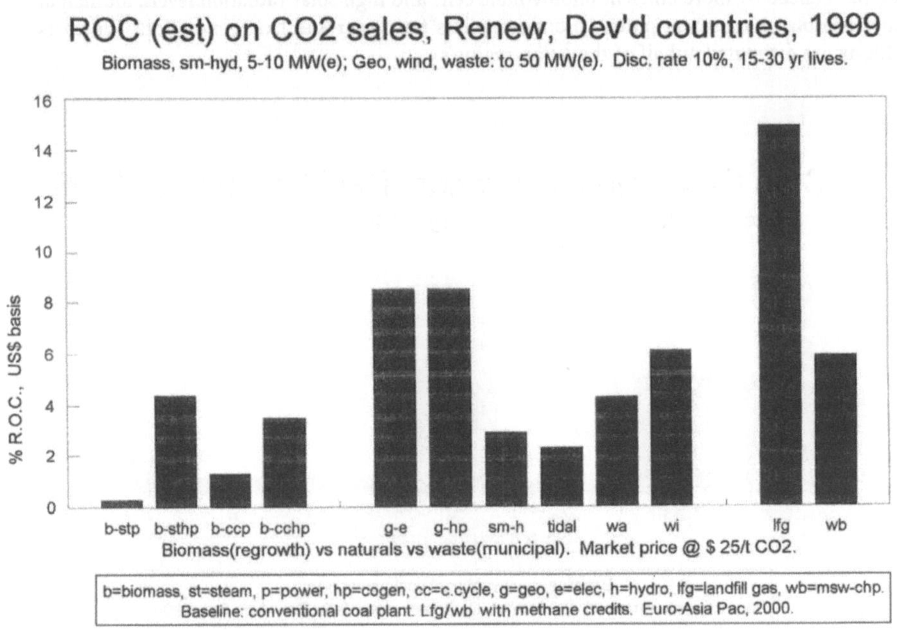

Fig. 8.14. ROC based on CO_2 sales, renewable energy, developed countries, 1999. In countries bordered or intersected by deep geological faultlines, geothermal energy provide relatively high returns on investments in dollar terms, as shown in the figure. Wind, wave and biomass cogeneration offer good returns, while methane credits may make landfill gas plants very attractive

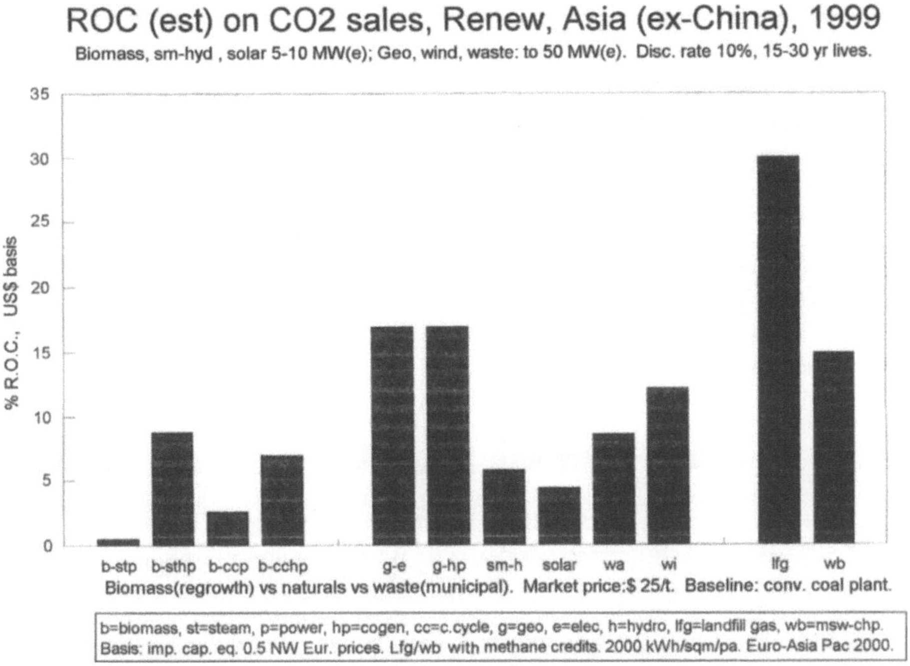

Fig. 8.15. ROC based on CO_2 sales, renewable energy, Asia, 1999. The benefits of renewable technologies exist along the "ring of fire" chain from the Russian Far East to the Philippine islands. Regions with high wind velocities, and small islands with embankments that can be retrofitted with wave energy devices should offer 10%–15% returns on capital, as the figure reveals. Solar energy is commercially viable in those locations that experience 2,000–3,000 kWh_p per square metre per annum

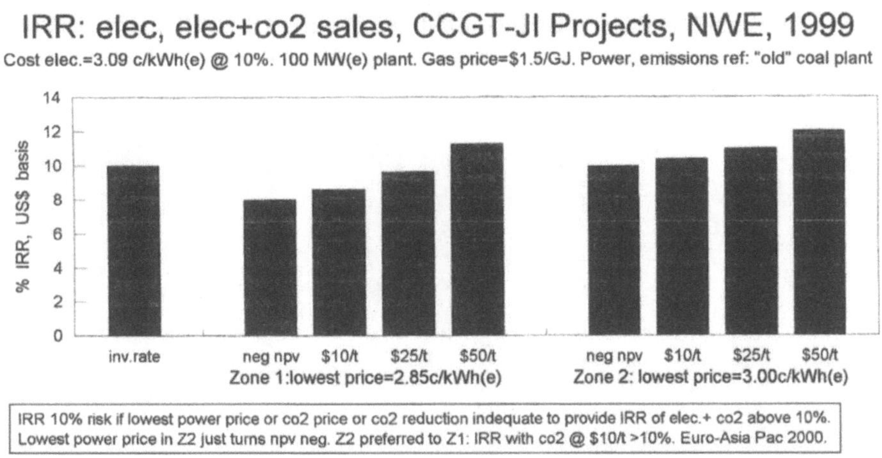

Fig. 8.16. IRR profile of a CCGT project in N.W. Europe, 1999. The financial impacts of the lowest or reference market price of electricity and the market or negotiated price of carbon dioxide credits in two different zones in North-West Europe are illustrated. At a project cost of electricity generation discounted at 10%, selecting Zone 2 instead of Zone 1 results in a higher project IRR from both electricity and carbon dioxide credit sales. Zone 2 offers a smaller electricity price difference than Zone 1 hence less price risk to the project company. However, the benefit from carbon credits is dependent on the market or negotiated price between the seller and buyer

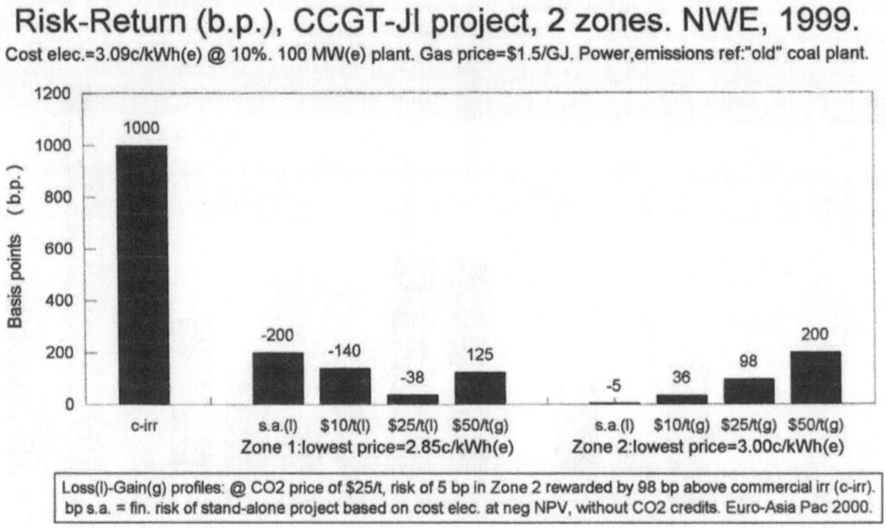

Fig. 8.17. Risk-return profile of the CCGT project (Fig. 8.16.) in N.W. Europe, 1999. The different rates of return for the same CCGT project in North-West Europe (see previous figure) is replotted in a comparative risk-return format. The vertical axis represents basis points (100 b.p. = one percent). At the market price of electricity, revenue in Zones 1 and 2 result in negative NPVs (negative NPV criteria of financial additionality). At a CO_2 price of US$ 25/t, revenue from CO_2 credit sales results in a 98 basis points return over the market price of electricity in Zone 2, but in Zone 1 revenue from CO_2 credit sales is insufficient to reverse a 38 basis point loss. The risk-return ratio in Zone 2 is hence more attractive for project investors and financiers

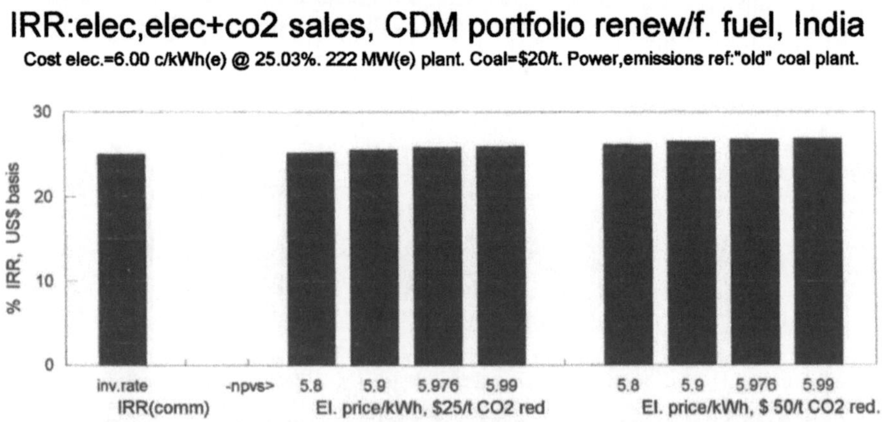

Fig. 8.18. IRR and risk-return profiles of a power projects portfolio in India, 1999. Assuming a superior return of about 25% on debt and equity capital, greenhouse gas emission reductions via the CDM enables an additional financial return of between 80 to 190 basis points (depending on market price of carbon dioxide equivalents), as shown. The portfolio consists of a number of plants using wind, biomass, solar, landfill gas, coalbed methane, coal and heavy oil gasification, pressurised fluidised bed combustion, CHP and CCGT technologies to replace the capacity of old conventional coal-fired plants

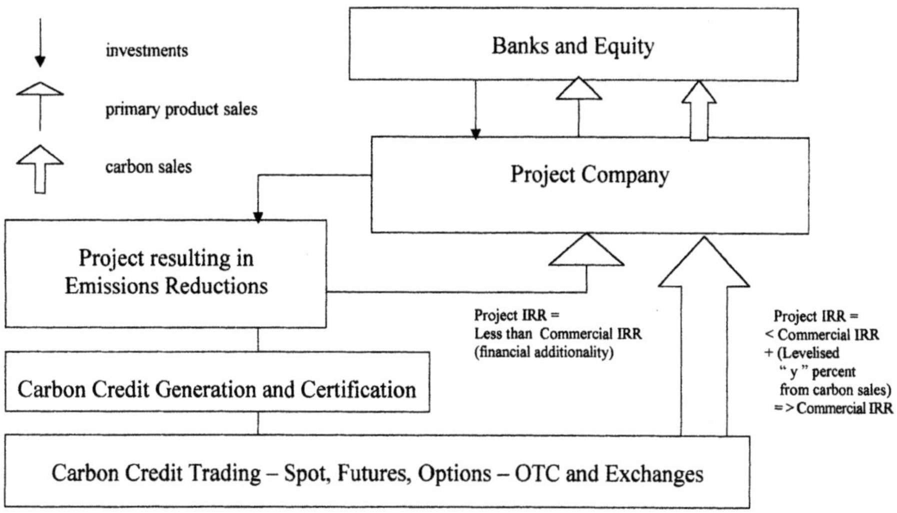

Fig. 8.19. Project financing flow diagram, additionality requirement.

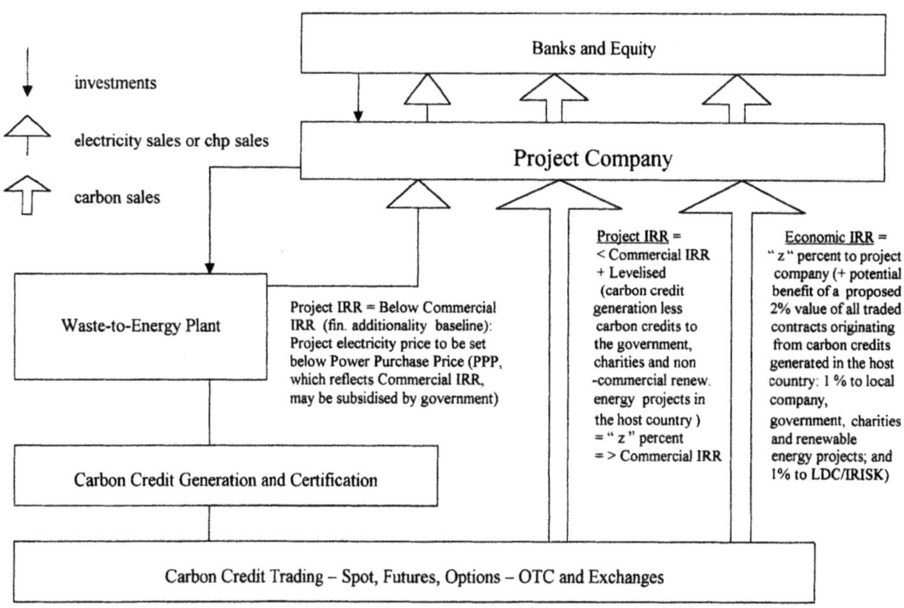

Fig. 8.20. Project financing flow diagram, proportion of carbon credits traded in secondary market is recycled to the host economy

9
Emissions Trading

This chapter covers a series of operations which are essential for the implementation of an efficient emissions trading market on the domestic and international level. An introduction to how a national emissions trading scheme may be structured in an Annex I country is firstly presented – in practice, consultations with UNFCCC are necessary so that the operation of the scheme complies with articles 3 and 17 of the Protocol.

The basics of the Value-at-Risk (VaR) technique and its application in emissions trading must be understood before one contemplates the sale or purchase of emissions reductions. VaR examples for energy and emissions contracts are provided in this chapter, which should assist the reader to appreciate the fundamentals of risk assessment in CDM and JI projects. The calculations are mainly based on power projects, but they can be similarly applied for cement, chemical and steel works; paper mills, smelting operations, heavy machinery and vehicle manufacturing plants.

Specifications for standard futures and options contracts (futures contracts would be priced using the forward price formula) are proposed next, followed by step-by-step examples for hedging price risk in the paper industry. The chapter ends with an evaluation of the impact of emissions trading on financial markets and credit rating of securities.

UNCTAD initiatives

UNCTAD (United Nations Conference for Trade and Development) has conducted a series of workshops on international emissions trading, but the current three initiatives of the Kyoto Protocol – CDM, JI and Emissions Trading – are likely to precede any allocation of global permits for emissions trading. Contracts on spot and derivatives trading of greenhouse gas emissions may require approval and authorisation of UNFCCC. An options market is expected to pre-empt the futures market, until the USA decides to ratify the Kyoto Protocol.

A framework for an emissions trading exchange and case examples in this chapter show how typical transactions may take place. Functioning of GHG derivatives markets assume that permits entitlements or emissions credits are allowed to be "banked" or sold forwards. If compulsory "sell-by dates" are imposed on permits, then it is also assumed that derivatives contracts are allowed to be made before the sell-by date of permits.

9.1
A National Pilot Scheme for Allocation and Trading of Carbon Permits

The fundamental of "emissions trading" is based on sale and purchases of units of reductions of specific gases between parties which have a surplus or deficiency of emissions allowances. Under the Kyoto Protocol, non-Annex I countries are not obligated to meet any reduction targets but they can be beneficiaries of inward investments through the Clean Development Mechanism or Joint Implementation initiatives.

In the absence or lack of CDM and JI investments, developing countries should consider implementing internally-regulated carbon emissions trading programmes if their own domestic environments are threatened. Carbon trading schemes, which may be preferred to energy or carbon taxes, "buy time" for emitters which would otherwise be financially obligated to invest in technology to meet a country's emissions reduction target.

A national permits scheme for an Annex I country: framework methodology

The United Kingdom could implement a scheme built on past experience of other countries, and improvise with factors that relate to its own political and economic environment. Factors which the UK government need to consider:

- A UK scheme could adopt the generic system based on USA's sulphur dioxide reduction programme, and include provisions to meet the UK's industrial and market environment. Consultation with firms will be needed to plan the pilot scheme.
- Market-based incentives in the air pollution sector in the USA are currently led by trading of emissions reductions. In 1992, participant companies realised cost savings of nearly 6% of compliance costs of US$ 29 billion.
- Emissions trading, however, does not represent a long-term solution to global warming, but allows time for companies to temporarily offset command and control costs. If the firm takes no positive action other than buying permits or credits indefinitely, then it is certain to incur huge financial losses as a result of higher production costs.
- The UK government hence must decide on "ceilings" for UK firms which may want to acquire emission reductions abroad through the CDM, JI and emissions trading initiatives of the Kyoto Protocol.

Plants and facilities pre-1990, 1990

(Basis of UK's emissions target for carbon dioxide of 20% below 1990 level by 2010)

- Plants, built before or in 1990, would be assigned permits equal to their level of emissions in 1990. Each permit would be an entitlement to emit, say, 1 tonne of carbon dioxide. This method of assignment allows permits to be credited to those firms which have invested in technology to reduce emissions.

- Case study – Year 2000. Company A, no additions to plants since 1990.

Emissions in 1990 = 100,000 t; 2000 = 90,000 t.
Reduction due to new technology investment. Allocation: 100,000 permits.
It is allowed only 80,000 permits by end 2010.

It is fined on the price of 20,000 permits otherwise.
If the plant reduced emissions to 70,000 t. in 2000 then it can sell 10,000 permits at the prevalent market price.

Plants and facilities post-1990 – considerations and issues

- Case study – Year 1991. Company B. In 2000, no additions to plant.

Emissions in 1991 = 100,000 t; 2000 = 90,000 t.
Reduction due to new technology investment.
Allocation: considerations

If a comparable plant can be identified, in the same industry and size, say, company A, then it would be reasonable to assume that B, if built in 1990, has not reduced emissions in 1991. Company A may not have made any reductions either in 1991. Company B may have used a higher carbon content fuel than Company A or the overall market economy in 1991 was stronger than 1990. A comparable plant owned by Company A may not exist for matching Company B's emissions.

A fair basis for permits allocation for a company such as B is provided next.

Plants and facilities post-1990: proposed system of permits allocation

After reviewing the different factors that influence emissions from plants built after 1990 compared to 1990, a fair system for the allocation of permits would be assign that number of permits equivalent to the plant's emissions in the year of operation.

- Case study (UK target of 20% below 1990 by 2010)

Company B is allocated 100,000 permits. It is allowed only 81,000 permits by end 2010.
It is fined on the price of 19,000 permits otherwise.
(B's obligation is the reduction of emissions of

$$\frac{\text{End 2010} - \text{end 1991}}{\text{End 2010} - \text{end 1990}} \times 20\% = 19\% \text{ (on 1991)}$$

If B reduced emissions by 19% at end 2000, then it has met the UK target. It can trade any further reductions of emissions.

- New plants

If expected emissions from a new plant or factory were known, it would need to have permits for these emissions.

Case 1.
Current and new emissions have been considered by the government in setting its target by 2012 - more importantly, on the 12.5% reduction on GHGs by 2012 on 1990 levels.
In this case, permits to the plant would be included in the emissions volume which is equal to 12.5% × 1990 level.

Case 2.
Emissions not envisaged by the government in setting its 2012 target. Permits to the plant would exceed the emissions volume as calculated above, and such issuances will increase UK's commitment to the EU bubble agreement, unless industry as a whole

manages to achieve cuts below UK's target. The target of 20% reduction below 1990 levels by 2010 is hence a fore-sighted move on the government's part. New plants should not be charged for their (notional) emissions.

Assignment and allocation of permits

1. Plants built before or in 1990

Allocation by historic emissions volume or "grandfathering" basis. Emissions permits are allocated according to the amount of GHG emissions in 1990 – emission levels before 1990 are ignored because they do not relate to the Kyoto Protocol. Its target emission level is less its mandatory reduction amount by the permits expiry year. It pays a fine equal to the number of permits for excess emissions relative to the expiry year or receives income from selling off any surplus reductions over its mandatory reduction amount.

2. Plants built after 1990, before permits allocation

Emissions target calculated by the number of months between plant operation and compliance year, divided by the number of years between compliance year and 1990, multiplied by the government's % target for the period.

3. New plants, current year of allocation, future years

New plants should not be penalised for "adding" to the country's total emissions, but treated fairly as other existing plants. New plants may not receive as many permits as they would like, unless they use high carbon fuels. They could be allocated permits based on an equitable distribution system or they could purchase permits by competitive auction. Annual permits with diminishing emissions rights to, say, 2003 could be allocated and auctioned.

Tradable economic instruments and parity issues: nationally issued permits vs. cross-border emission credits

Parity of GHG permits and "overseas-generated" credits represents a topical issue in the Kyoto Protocol agreement – whether one unit of carbon dioxide equivalent reduction unit purchased from CDM or JI projects or Annex B sellers is equal to one unit of mandatory permits issued or sold to emitters in Annex I countries. The rationale and admissibility of different baselines, discussed in Chapter 7, may enable project stakeholders to assign equivalent values of credits to permits.

JI credits, however, may have to be re-evaluated against CDM credits for projects that are implemented in non-Annex I host countries – this could be done by discounting JI credits if JI project criteria does not match the more stringent CDM project requirements.

9.2 Derivatives for Hedging Price Exposures of Carbon Credits

Financial instruments are imperative for the hedging of prices at a future date. Carbon credits can be regarded as tradable commodities with fluctuating or static prices as determined by market conditions. The most basic tools employed by financial institutions – futures, options, swaps – would be the most appropriate techniques of hedging for a buyer or seller of carbon credits in the marketplace, assuming perceived risks of their intrinsic values with time.

In the absence of actual prices of traded carbon contracts at the time of publication of this book, hedgers can only formulate a view of predicted prices on the basis of economic models, past prices in the Joint Implementation market, and settled prices in internal trading schemes of major energy companies. The next section deals with Value-at-Risk evaluation of the price risk of carbon credits based on standard forwards and options formulae.

9.3 Value-At-Risk of CDM/JI Investments

The maximum expected loss of a CDM/JI project may be estimated by a combination of (i) the loss as calculated by the Riskmetrics method – developed by J.P.Morgan to assess potential losses in financial markets under normal market conditions, and (ii) intrinsic losses which may occur as a result of project compliance to Kyoto Protocol criteria.

Value-at-Risk for a project is usually abbreviated as VaR @ the level of confidence interval required, e.g. VaR @ 95%, VaR at 97% etc. A VaR @ 95% represents a probability of 5% that the value of an observation will be at least 1.645 standard deviations below the mean (the number of standard deviations can be obtained from normal distribution tables). In simple terms, a worse-case scenario of 95% losses can be characterised by 1.645 multiplied by the standard deviation from the mean of the normal distribution.

In evaluating VaR, a host of numerical techniques which involve simple to highly complex calculations can be employed – these include inputs from Lognormal, Variance Covariance, Monte Carlo and Mean-Reverting methods. Simulations have been performed in energy and electricity markets in North America, UK and Germany but the scope for application to the emissions market remains a real challenge.

In recent years, particularly during very hot summers in the USA, electricity prices have exhibited day or weekly peak volatilities between 10% to 200% due to power demand surges and supply shock constraints. Riskmetrics and other financial models have not been able to predict such volatilities because energy prices do not behave like stock prices across a given time interval. Energy producers rarely offer long-term fixed price contracts on fuels, and higher surface temperatures as a result of global warming may dramatically increase VaR for both generators and suppliers.

In the power market, there are three primary risk elements which affect Value-at-Risk in CDM and JI projects – the mainstream product revenue (electricity), fuel price (e.g. gas, oil, coal, biomass) and the emissions trading price. Other major concerns are political, economic and currency risks, and risks normally associated with project financing.

The most basic VaR equations and present value terms which relate to quantifying total losses of CDM/JI projects are employed by the author. More complex energy models exist but they can only provide misleading results as no reliable database on emissions prices or volatilities were available at the time of manuscript finalisation.

There are numerous good references in the market which offer econometric models to predict VaR under different conditions and assumptions – McGraw-Hill's *Energy Risk Management*, P. Fusaro, 1998, and Risk Publications' *Managing Energy Risk*, 1999 are recommended.

Futures and Forward price in Emissions Trading: contract differences and price parity

Case examples of carbon dioxide trading, provided later in this chapter, refer to futures prices and not forward prices. Futures contracts are obligatory agreements between parties to pay or receive any differences between the futures price after contract is made and the futures price as quoted in the contract. The differences are marked-to-market daily and settlements are cleared through an exchange by margin-calls on the parties. A forward contract does not require daily settlements of price differences, but parties are obligated to trade the underlying asset at the forward price at the maturity date of the contract. Although the cash flows may be different for forward and futures contracts, the futures and forward price may be used interchangeably as both reflect the same value.

Value-at-Risk of CDM/JI power projects

The Value-at-Risk of power projects under the Kyoto Protocol framework, notwithstanding any unexpected amendments or additions to current articles of the Protocol, will principally depend on (i) fuel price volatility, security and duration of purchase contracts, (ii) power purchase agreements, power price volatility in the market, shock supply and demand factors, and (iii) choice of baselines to meet financial and emissions additionality criteria. The last factor could represent the highest risk to investors, as initial estimates of emissions abatement costs of Activities-Jointly-Implemented projects exhibit large standard deviations. However, the market value of emission credits is expected to stabilise after the first periods of active trading as a result of sellers' discounts on actual costs to meet market prices.

According to the author's calculations, the bulk of returns on most CDM/JI power projects will be based on the cost or selling price of electricity and process steam or district heating. The returns on emission credits may become significant – up to 10% in present value terms -in highly efficient and cost effective power projects.

Nominal formula for mainstream power product (electricity)

Single Power Contract: VaR based on current unit price of electricity

VaR @ 97.5% (VaR formula based on current unit price of electricity)

$$= 1.96 \times q\ (\text{power}) \times P \times \sigma$$

where

q = power contract value in MWh_e

= number of days × number of hours/day × number of MW_e/hour

P = current or spot unit price of electricity underlying the contract, US$/$MWh_e$

σ = Black-Scholes implied volatility of P if current call option price is known (assuming the contract has embedded options). Standard deviation of continuously compounded period rate of return, i.e. stock price volatility, if current call option price unknown

VaR @ 97.5% = VaR formula based on 97.5% potential loss of electricity contract, based on observations by P. Fusaro (* pp 163, Energy Risk Management, McGraw-Hill) and not 95% potential loss, the industry standard adopted by J.P.Morgan

1.96 = number of standard deviations from the mean (obtainable from normal distribution table) at which 97.5% of losses can be incurred – based on daily position fluctuations of electricity prices (see P. Fusaro *)

Single power contract – case example

3 year power contract, around-the-clock (ATC) delivery, power supply rate 50 MW_e/hour, current unit price of electricity US$ 31/MWh_e, daily price volatility 1.8%.

$$\begin{aligned} \text{VaR@97.5\%(daily VaR)} &= 1.96 \times q\ (\text{power}) \times P \times \sigma \\ &= 1.96 \times (3 \times 365 \times 24 \times 50) \times 31 \times 0.018 \\ &= \text{US\$ 1.43 million} \end{aligned}$$

Portfolio Power Contract: VaR based on current unit price of electricity

The classical portfolio theory for financial markets can be extended to the power market – instead of assessing the variance of a portfolio of financial assets, the VaR formula evaluates the variance of power contracts. The portfolio variance is the square of the standard deviation of the portfolio; the lower the portfolio variance, the smaller the VaR. The objective of risk managers in the power market is to reduce portfolio variance to as close to zero as possible with risk diversification strategies.

$$\text{VaR@97.5\%} = 1.96\sigma_p$$

where

$$\sigma_p = \sqrt{}\ \text{Var}\ (\Delta\ \text{Portfolio})$$

Var (Δ Portfolio), variance in the change of portfolio (or variance of the daily fluctuations in the portfolio value in the VaR @ 97.5% formula above)

$$= \Sigma q_i^2 P_i^2 \sigma_i^2 + q_j^2 P_j^2 \sigma_j^2 + 2\ \Sigma q_i q_j P_i \sigma_i P_j \sigma_j \rho_{ij}, \text{ for a two asset portfolio}$$

q_i = total number of MWh_e left in the *i*th contract
q_j = total number of MWh_e left in the *j*th contract
P_i = current unit price of electricity in the *i*th contract
P_j = current unit price of electricity in the *j*th contract
σ_i = Black-Scholes implied volatility of the current unit price of electricity underlying the *i*th contract if current call option price for the ith contract known; or the standard deviation of the continuously compounded annual rate of return of historical electricity price underlying the *i*th contract if call option price not known
σ_j = same definition as above applied to the *j*th contract

ρ_{ij} = correlation coefficient between ΔP_i and ΔP_j
Notes: VaR (Value-at-Risk) is different from Var (variance). For multi-asset portfolios, more complex methods are available which involve variance-covariance and matrix multiplications.

Two asset power portfolio – positive, zero and negative correlations

1. Correlation coefficient of +0.8.

Two power contract portfolio. The first contract and second contract with 3 and 6 years to expiry; power supply rate of 35 MW_e/hour, 30 MW_e/hour; respectively; around-the-clock (ATC) power delivery. For first and second power contracts, current unit price of electricity at US$ 26/MWh_e, US$ 28/MWh_e; price volatilities of 45%, 60% per annum, respectively.
Correlation coefficient of electricity prices between the first and second contract is +0.8.

- Risk evaluation required: Monthly VaR @ 97.5%

Solution

VaR @ 97.5% = 1.96 σ_P

$\sigma_P = \sqrt{[q_i^2 P_i^2 \sigma_i^2 + q_j^2 P_j^2 \sigma_j^2 + 2\, q_i q_j P_i \sigma_i P_j \sigma_j \rho_{ij}]}$, where

σ_i = Standard deviation of the continuously compounded annual rate of return of historical electricity price underlying *i*th contract = annualised volatility of prices in first contract. Equivalent monthly price volatility = annual volatility/$\sqrt{12}$ periods.

σ_j = Same definition as above, applied to the second power contract. Equivalent monthly price volatility = annual volatility/$\sqrt{12}$ periods.

$q_i = 3 \times 365 \times 24 \times 35 = 919{,}800$; $q_j = 5 \times 365 \times 24 \times 30 = 1{,}314{,}000$; $P_i = 26$, $P_j = 28$;

$\sigma_i = 0.45/\sqrt{12} = 0.129$; $\rho_j = 0.60/\sqrt{12} = 0.173$; $\rho_{ij} = 0.8$

$\rho_P = \sqrt{[(919{,}800 \times 26 \times 0.129)^2 + (1{,}314{,}000 \times 28 \times 0.173)^2 + (2 \times 919{,}800 \times 1{,}314{,}000 \times 26 \times 0.129 \times 28 \times 0.173 \times 0.8)]}$

$= \sqrt{[(9.517 \times 10^{12}) + (40.513 \times 10^{12}) + (31.417 \times 10^{12})]} = 9.024 \times 10^6$

Hence, monthly Value-at-Risk @ 97.5% confidence interval = 1.96 ρ_P = US$ 17.68 million.

2. Two asset power portfolio, zero correlation: correlation coefficient of 0.

Using the formula and data based on the 2 asset positively correlated portfolio, substituting σ_{ij} as 0, the monthly VaR @ 97.5% confidence interval = US$ 7.07 million.

3. 2 asset power portfolio, inversely correlated assets: correlation coefficient of –0.5.

Using the same procedure, monthly VaR @ 97.5% confidence interval = US$ 5.51 million.

Risk diversification of power portfolio:
For the 2 asset power portfolio, the monthly Value-at-Risk @ 97.5% confidence interval is diluted from US$ 17.68 million to US$ 5.51 million – representing a reduction of almost 70% in potential capital losses to the power supplier.

Single power contract: VaR based on forward price

The VaR formula for a single power contract based on a forward price is almost similar to the earlier formula based on current unit price of electricity, except that an exponential factor raised to the negative value of interest rate multiplied by time to expiration is included.

$VaR@95\% = 1.645 \times q\ (power) \times F \times e^{-rT} \times \sigma \times \Delta t$

where

q = power contract value in MWh_e
F = current forward price
r = risk-free discount rate
T = delivery at time T from present
σ = volatility of forward contract (standard deviation of continuously compounded period rate of return on forward price if option call price is not known); the Black option pricing model implied volatility if call option price based on forward price is known.
Δt = time to expiration
Note: the current forward price can be calculated by a formula introduced in the section on customising derivative products for CDM/JI power projects (page 189).

Case example:
Power contract value of 500 MWh_e to be delivered in one year. The current forward price is US$ 33/MWh_e, annualised volatility of a 1 year forward contract is 40%, and the risk-free rate is 6% per annum.

$VaR\ @\ 95\% = 1.645 \times 500 \times 33 \times e^{-0.06\times1} \times 0.4 \times \sqrt{1} = US\$\ 10{,}224$

$VaR\ @\ 95\%/power\ contract = 10{,}224/16{,}500 = 61.96\%$

VaR for CDM/JI power projects; assuming electricity, fuel and emission price risks

Basis of Single Power Contract: VaR based on current price of power, VaR 95% (not 97.5%)
In estimating contingent VaR for a combination of electricity, fuel and emission prices, a common confidence interval for assessing losses from these markets is desirable. The standard industry confidence interval of 95% is hence chosen for this purpose. The error contribution from not using a 97.5% confidence interval based on daily electricity price fluctuations (refer to P. Fusaro's observations, *Managing Energy Risk*, McGraw-Hill) is expected to be minor.

$VaR\ @\ 95\%\ (electricity\ price) = 1.645 \times q\ (electricity) \times P\ (electricity) \times \sigma\ (electricity)$

- Fuel price: VaR based on current unit price of fuel

Escalation of fuel price over the life of the project will increase the average cost of electricity unless hedged with futures or options contract. The potential increase in fuel price should be discounted to the present value.

$VaR\ @\ 95\%\ (fuel\ price) = 1.645 \times q\ (fuel) \times P\ (fuel) \times \sigma\ (fuel)$

- Emissions traded price: VaR based on current unit price of emissions reduction

VaR @ 95% (emissions) =
1.645 × q (emissions reduction) × P (emissions reduction) × σ (emissions reduction)

9.3.1
Case Example: VaR of CDM/JI Power Project, China, 100 MWe CCGT Plant

Basis: estimation of VaR @ 95% in the year when emission credits are realised and traded.
Assumptions – deregulated electricity market in region where project is implemented; baseline for carbon dioxide emissions is conventional coal plant; electricity price at low discount rate (agreed with UNFCCC) which results in negative NPV (generation), US$ 9/MWh_e.

Data
Annual load factor 90%, i.e. 90 MW_e; thermal efficiency 54%; power supply 16 hours/day to the market; current unit price of electricity (generation) US$ 10/MWh_e (1 cent/kWh_e), average electricity price volatility 10% per annum; price volatility at US$ 9/MWh_e estimated at 2% per annum; gas price US$ 1/GJ, gas supply requirement 3500 × 10^{12} J/per annum; gas price volatility 20% per annum; emissions reduction based on (964–349) tonnes of carbon dioxide per GWh_e. Other greenhouse gas emission reductions not taken into account.

- VaR based on (negative NPV) electricity price: VaR if emission credits are certified but not sold in the year when credits are realised – company decides to defer sales to next or future years because of low emission prices in the market. Decision may be made jointly by project investors and financiers.

VaR @ 95% (negative NPV) electricity price = 1.645 × q (electricity) × P × σ
= 1.645 × (90 MW_e/hour × 16 hours × 365 days) × US$ 9/MWh_e × 0.02
= US$ 155,630

VaR @ 95% of carbon dioxide emissions traded price for following year, assuming price volatility will increase with longer duration, market uncertainty, and any UNFCCC decisions to review or change certification procedures.

VaR @ 95% (postponed emissions trade to following year, price volatility 75%)
= US$ 7.97 million (1.645 × 323,244 × 20 × 0.75, see sub-section on emissions traded price)

- VaR based on electricity price: VaR if credits are not certified or if Protocol is not ratified

Net loss = Market price – unit price of electricity which results in negative NPV
= US$ 10-9/MW_e = US$ 1/MWh_e × (90 × 16 × 365 = 525,600 MWh_e/year)
= US$ 525,600.

- VaR based on current unit price of fuel (fuel VaR exists regardless of CDM/JI)

Escalation of fuel price over the life of the project will increase the average cost of electricity unless hedged with futures or options contract. The potential increase in fuel price should be discounted to its present value.

VaR @ 95% (fuel price)
= 1.645 × q (fuel) × P (fuel) × σ (fuel price) = 1.645 × 3,500,000 × 1 × 0.2
= US$ 1.15 million.

- Emissions traded price: VaR based on current unit price of emissions reduction, credits are sold in the same year when reductions are realised

VaR @ 95% (emissions) =
1.645 × q (emissions reduction) × P (emissions reduction) × σ (emissions reduction)
Amount of carbon dioxide reduction = 615 tonnes/GWh_e per year
Annual power output = 90 × 16 × 365 = 525,600 MWh_e/year = 525.6 GWh_e/year
Total amount of carbon dioxide reduction = 615 × 525.6 = 323,244 tonnes
Market value of carbon dioxide reduction = US$ 20/tonne (predicted upper band)
Estimated volatility of price of carbon dioxide reduction
= 50%, if US$ 10/tonne represents the lowest market threshold.
VaR @ 95% (emissions) = 1.645 × 323,244 × 20 × 0.5 = US$ 5.31 million.

9.3.2 Total VaR Analysis, CDM/JI Power Project, China – Year in Which Emissions Reduction is Achieved

VaR analysis under different conditions:

1. Case if CDM/JI credits are certified and sold in the same year

Total VaR =
VaR @ 95% of revenue from selling electricity at US$ 9/MWh_e (US$ 155,630) +
VaR @ 95% of revenue from selling carbon dioxide credits (US$ 5.31 million) +
VaR @ 95% of potentially higher fuel price (US$ 1.15 million)
= US$ 6.61 million.

2. Case if CDM/JI credits are certified but not sold in the same year

Total VaR =
VaR @ 95% of revenue from selling electricity at US$ 9/MWh_e (US$ 155,630) +
VaR @ 95% of revenue from not selling emissions in current year (US$ 7.97 million) + VaR @ 95% of potentially higher fuel price (US$ 1.15 million)
= US$ 9.27 million.

3. Case if CDM/JI credits are not certified or Kyoto Protocol is not ratified

VaR @ 95% is not applicable in this scenario; total loss results from charging electricity below the market rate (US$ 525,600) + VaR @ 95% of potentially higher fuel price (US$ 1.15 million) = US$ 1.67 million. This amount is indicative of the maximum possible loss incurred by the company if credits are not certified or if the Protocol is not successfully ratified. There would be no subsequent VaR of emissions assets as these do not exist *de facto*.

Payoffs vs Total VaR @ 95% for CDM/JI power project, China:
If the power project in central China resulted in certified emissions and these are traded in the same year reductions are realised, the payoff assuming a carbon dioxide credit price of US$ 20 per tonne is 323,244 × 20 = US$ 6.46 million. This potential gain,

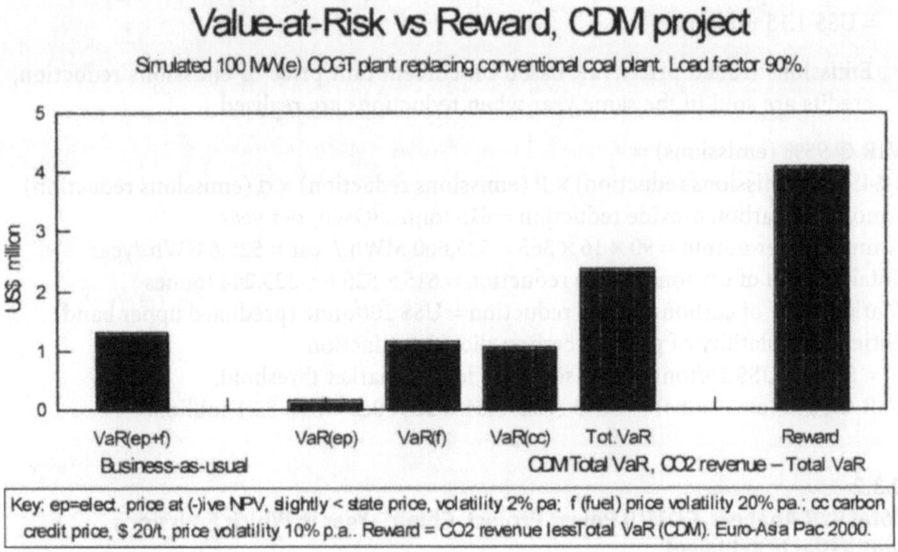

Fig. 9.1. Value-at-Risk vs. reward, CDM project
A low price volatility and a high carbon price results in a financial reward which exceeds the total Value-at-Risk of the project. If VaR of fuel is secured in a fixed price contract, and annual increments of the power price is agreed, then the reward escalates.

compared to a total VaR risk of US$ 6.61 million, represents a fair risk-return trade-off and a reasonable financial hedge for the project company and his investors.

Minimising Value-at-Risk (VaR) of CDM/JI power projects:
VaR can be potentially reduced by diversifying singular deals with a portfolio of multiple deal contracts – the risks could be further diluted by non or negatively correlated electricity, fuel or emission prices. The mitigation of VaR for a 2-asset portfolio with different correlation coefficients (+0.8, 0, –0.5) has been shown earlier. International power companies effectively utilise portfolio risk diversification to lower the overall VaR of projects in different countries with different political, economic and currency risks.

9.3.3
Risk Erosion Strategies for the Electricity Market – Generator and Consumer

In the case studies for the generator and consumer, it is assumed that an efficient, deregulated electricity market exists – where prices are determined by demand and supply, and in which numerous sellers, buyers and traders are able to operate competitively.

Electricity generator – Anxious that electricity prices in the market may fall in the future, hence reduce expected revenues to finance the average cost of electricity generation.

Strategy: Sell (short) a forward contract. If market price at maturity is below forward price, then the counterparty has to pay the difference to the electricity generator. Contracts are often settled before the maturity date.

If the market price rises above the forward price, the generator has to pay the difference to the counterparty. To minimise this risk, the generator may decide to buy an option.

Long (buy) a call option. Generator pays the call premium. If the market price exceeds the strike price, then the generator will exercise the option and receive the difference between the higher market price and the lower strike price.

Electricity consumer – Anxious that electricity prices in the market may rise in the future.

Strategy: Buy (long) a forward contract. If market price at maturity is above forward price, then the electricity consumer will be compensated by the counterparty with the difference between the higher market price and the lower forward price.

If the market price falls below the forward price at the forward maturity date, the electricity consumer is obligated to pay the difference in the two prices to the counterparty. To minimise this risk, he may decide to sell an option.

Sell (short) a call option. Electricity consumer receives the call premium. If market price falls below the strike price at the strike date, the counterparty has no incentive to exercise the option. If counterparty did exercise the option, he would notionally be buying at the higher strike price and selling at the lower market price, effectively paying the difference in prices to the electricity consumer.

Electricity traders or speculators would adopt the buyer or seller positions above with a "buy low, sell high" strategy to profit from any favourable movements on the underlying market.

In more complex risk management or trading strategies, "delta" hedging is used to reduce the risk of the most volatile contracts. Delta, a measure of price risk, represents the relative movement of the futures or option price with the underlying market price. For example, long futures, long calls and short puts have deltas of 0 to +1.0; short futures, short calls and long puts, 0 to −1.0. The net effect of delta hedging is to buy and sell the right amount of futures and options contracts to achieve overall delta neutrality.

Customising derivative products for CDM/JI power projects

A selection of forward and options pricing models for the energy and emissions markets is presented in this section. The derivative products include generic financial models as well as recently developed models for the energy market. Although the emissions trading market is embryonic at the time of manuscript completion, carbon credits ranging from thousands to ten thousands of tonnes were transacted in the North American market. The scope for Europe and the Asia-Pacific region is potentially enormous in the Kyoto Protocol framework.

Derivatives to hedge price, basis and credit risks in global emissions trading will be heavily sought by both sellers and buyers in what could be a highly unpredictable and volatile market. The author hopes that the models and case examples introduced in this section, which attempt to illustrate the risks in the power market and power-related emissions – and how these risks may be effectively managed – will offer some degree of comfort to both the novice as well as the experienced practitioner of financial instruments.

9.4 Pricing Techniques for Futures and Options

This section deals with the basic fundamentals of futures and options, which are applied to power assets (electricity) and emissions reductions using case examples. At mid-2000, there are no substantive reports or publications on emissions trading derivatives, although some boutique financial houses in USA and Australia may have developed OTC futures contracts.

9.4.1 Forward Pricing for Electricity Contracts

D. Pilipović (*Energy Risk*, McGraw-Hill, 1998) provides a convenient method for calculating the forward price of an energy product –

Forward price = $E_t [S_t] e^{-\lambda\sigma(T-t)}$

where $E_t [S_t]$ = expected price of electricity at expiration time T
λ = market cost of risk
σ = volatility of asset price
T = time of expiration
t = time of observation

Case example

Expected price of spot electricity (US\$ 30/MWh$_e$) in three months = US\$ 35/MWh$_e$
Market cost of risk = 10% per annum = 2.5% over three months
Volatility of electricity price = 25% per annum = 6.25% over three months
Time to expiration = 3 months or 0.25 year

Forward price US\$ 35/MWh$_e$ $e^{-0.025\times0.0625\,(0.25)}$ = US\$ 34.9863/MWh$_e$.

If electricity price volatility = 100% pa, 3 month forward price = US\$ 34.9453/MWh$_e$.

9.4.2 Forward Pricing for Emissions Reduction Contracts

Although there is no reliable historical data on emission prices or emission traded prices at the time of completing the manuscript, the price behaviour of emission assets may assume a normal distribution characteristic and experience volatilities typical of the electricity markets in which the emission reductions are achieved. If so, Pilipović's formula as described above may provide a first approximation of the forward price of emission assets.

Forward price = $E_t [S_t] e^{-\lambda\sigma(T-t)}$

where $E_t [S_t]$ = expected price of emissions credit at expiration time T
λ = market cost of risk
σ = volatility of asset price
T = time of expiration
t = time of observation

Case example

Expected price of spot CO_2 reduction (US$ 20/t) in three months time = US$ 23/t
Market cost of risk = 8% per annum = 2% over three months
Volatility of CO_2 asset price = 30% per anum = 7.5% over three months
Time to expiration = 3 months or 0.25 year

Forward price = US$ 23/t $e^{-0.02\times 0.075\ (0.25)}$ = US$ 22.9914 per tonne.

If CO_2 price volatility = 70% pa, 3 month forward price = US$ 22.9799 per tonne.

9.4.3 Option Pricing for Electricity Contracts

The Black-Scholes Option Model

The Black and Scholes Option Pricing Model, based on current price of a stock, can be used to calculate an approximation of the unit price of electricity in the future assuming the following conditions; (i) prices change continuously and display a log normal distribution, (ii) prices adjust constantly to prevent arbitrage, (ii) no dividends are paid on the stock, (iii) option can only be exercised at the expiry date, i.e. European stock option, (iv) transaction costs and taxes are excluded, and (v) the risk-free rate is the same throughout the period.

Provided that these assumptions hold, the current call option price, C_o is expressed by

$$C_o = S_o \,.\, N\, d(1) - X \,.\, e^{-rt} \,.\, N(d2)$$

where:

S_o = current stock price
X = stock exercise or strike price
e = base of the natural logarithm = 2.7128
r = continuously compounded annual riskless rate of interest over the option life
t = the amount of time remaining to the expiration of the option, expressed as a fraction of a year

N (d1), N (d2) = values of the cumulative normal distribution at the points d1 and d2 respectively, where:

$$d1 = \frac{\ln(S_o/X) + (r + \sigma^2/2)t}{\sigma\sqrt{t}}$$

$$d2 = \frac{\ln(S_o/X) + (r - \sigma^2/2)t}{\sigma\sqrt{t}} = d1 - \sigma\sqrt{t}$$

σ = standard deviation of the continuously compounded annual rate of return, which reflects the volatility of the stock price.

Note: X. e^{-rt} is the present value of the stock exercise price (quoted at expiry date)

The Black Option Model for Forward Contracts

The Black Option Pricing Model enables a call option to be valued if settlement is based on a forward price and not the spot price. In this case, the forward price is used to hedge the option, and the forward call option price, C_f, is expressed by

$$C_f = F \cdot e^{-rt} \cdot N(d1) - X \cdot e^{-rt} \cdot N(d2)$$

where:

F = current forward price
e = base of the natural logarithm = 2.7128
r = continuously compounded annual riskless rate of interest over the life of the option
t = the amount of time remaining to the expiration of the option and the forward price, expressed as a fraction of a year
X = stock exercise or strike price
N (d1), N (d2) = values of the cumulative normal distribution at the points d(1) and d(2) respectively, where

$$d1 = \frac{\ln(F/X) + (\sigma^2/2)t}{\sigma\sqrt{t}}$$

$$d2 = \frac{\ln(F/X) - (\sigma^2/2)t}{\sigma\sqrt{t}} = d1 - \sigma\sqrt{t}$$

σ = standard deviation of the continuously compounded annual rate of return, which reflects the volatility of the stock price.

Note: The Black Option Model can be applied to projects where the forward price is known – values of the forward price in electricity sector (earlier section on forward electricity pricing) could be used as data input in the model.

9.4.4 Option Pricing for Emissions Reduction Contracts

The Black-Scholes Option Model, and the Black Option Model, based on spot and forward prices respectively, could be used for the pricing of call options if the spot and forwards prices of emissions reductions are known.

The Black and Scholes Option Model, based on spot price of CO_2 reduction

Provided that the assumptions on which the Black-Scholes Option Model was formulated are satisfied, the current call option price, C_o is expressed by

$$C_o = S_o \cdot N(d1) - X \cdot e^{-rt} \cdot N(d2)$$

Data

S_o = current stock price = US$ 20/tonne
X = stock exercise or strike price = US$ 22/tonne (3 month option)
e = base of the natural logarithm = 2.7128
r = continuously compounded annual riskless rate of interest over the life of the option = currently 6 percent per annum

t = the amount of time remaining to expiration of the option, expressed as a fraction of a year = 0.25 (3 months)

N (d1), N (d2) = values of the cumulative normal distribution at the points d1 and d2 respectively, where:

$$d1 = \frac{\ln(S_o/X) + (r + \sigma^2/2)t}{\sigma\sqrt{t}} = -0.46$$

$$d2 = \frac{\ln(S_o/X) + (r - \sigma^2/2)t}{\sigma\sqrt{t}} = d1 - \sigma\sqrt{t} = -0.61$$

σ = standard deviation of the continuously compounded annual rate of return, which reflects the volatility of the stock price = quarterly standard deviation multiplied by √ 4 (periods) = 15 percent (observed) √ 4 = 30 percent per annum.

Note: X. e^{-rt} is the present value of the stock exercise price (quoted at expiry date)

Solution

$$C_o = 20.\ N(-0.46) - 22.\ e^{-0.06\times 0.25}\ .\ N(-0.61)$$
$$= [20 \times 0.3228] - [22 \times 0.985 \times 0.2709]\ \text{(using normal distribution tables)}$$
$$= 6.456 - 5.870 = \text{US\\$ } 0.586 \text{ or about 3\% of current stock price}$$

Notes

If the observed stock volatility is higher at 60% per annum, the call option price increases to US$ 1.73 or nearly 9% of the stock price.

On examining the call option price formula, the call option premium increases with higher current stock prices, higher risk-free rates of interest, lower exercise prices, longer maturity periods to exercise date, and the product of the stochastic element (randomness of price or price volatility) and the square root of time to expiry, σ √ t, also known as variability.

The value of the exercise price is quoted by the seller or writer of the option. One possible explanation of how the exercise price may be determined by the seller is strategic reduction by a percentage of the calculated forward price (using a forward pricing formula) to provide an exercise price which appear atttractive to an option buyer, but hedged by the seller with an option premium (Black-Scholes Option Model) which reflects the true volatility of the underlying asset. Other exercise pricing methods are likely to adopt the same principle.

The Black Option Model, based on forward price of CO_2 reduction

Provided the assumptions on which the Black Option Model was formulated are satisfied, the forward call option price, C_f, is expressed by

$$C_f = F.\ e^{-rt}.\ N(d1) - X.\ e^{-rt}.\ N(d2)$$

Data

F = current forward price = US$ 26/tonne (3 month)

e = base of the natural logarithm = 2.7128

r = continuously compounded annual riskless rate of interest over the life of the option r = currently 6 percent per annum

t = the amount of time remaining to the expiration of the option and the forward price, expressed as a fraction of a year = 0.25 (3 months)

X = stock exercise or strike price = US$ 24/tonne (3 month forward option)

N (d1), N (d2) = values of the cumulative normal distribution at the points d(1) and d(2) respectively, where

$$d1 = \frac{\ln(F/X) + (\sigma^2/2)t}{\sigma\sqrt{t}} = \frac{\ln(26/24) + (0.5^2/2)0.25}{0.5\sqrt{0.25}} = 0.444$$

$$d2 = \frac{\ln(F/X) - (\sigma^2/2)t}{\sigma\sqrt{t}} = d1 - \sigma\sqrt{t} = 0.444 - 0.50\sqrt{0.25} = 0.194$$

σ = standard deviation of the continuously compounded annual rate of return, which reflects the volatility of the stock price = quarterly standard deviation multiplied by √ 4 (periods) = 25 percent (predicted) √ 4 = 50 percent per annum

Solution

$$C_f = F.e^{-rt}.N(d1) - X.e^{-rt}.N(d2)$$
$$= e^{-0.06\times0.25}[(26 \times N(0.444)) - (24 \times N(0.194))]$$
$$= 0.985[(26 \times 0.6716) - (24 \times 0.5769)] \text{ (using normal distribution tables)}$$
$$= 0.985[17.461 - 13.845] = \text{US\$ 3.56 or 13.69\% of current forward price.}$$

Notes

The Black Option Model can be applied to projects where the forward price is known – values of the forward price in electricity sector (earlier section on forward electricity pricing) could be used as data input in the model.

On examining the Black Option Model formula, the forward options price increases with higher forward prices, lower exercise prices, higher expected forward price volatility, longer periods to contract maturity but independent of the risk-free rate of interest.

Relationship between price of call and put options

For European options, there is higher expectation of a price rise for a call option buyer than a put option buyer if both options have the same exercise price. A put option conveys the right but not the obligation to the option buyer to sell the underlying asset at a fixed price in the future. The following formula shows the relationship between the call and option price:

Call option price + PV (exercise price) = Put option price + stock price

If an investor buys a call option and sets aside money to purchase the stock at the exercise price (at risk-free rate of interest), that amount of investment would be equivalent to buying the stock at the current price and buying a put option to sell the bought stock at the exercise price in the future – in both cases, the pay-off is the same, i.e. owning the stock at the same exercise price or receiving the same cash payment equal to the exercise price. If the market price exceeds the exercise price, the investor will realise a gain; if it falls below the exercise price, then the investor has achieved a hedge.

9.5 Pricing of Packaged Electricity and Emissions Contracts: A CDM/JI Case Example Based on Export of Power and Emission Credits, China to Japan

Project specification and data

Construction of a 500 km power line from Jiangsu province in north-east China to either Kyushu or southern Honshu, carrying a load of about 600 MW_e – a green project based on the framework of the Kyoto Protocol. The project may be a CDM or JI project, depending on the origin of the investor country. China is a non-Annex I country and Japan, an Annex I country. If the project is commercially successful then additional funds could be sourced from capital markets to finance a second power line.

According to BICC General Cables, the subterranean cable would cost US$ 100–150 m., current converters at generation and reception terminals US$ 200 m., and installation work of about US$ 50 m.

High Voltage Direct Current (HVDC) transmission is the preferred mode, as it requires a thinner cable and would incur a lower overall capital cost compared to alternating current. The power losses using DC would also be small, in the range of up to 5%.

Jiangsu and its environs has substantial reserves of coal, but the enabling technology would have to reduce present levels of emissions – pressurised fluidised bed combustion or integrated gasification combined cycle using coal as feedstock would be the likely candidates.

The author's 1999 estimates of the cost of electricity from 100 MW_e-rated PFBC and IGCC plants in China range from 1.5 to 2 cents/kWh_e for each plant (discounted at 10%) – the cost may fall to 0.5 cent/kWh_e for 600 MW_e capacity plants. If the estimated cost of electricity transmission is US$ 400 m. for the Jiangsu-Kyushu power cable project, then the additional cost of financing at 10% discount (annuitised method approximation, inclusive O&M costs), of 1.5 cents/kWh_e, results in a cost of about 2 cents/kWh_e to transmission terminal off-takers. This sales price is below the average market price of electricity at transmission line terminals in Japan, and should hence meet the financial additionality criteria of the Kyoto Protocol. The upside of emission credit sales to either Japanese companies with emissions targets or via exchange trading increases the internal rate of return to project financiers and stakeholders.

Calculation methodology

Plant: 600 MW_e PFBC, 3,600 GWh_e per annum, author's estimated forward prices
Note: individual project calculations should be made using the basic forward pricing formulae for electricity and emissions, introduced earlier in the current chapter, or by applying other suitable models or methods available in the marketplace.

Pricing of electricity, forward price (as delivered to Kyushu terminal)

Annual forward price = US$ 20/MWh_e, for year 2003.
Total forward price for year 2003 = 3,600 GWh_e × US$ 20/MWh_e = US$ 72 million.

Pricing of emission reductions, forward price (PFBC, coal; hypothetical, as capacity of the largest commercial plants is limited to 170 MW_e at end 1999)
Forward sale of emissions, assuming 30% capital cost reduction by project company in the PRC on a scaled-up 600 MW_e PFBC plant – basis of estimated abatement cost in the PRC of US$ 80/tonne, 100 MW_e PFBC plant at 10% discount. The cost reduction achieved by lower capital cost per MW_e on procurement of scaled-up 600 MW_e plant and higher plant thermal efficiency at 45%. Measurement of emissions reduction of CO_2 based on conventional coal fired plant. Author's theoretical estimate of CO_2 emissions/GWh_e based on 600 MW_e scaled up PFBC plant, 45% efficiency, Lower Heating Value of coal, is about 700 tonnes.

Total tonnage = (964 – 700) tonnes/GWh_e × 3,600 GWh_e × (70% × US$ 80/tonne)
= US$ 53.22 million.

Packaged forward price of electricity and emissions, with price discount (e.g. a five percent discount for combined purchase of electricity and emissions could be offered)

Package price for delivery of electricity in year 2003 and forward sale of emission reduction
= US$ (72 + 53.2) million × 95%
= US$ 118.94 million.

If the project company was not able to sell the emissions credits at US$ 56 per tonne, when the forward market price is, say, US$ 20/tonne, then the packaged forward price will be revised to US$ [72 + {(964 – 700) × 3,600 × 20)}] × 95% = US$ 86.45 million.

9.6 Factors Influencing Pricing of Futures, Options and Swaps of CO_2 Assets

The traditional methods for pricing forward contracts, and options using the Black-Scholes and Black pricing models were introduced earlier in this chapter. This section reinstates the primary risk factors and other risk factors which investors and trading parties are likely to encounter in the emissions trading market.

Factors influencing futures prices

1. Current costs of CO_2 reduction (issues that are likely to affect pricing were raised in the previous section).
2. Real interest rates, which reflect asset value if "banked".
3. Market volatility.
4. Dominance of large players in the market (industrial companies with emissions targets, as well as industrial and financial firms with speculative motives).
5. Discounting of credits or permits that are earned or purchased before the first commitment period of 2008–2012. The cost of technology are likely to fall, say, from 2005 compared to costs incurred over 1990 to 2005, hence buyers would expect lower trading prices compared to the alternative of, perhaps, investing in technology at costs lower than trading prices. This discounting rationale may be counteracted on the probability that emissions baselines are also likely to fall in the 2008–2012 period, hence companies which decide to install CO_2 reduction equipment in

this period will need to spend as much as those companies with higher emissions baselines over the 1990–2005 period.
6. A shortage or surplus of permits and credits in the market.
7. Credit risk of buyers and sellers of futures contracts.
8. Risk of settlement or delay of settlement. Market factors, which may include sovereign risk in developing and LDC countries.
9. Adjustment of futures price with Value-at-Risk assessments; this would inflate the normal futures price calculated by non-VaR methods.

Factors influencing option premiums

Option premiums depend on

1. Benchmark, e.g. forward or futures prices.
2. Intrinsic value, that is, present value of difference of the strike price and the forward price of the CO_2 permit or credit.
3. Time value, which depends on volatility, time to expiry, and the "cost of carry" or cost of borrowing to finance the option purchase.
4. As for futures prices, addition of VaR to option premiums will inflate the value of the option for sellers.

Factors influencing swap rates

Swap prices would depend on the swap rate that sellers of contracts consider as "fair risk" to both parties of the contract. The factors described for futures and options will also affect the pricing of the swap. Value-at-Risk techniques could be applied to swap contracts using period volatilities of emissions prices.

9.7 Emissions Trading Contract

For project sponsors or stakeholders wishing to sell carbon credits, information on the project must be provided with as much detail and transparency if buyers are to be attracted. This section provides guidelines on how project data should be presented to the emissions trading market. To conduct paper or physical trading, a proposed format for the specification of standard contracts is presented.

9.7.1 Data Input

Part A: Data input from project sponsors (Data to determine Quality of Asset to Investors – this would form part of the offer for sale)

1. Asset type

Carbon dioxide, methane, nitrous oxide, hydrofluorocarbons, perfluorocarbons, sulphur hexafluoride (for halocarbons, refer to IPCC, GHG Inventory Manual 1997).

2. Asset quality

- Does project meet

a) Emissions additionality criteria
b) Financial additionality criteria
c) Economic, social and environmental sustainability criteria
d) Certification of GHG reduction requirement
e) Investor risk aversion requirement
f) CDM eligibility requirement by UNFCCC

- Government's information input to assist project appraisal by investors

a) Accurate data is required to determine the emissions baseline. How reliable is data on the existing emissions source? Does the existing source form a small percentage or dominant type of plant or facility in the project location? What are the percentage contributions of manufacturing and energy technologies to total GHG emissions, GWh_e and GWH_t production, and GHG reductions in the current period and the next five years? If more renewable energy technologies are to be introduced, national baselines will fall (i.e. lower levels of GHG production per year). Is the economy predicted to expand or regress in the future? If the source emits higher emissions (from baseload to peakload, say, 600 GWh_e to 650 GWh_e in the power generation industry) as a result of higher economic demand, is there an adjustment factor to revert to the original emissions benchmark level? Fixed emissions baselines, irrespective of unfavourable changes in economic circumstances or adopted technologies, are likely to represent a non-negotiable term for project financiers as they affect the amount of carbon credits generated to stakeholders.

b) Financial additionality requires the investment to be made in projects which exhibit a negative net present value of cash flows at a low discount rate – this requirement ensures that the investment is additional in terms of projects which do not appear financially viable in the market – but could be attractive with GHG credits. Projects which have high entry costs, normally ignored by investors, would be considered additional – hence CDM projects may ultimately outperform non-CDM projects.

   Is the country able to afford this technology over the period of pre-financing to the project commissioning phase?

   What are the current and future official aid projects which result in the reduction of GHG emissions (World Bank, GEF, official development aid) – this answers, in part, the financial additionality requirement. CDM projects should be financially additional to official aid-linked projects to be implemented in the host country. However, it is up to UNFCCC to decide if such official aid projects, although forthcoming, only form a small percentage of CDM-linked investments. UNFCCC should encourage further (CDM) investments from the private sector.

   The combination of negative NPV at low discount rates and private sector investment should strengthen "additionality" characteristics of the project.

c) What is the socio-economic-environmental impact of CDM projects on the investment location? What is the socio-economic-environmental impact on neighbouring locations and the host investment region? These questions address the sustainability requirement.

d) What inspection mechanism is offered with regard to measuring and monitoring GHG reductions? These tasks should be undertaken by the Environmental Department or reputable environmental emissions testing body approved by UNFCCC.
e) Is the government able to enforce international laws which pertain to the protection of foreign investments, and agreed terms which relate to CDM projects in the host country?
f) Has the potential CDM project been referred to UNFCCC for their comments or decision on project eligibility? If so, what is the response or recommendation?

3. Baseline emissions reference for project
GHG emissions in mass or volumetric units, in average terms in 1999, and the average for past three years, if possible. Volumetric units should be expressed or converted to standard temperature and pressure conditions of 0 C and 101.325 kNm^{-2} (1 bar).
4. Risks undertaken by CDM project investors
Investors bear the risks of non-ratification of Kyoto Protocol by parties to the Protocol – their investments shall be subject to country, financial and force majuere risks and risks normally encountered by foreign investments, even with provision of sovereign guarantees. In the event of non-ratification of the Protocol, subject to agreement by parties involved:

a) CDM credits already traded and financially settled, may be returned in full in good faith, part return, or deemed non-returnable by and without liability to the seller;
b) Goods and services ordered and/or paid for potential CDM or CDM projects may be deemed non-returnable.

5. Non-performance of contract: seller's obligations to buyer
If the quantitative emission reduction is not achieved and certified as stated in the contract then the buyer has the right to claim compensation. Compensation may be settled by one or more of the following remedial actions:
Reimbursement of that part or total contract value which is equal to the shortfall or absence of emission reduction at the spot date (if reductions are made, the reduction on the day of spot trading may not be correctly and officially certified), options and futures pre-maturity or maturity dates of the contract – plus the interest at a rate agreed between the seller and buyer: for spot trades, interest accrued between the spot transaction day and the date partial or non-performance of contract is attested by an approved authority; for futures contracts, interest accrued between the date money is invested and the date of contract performance, partial or non-performance conditions to be applied. Any disputes, cross border or within the host country, shall be settled under International Chamber of Commerce or UNCITRAL rules of arbitration.
6. Non-performance by signatory/ratifying country: Implications and remedies for traders
If the host country in which the CDM (or JI) project does not fulfil its obligations in any part to the Kyoto Protocol before or after its signing or ratification of the agreement, then the seller nor buyer may not claim any compensation from the host country – any losses in value of traded emission reduction assets would be undertaken by buyer, shared in such proportions as agreed between buyer and seller, or recoursed to the seller.

Part B: Draft listing of potential CDM/JI projects in contract document

Industry Type	Project Type	Energy Technology Type	Approx. capital cost (Total of equipment + construction), $ m	Greenfield (G) or Retrofit (R)

Guide to classifying Industry and Project types:
(types are arranged in alphabetical order; rows do not represent industry and project matching)

Industry Type	**Project Type**	**Energy Technology Type**
Agriculture	Energy/fuel saving	Renewable energy:
Cement production	Fuel switching	Biomass – electricity
Chemical, petrochemical	Fugitive gas capture	Biomass – heat + electricity
Electricity generation	Process improvement	Biomass – transport
Energy savings in buildings	Technical improvement	Geothermal – electricity
Fuel extraction		Geothermal – heat + electricity
Fuel flaring, venting		Landfill gas
Fuel transformation		Municipal Solid Waste – incineration
Fuel transport, distribution		Small hydro
Heat + electricity generation		Solar
Halocarbon production		Tidal
Halocarbon, SF_6 utilisation		Wave
Iron, steel and alloys		Wind
Metal production		Fossil fuel:
Paper mills		Coal-oil dual fuel, Oil-fired
Pharmaceutical		Combined Heat and Power
Transport vehicles		Fluidised bed combustion
Waste and sewage operations		Gas Turbine, Combined Cycle Gas Turbine
		Integrated Gasification Combined Cycle

Notes

Biomass – short rotation coppice; energy crops; refuse, MSW municipal solid waste; commercial/industrial wastes; agricultural residues, temperate/tropical crop wastes; animal wastes; landfill gas; biogas/biofuels.

Biomass steam cycle – power, heat + electricity
Biomass IGCC cycle – gasifier, steam and gas turbine – power, heat + electricity
Biomass – steam and IGCC cycles – short rotation coppice regrowth fuel
Biomass – biofuels for vehicle transport
Combined Cycle Gas Turbine (and steam turbine) – Cogeneration: heat + electricity
Combined Cycle Gas Turbine (and steam turbine) – electricity only

Combined Heat + Power, or Cogeneration
Fluidised bed combustion – circulating, pressurised
Gas turbine – open cycle, combined cycle
Integrated Gasification Combined Cycle – coal heavy oil residue hydrocarbon feed
Landfill gas – electricity, heat + electricity
Solar photovoltaic, Solar thermal
Solid Municipal Waste Incineration – electricity, heat + electricity
Tidal barrage
Wave power – onshore, offshore
Wind – onshore, offshore

9.7.2 Emission Trading Contract – Proposed Format for Spot/Futures/Options

FUTURES/SPOT/OPTIONS FORMAT
FOR GREENHOUSE GAS EMISSIONS REDUCTION TRADING:
EXEMPLARY CONTRACT FOR CARBON DIOXIDE

CDM VERSION
(Replace "CDM" with "JI/Annex B" if contract is used by Annex I and Annex B parties in the context of the Kyoto Protocol and assuming additionality and sustainability criteria apply)

Embodiment of contract

1. Project title.
2. Industry type.
3. Project type.
4. Greenfield or retrofit.
5. Asset type.
6. Asset quality.
7. Project Details

Capital cost
O&M cost
Fuel cost (n/a to natural renewable).

Technology cost if project investment does not require any fuel to be involved (e.g. process or structural improvement, retrofit, or materials used to reduce energy loss)

Project sponsor/developer, operator, financiers, syndicated capital, technology providers

Recommended for inclusion:
Overnight cost
Discounted cost, discount rate used.

Other project information, e.g. project life, project gross and net efficiency, gross or net heating value of fuel used, carbon content, GHG compositions, load factor, availability; other relevant factors like sensitivity of fuel price and security of supply.

8. Baseline reference emissions level.
9. Statement of risks facing investors.
10. Terms and conditions of contract.
11. If emissions are achieved through economic inactivity, state baseline year and amount (applicable if contract is used for JI initiative and Annex B emissions trading).

9.7.3 Proposed Specifications for Futures and Spot Contracts

Note: *unit subject to revision; **not applicable to spot contracts

1. Trading units*

Trading unit = 100 one-tonne emissions reductions in CO_2

2. Standards, certification and legality of trading parties

Verification and certification of any emissions reductions must be made according to such procedures acceptable by both the UNFCCC and relevant governmental environmental agency (or agency empowered by the agency).

The certification requires the services of approved inspection agencies or entities – such costs may be proposed by UNFCCC to be shared or absorbed by the seller and/or buyer.
Legality of trading parties:
Three issues arise at the time of drafting this report (December 2000) – on whether:

a) Non-signatory or non-ratifying parties to the UNFCCC and the Kyoto Protocol are able to participate in spot or futures emissions trading.
b) Speculators, financial firms and institutional investors – which may or may not have any connection to environmental emissions or obligatory targets, are allowed to trade emissions. Perhaps two pre-conditions for financial institutions or entities to trade emissions could be their appointment by industrial companies that are seeking the best prices as buyers or sellers, and for investment funds to take an active part in providing needy capital to projects. Trading for purely financial motives may not be acceptable to UNFCCC.
c) Purchasers, owners or traders of carbon credits should also include governments.

3. Quotation
Price quotation of entitlements or emissions permits or emissions credits in US$ and cents per tonne of CO_2.

4. Minimum price movement**
Minimum price fluctuations of, say, multiples of US$ 0.25 per tonne, on futures prices. The minimum price movement per contract (of 100 tonnes) or "tick" is therefore, US$ 25.

5. Initial and variation margins**
Initial margins are required to be deposited at the exchange or clearing house by both parties of the trade as a measure of good faith. Interest collected in the interim shall be returned to the parties at the settlement date. The initial margin for CO_2 futures is estimated between 2% and 4% of the contract value, based on sulphur dioxide and energy futures contracts.
Variation margin payments shall be paid or received at the close of business daily, based on the difference between the futures price and the day's official settlement price.

6. Trading limits
Daily trading limits, that is, minimum price per tonne of CO_2, shall be set at US$ 25/tonne or US$ 2,500 per contract**.
Variable limits shall be set at say, 100% to 200% of the daily trading limit. No limits shall be in effect for the current month (as to establish kick-start market price for CO_2 trading).

7. Delivery
Delivery or settlement of contracts or spot trades shall be recorded via book entry in both the registers of the national environmental agency register and UNFCCC, and registers of the relevant exchanges (ex-exchange OTC deliveries to be accounted for in the same manner).
Settlement of trades to be enforced by the exchange authority.

8. Trading periods**
Current year, and quarterly futures quotations for current and next two years in March, June, September, December.

9. Trading hours
9 A.M. to 3 P.M. per working day, or according to agreed hours in the country concerned.
For CDM and Annex B emissions trading, a 24-hour electronic system has been advocated by interested parties in these initiatives.

10. Last day of trading**
Any trades on CO_2 entitlements or CO_2 futures in the current month must take place before the last three business days of the month.
Alternatively, the last trading day of the month could be the last Thursday, or last Wednesday if the Thursday falls on a public holiday.

11. Limits for positions*
To reduce investment risk in the market by highly speculative trading, position limits shall be set on any trader on say, 1,000 contracts in the current month and 2,000 contracts overall.

12. Allocation of part of proceeds
The author recommends the following proportions to be allocated from the proceeds of the sale of the contract, i.e. final settlement price of the contract.
Direct proceeds to host country, certification and costs of adaptation (proposed)

a) A surcharge of 3% to 5% to host government – This surcharge fulfils the obligation of investors – according to the Protocol – to reward the host country of investment and the local community in which the investment is conducted.
This financial allocation would be additional to any stakeholding in the contract by the host government – e.g. land offered as collateral; materials, fuel or labour supply as part of the total cost of investment.
On the secondary trading of carbon contracts, a 1% mandatory contribution of the settled value of each and every contract originating from the host country to be credited to the host government. This ensures that the host country will receive more than the fixed amount of financial benefit from its first entitlement of carbon units, compared to the benefits of cost savings to foreign investors and commissions to traders.
b) A surcharge of up to 2% on contract trading costs – this would include consultancy and advisory services to exchange or OTC transactions, development of software, on-line and telecommunications systems, computing, personnel, fixed and variable costs. The costs of matching/advising sellers and buyers, and negotiations would be additional.
c) A surcharge of 1% to renewable energy development in the host country – particularly in the purchase of foreign capital equipment.
Renewable energy investments – capital intensive with long term payback periods – may be considered viable in those remote locations where the cost of grid-connected electricity is prohibitively high but offer best conditions for renewable energy to be harnessed.
d) A surcharge of 1% to SMEs to reduce their project costs for consideration by foreign investors – this would encourage more SMEs to participate and compete for foreign capital with large scale enterprises.
e) A contribution of 1% of the contract value of emissions reductions from first trade and secondary trading of each and every contract to a proposed LDC/IRISK Climate Change Fund (48 Least Developed Countries and 2 islands-at-risk from sea-level rise)- which could act as a precursor to the proposed "Adaptation Fund". The beneficiaries of the fund may choose to receive benefits on the basis of contract value to be traded, or basis of actual emissions reduction units (CERs for CDM projects and ERUs for JI projects).
70% of the contribution to be directed at the humanitarian sector, the balance to the energy sector with preference to renewable energy, and environmental pollution control. For small islands-at-risk, climate change adaptation shall be the primary objective.
f) Cost of final certification of emissions reduction to inspection body or agency. This may entail measurements of emissions from changing baseline sources as CDM rules are revised. An allocation of a quarter of 1% is recommended.
g) A success fee of a quarter of 1% is proposed to parties responsible for initial feasibility and due diligence studies, project sponsors and contractors, and intermediaries which are directly responsible for raising or arranging finance – but may exclude banks, equity funds and entities whose core business is the provision of capital financing.

A proposed 2% contribution in surcharges for developing renewable energy resources and assisting SMEs to compete for foreign investments (on level footing with large enterprises) may be applied to the investment capital instead of emissions contract value/entitlement.

The author proposes that the total cost of direct proceeds should not exceed 10% (excluding stake holdings by the host government) of carbon credits are created from the project.

Legal fees for the drafting of the offer document for sale in the English text, native language of the host country, and other major business languages, is not included.

Notes:
Secondary trading of CER contracts

1. 1% or more of CER contract value when settled to CER country (government) of origin, of each and every trade. Allocation as a standard term in all emissions contracts, or as a unilaterally (individual country initiated) embedded term in the contract regardless of COP decisions to the host CDM or low GNP/capita Annex I country is expected to be discussed at COP VI in the Hague, November 2000.
2. 1% or more of CER contract value when settled to the LDC/IRISK Climate Change Fund; to end of the first commitment period of the Kyoto Protocol, that is, the year 2012.

The rationale for these proposed contributions is provided in the section "maximum threshold of CER benefits to CDM host countries" and "North-South" benefits comparison – via two case studies on North-West European investments in India, and emission trading commissions of CERs orginating from India in Chapter 7.

Item 7 in Embodiment of Contract – Investment capital
(Project initiation costs – part of investment capital)

The author recommends the following costs to be charged as part of the investment capital – it should be emphasised that no investment would ever occur without due diligence work.

1) A surcharge of 1% for site inspection, measuring and monitoring emissions reductions over period of plant or site operation to obtain a value for comparison against the emissions reference source, and certification of reductions.
2) A surcharge of 2% for evaluating feasibility studies and viability of CDM or potential CDM project. Such studies may include due diligence research and site work conducted to establish the nature and ideal location for the project, training requirements, current emission levels, socio-economic-environmental impacts and assessment, and operational factors.

9.7.4 Proposed Specifications for Options Contracts

1. Trading Unit

100 tonnes of CO_2 = One contract.

2. Quotation
US$ and cents per tonne.

3. Minimum price movement
US$ 0.25 per tonne.
Tick value per contract is US$ 25.

4. Strikes/strike price
The ATM (At-The-Money) strike is that value which is nearest the previous day's closing price on the futures contract.
The strike price is that futures price at which the client can exercise his right to physical or financial settlement – usually specified as ATM, ITM or OTM compared to the futures price.

5. Exercise/settlement
American style. At any time at or before expiry of the option.
Option should be exercised not later than 16.00 hours, or 60 minutes after the underlying futures settlement price is quoted, on any business day.

6. Premium
The price of the option, as determined by an appropriate option pricing model. It should reflect all the investor's risks as itemised in the greenhouse gas trading contract.

7. Intrinsic value
The strike price minus the current market price of the option.

8. Time value
Option premium minus intrinsic value of the option, which reflects the time to expiry and expected market volatility.

10. Delivery
Physical delivery of CO_2 entitlement, permits or credits; re-sale of option; exchange of futures for physicals (EFP), or according to terms of settlement between option buyer and seller, in which case the exchange must be notified.

11. Trading periods
Current year, and quarterly quotations in March, June, September, December.

12. Trading hours
9 A.M. to 3 P.M. per working day.

13. Last day of trading
Friday preceding the underlying future contract's expiry (if three trading days remain until the future's expiry date), or second Friday before the contract expiry date.

14. Limits for positions
Position limits shall be set an any trader on say, 1,000 contracts in the current month and 2,000 contracts overall.

9.8
Case Examples using Spot, Futures and Options

Project stakeholders and traders wishing to minimise price risk as a result of external factors may wish to hegde their risk exposure using derivatives such as futures, options and swaps. The emissions trading market will provide a new stimulus to global financial flows, and the use of financial instruments will raise market liquidity as well as capital for projects. The objective of this section is provide the reader with an understanding of the risk management concepts of derivatives as applied to emissions trading, starting with a spot market example. Further case examples are provided to acquaint the reader with more advanced derivatives.

Spot market – An illustrative example

Unit of trading = 100 tonnes of CO_2.

Date: say, a business day in February 2003, traded on an exchange.
Say, a seller wants to sell 100,000 tonnes of CO_2 (1,000 contracts).

Seller's price	US$ 30/t, plus cost of certification of emissions reduction*. *certification cost to be negotiated between the parties; may include negotiation costs incurred by the negotiator.
Seller's quotation	In US$ and cents per tonne of CO_2.
Buyer's price	Accepts seller's price.
Delivery	Transfer of title Settlement shall be made on agreed terms at the close of business of the trading day, according to the method agreed with the exchange.
Exchange role	The exchange shall supervise the trade, and enforce payments as agreed between the parties concerned.

9.8.1 Futures Market – Pricing Methodology, 3 Month Futures Contracts, Case Examples

The pricing of 3 month carbon dioxide futures, options and swaps in the case examples in this section of the book is principally based on historic forwards price and options price curves of crude oil traded over selected periods at IPE and NYMEX. Such simulations should not be regarded as models for future carbon dioxide contracts as the factors which influence emissions reductions are determined by a host of market and non-market conditions. The absence of a liquid market in carbon dioxide trading at end year 2000 provides no reliable data for forecasting futures prices, hence the emissions trading market is expected to be highly volatile in JI and CDM countries.

The most significant non-market factors are price constraints set by financial and emissions additionality criteria of the Kyoto Protocol – financial additionality "caps" on electricity prices of power generators, transmission and distribution companies at values below previous average costs of these services at discount rates recommended by UNFCCC. The electricity price caps may deter some companies to invest in fuel-switching or energy efficiency projects because the difference in prices before and after project implementation may be too low for the company to recover revenue at pre-project prices. The alternative would be to increase carbon dioxide prices to meet expected levels of return demanded by investors.

In the case examples, the gradient of forwards and options price curves are designed by the author to reflect the risks perceived by the seller and to offer hedging opportunities to buyers. For options examples, futures prices are re-used to simulate options strike prices at quarterly settlement dates to simply calculations. The net effect of additionality factors of the Kyoto Protocol would be to maintain a price check on electricity prices, but inflate carbon dioxide prices to keep sellers in business.

9.8.2 Futures Market – CDM Example

Case: a Hedging Strategy, based on a CDM project

In February 2003, an Annex 1 company wishes to buy CO_2 reductions to comply with its emissions target by May 2003.

Market scenario, in February 2003

Say, a day in February 2003. Market data:

February 2003
Futures market prices, US$/tonne CO_2
(will change daily after February 2003)

Mar 2003	June 2003	Sept 2003	Dec 2003
30.00	30.50	30.75	31.25

Spot CO_2 Index price at the exchange: US$ 30/tonne CO_2

Buyer's view: Fears that prices in May 2003 will rise above current price of US$ 30/t, as reflected by futures prices for June 2003, and perhaps by an additional amount determined by market conditions in the future.
Buyer has three alternatives:

1. Buy the physical at the Spot CO_2 Index rate quoted at the exchange in May 2003, in which case he would be paying the market price in May 2003, from a seller in the physical market (say a project company that has achieved a surplus of emissions reductions in a developing country).

Two scenarios arise – the trade is:

1a. Made in May 2003, but the buyer may face the risk of low stocks of CO_2 credits in the host country or as held by the seller.
1b. Agreed contractually between the seller and the buyer in February 2003. In this case, the buyer has secured the CO_2 credits at the spot rate in May 2003 and averted the risk of non-supply in (1a).

2. Buy the physical at the Spot CO_2 Index rate quoted at the trading exchange in May 2003, and purchase a June 2003 futures contract at the futures price that is quoted in February 2003 (US$ 30.50/t); any further futures price increase in the June 2003 futures contract – say, in May 2003 – will result in a gain equal to the difference in the June 2003 futures price quoted in February 2003 and the June 2003 futures price quoted in May 2003, when the buyer closes out his position (sells the underlying futures contract). Buyer becomes a speculator in this case – he may also incur a financial a loss if the June 2003 futures price in May 2003 should fall below US$ 30.50/t.
3. Buy the physical at an arranged pricing mechanism with the (physical) seller, at a rate that is calculated, say, on the Spot CO_2 Index in May 2003, plus difference

(+/–) between the Spot CO_2 Index in June 2003 and the Spot CO_2 Index in May 2003. In this scenario, the seller in the physical market assumes a financial risk if the Spot CO_2 Index in May 2003 falls below the Spot CO_2 Index in February 2003, but he is willing to accept a return dictated by the level of the risk.

Seller (physical market) analysis
In a CDM project, the seller, which could be a corporate investor from a developed country, or a co-investor from the host country, may want to sell the emissions credits as soon as they are realised. A co-investor from the host developing country is likely to sell his share of emissions credits based on the method described by (3) if he needs hard currency as a priority.

Hedging Strategy by the Buyer

The Buyer chooses alternative (3) because this strategy offers:

a) Security of emissions credits to be delivered in May 2003, rather than risking low stocks or inavailability of these credits in May 2003.
b) The upside of lower spot prices in May 2003.

However, it is exposed to any increases in the Spot CO_2 Index rate in May 2003 over the Spot CO_2 Index rate in February 2003.

Available to Buyer:	- A futures market in CO_2 emissions trading - A cash market based on a CO_2 Spot Index at the exchange (or another cash settlement market in CO_2 emissions trading)

Buyer is interested in buying emissions reductions from a seller in the *physical* market in May 2003, but the buyer may incur a loss in the cash market if prices rise above the US$ 30/t as agreed with the seller. He decides to hedge any losses in the *futures* market.

Seller in the futures market wants to sell 100,000 tonnes of CO_2 = 1,000 contracts.

Futures contract based on, say, settlement on last Thursday of the month.

Buyer's action: Takes out a June 2003 contract, to be settled, in theory, on 29 June 2003. Buyer, in theory, agrees to pay seller US$ 30.5 × 100,000 tonnes = US$ 3,050,000.

Market scenario in early May 2003

Cash price in early May 2003 = US$ 31.00/t CO_2

Early May 2003
Futures market prices, US$/tonne CO_2

June 2003	Sept 2003	Dec 2003	Jan 2004
31.25	31.50	31.75	32.00

Spot CO_2 Index at the exchange: US$ 31.00/tonne CO_2

Buyer's position in the futures market (buyer with CO_2 emissions target)

Buyer's view: His June 2003 futures contract purchased in February 2003 (US$ 30.5/t CO_2) will secure a futures gain in June 2003 provided the cash market price in May

2003 dictates the futures market price in June 2003, or he is lucky, the futures price for June 2003 may rise above US$ 31.25/t CO_2. However, the buyer also accepts the reality that future economic as well as market conditions are beyond his control, and hence, the futures price for June 2003 may fall below the quoted US$ 31.25/t CO_2 or, at worse, fall substantially below US$ 30.5/t – if so, he would incur a loss under the futures contract (net gain or loss to the buyer is based on the difference between the futures price quoted for June 2003 in February 2003 and the futures price for June 2003 provided the contract is held until the maturity date. If the buyer closes out his position before the maturity date, say in early May 2003, the net gain or loss is based on the difference between the futures price quoted for June 2003 in February 2003 and the futures price for June 2003 quoted in early May 2003, neglecting transaction costs). Buyer's action: Buys CO_2 emissions reductions from a physical seller at agreed Spot Index rate at May 2003 plus the difference between this rate and the Spot Index rate in February 2003, and decides to close out his position in early May 2003.

Buyer's position in the physical market (same buyer with CO_2 emissions target)

Buyer purchases CO_2 emissions reductions from the Spot CO_2 Index market at the exchange, at the rate of US$ 31.00/t, as agreed with the physical seller from the developing country. He incurs a loss of US$ (31 – 30.50) per tonne of CO_2.
Futures market transaction details, early May 2003

Seller (exchange) pays Buyer	= US$ (31.25 – 30.50) per tonne of CO_2
Buyer's gain	= US$ 0.75 per tonne of CO_2
Buyer's gain/contract of 100 tonnes each	= US$ 0.75
Tick value per contract	= US$ 25
Buyer's gain per contract	= 3 ticks
Number of contracts	= 1000
Overall gain to Buyer	= Tick value per contract × Number of contracts
	= 3 ticks × 1000
	= (3 × US$ 25) × 1000
	= US$ 75,000

Net position in futures market

The seller (exchange) in the futures market makes a net credit of US$ 75,000 to the buyer.

Cash/physical market transaction details, early May 2003

Buyer pays additional cost to seller	= US$ (31.00 – 30.50) = US$ 0.5/t CO_2
Total loss in cash market	= 100,000 tonnes × US$ 0.5 = US$ 50,000

Net position in cash market
The buyer in the cash market makes a net loss of US$ 50,000 to the seller (physical market).

Overall hedge position:

The buyer has achieved a successful hedge, and managed to realise a net gain of US$ 75,000 – US$ 50,000 = US$ 25,000 in the emissions trade using a futures contract.

Comments on the case example

1. Buyer has hedged his price exposure *and* profited from the futures trade.

Notes:
However, futures contracts may have the reverse effect on the buyer. If the futures price for June 2003 quoted in early May 2003 is lower than the (June 2003) futures price quoted in February 2003 – say, US$ 29/t CO_2 – then the buyer, if he decides to settle in May 2003 because he fears that the futures price may drop even further below US$ 29/t CO_2 – will incur a loss per tonne of CO_2 equal to US$ 30.5 – 29 = US$ 1.5. In the cash market, the buyer will incur an additional loss per tonne of CO_2 equal to the difference in the Spot CO_2 Index in May 2003 and the agreed rate of US$ 30.50 per tonne.

2. Seller has made a loss.

Notes:
In February 2003, the seller has taken the view that market prices of CO_2 would fall below the indicative June 2003 futures price of US$ 30.50, hence hoping that by selling at the higher futures price, he would realise a gain in June 2003 by buying at the lower futures price. Unfortunately, the events turned out as the market predicted in February 2003.

3. Closing out.
Seller or buyer can close out his position at any time prior to the contract maturity date, under the terms of the trade, which is facilitated by the "counterparty" role assumed by the exchange or clearing house.

4. Exchange of positions in futures and physical market (this trading strategy is abbreviated in the market as EFP – Exchange of Futures for Physicals).
Positions in the futures markets could be exchanged for equivalent positions in the physical market, or could be exchanged for swaps.

9.8.3 Futures Hedging Example, Firm with Net CO_2 Emissions Reductions

Following the analogy of the previous example, another example of hedging based on a firm with a surplus of CO_2 emission reductions.

Case: Hedging strategy for a firm with net CO_2 emissions reductions

Market scenario, early March 2003

In March 2003, Megapaper Plc, a UK manufacturing company has a surplus of 20,000 tonnes of CO_2 permits which it intends to sell to the market by December 2003. The market value of these permits in March 2003 is US$ 29.50 per tonne.

Early March 2003
Futures market prices, US$/tonne CO_2

Mar 2003	June 2003	Sept 2003	Dec 2003
29.50	29.50	29.25	29.00

Spot CO_2 Index at the exchange: US$ 29.50/tonne CO_2

The futures market indicates a gradual fall of prices of petrmits in the next nine months. Megapaper finds a buyer in the same industry, Ultrapulp and agrees to sell all these permits at the Spot CO_2 Index price (quoted at the exchange) in December 2003. As Megapaper will incur a loss in the physical market if the futures market trend turns out correct over the next nine months, it decides to adopt a hedge using a futures contract.

Market scenario, November 2003

November 2003
Futures market prices, US$/tonne CO_2

Dec 2003	Mar 2004	June 2004	Sept 2004
28.50	28.50	28.25	28.25

Spot CO_2 Index at the exchange: US$ 28.50/tonne CO_2

Action by Megapaper in November 2003

Cash market: Sells CO_2 permits to Ultrapulp at US$ 28.50/t in November 2003, a month before the original date agreed, which is accepted by Ultrapulp.
Futures market: Close out futures position by buying back futures contract at US$ 28.50/t.

Details of actions by Megapaper in November 2003

- Futures market transaction details, November 2003

Buyer (exchange), pays Megapaper	= US$ (29.00 – 28.50) per tonne of CO_2
Seller's gain	= US$ 0.5 per tonne of CO_2
Seller's gain/contract of 100 tonnes each	= US$ 50
Tick value per contract	= US$ 25
Seller's gain per contract	= 2 ticks
Number of contracts	= 200 (based on 100 tonnes each)
Overall gain to Megapaper	= Tick value per contract × Number of contracts
	= 2 ticks × 200
	= (2 × US$ 25) × 200
	= US$ 10,000

Net position in futures market:
The buyer in the futures market makes a net credit of US$ 10,000 to Megapaper.

- Cash/physical market transaction details, November 2003

Ultrapulp buys CO_2 permits from Megapaper at the agreed Spot CO_2 Index rate of US$ 28.50 per tonne; US$ 1.00 less per tonne than the market value at March 2003 (US$ 29.50/t).

Net position in cash market:
Megapaper incurs a financial loss of US$ 20,000 (US$ 1.00 × 20,000 tonnes) in the permits sale versus the spot market.

Overall hedge position

Megapaper has achieved a semi-perfect hedge in this case example; losing in the cash market to Ultrapulp by US$ 20,000 but gained in the futures market by US$ 10,000; resulting in a net loss of US$ 10,000.

Notes:
Megapaper could have swapped its futures obligation (Exchange of Futures for Physicals) with a CO_2 emissions reductions buyer, who would pay the swap rate to Megapaper for a guaranteed purchase price for the emissions credits. Its overall position with the swap would have been better than the trade with Ultrapulp.

9.8.4
Options Market – Principles

In the previous case examples, options could have been taken out by the seller and buyer – *only if they were uncertain of the market price outcome in the future.*
The options market would offer the buyer the right, but not the obligation, to purchase the underlying asset or commodity at a price if it is favourable to him, or terminate (expire) the option if the converse occured. Options carry a fixed premium charge, normally a percentage of contract price, which must be paid upfront. While an options buyer can enjoy unlimited gains, the seller or writer of options does not have rights, but carry the obligation to settle options contracts with potentially unlimited losses. Options, when exercised, are cash settled.

Circumstances for considering options in the previous case examples

- CDM case example – uncertainty of predictions or contrarian views

Uncertainty
If a different set of futures prices were quoted over the period in the example – which did *not* show any degree of consistency on whether market prices in the future will assume an upward or downward trend, then the corporate buyer from the developed country would be hesitant to take out a futures contract.

Contrarian views
If the same set of futures prices as shown in the example were indeed listed, the buyer from the developed country may be indifferent to these figures and perhaps, form his own opinion as to what market prices would be over the period. The buyer may be only willing to buy emissions credits at, say, no more than US$ 30.00/t in June 2003 (buy a call option, that is, receiving the right to buy at the strike price at or before the option's expiry date) but would have to pay a premium to secure this option.

- Megapaper/Ultrapulp case example

If Megapulp, the seller of emissions permits, held two views – that while a futures contract may hedge any CO_2 price falls from March to December 2003, any increases in futures prices will result in financial loss equal to the difference of futures prices over the period. Megapulp hence could have taken out an option contract if it was ambivalent on how market prices would change from March 2003. An example of Megapulp hedging with a option on CO_2 permits is provided next.

9.8.5 Options Market – CDM Example

The case example on the Clean Development Mechanism is re-visited.

Case: Hedging with an option strategy – a firm wishing to purchase CO_2 credits

In February 2003, Alphatech Plc, an Annex I company, wishes to buy 10,000 tonnes of CO_2 credits to comply with its emissions target by May 2003. It does not have sufficient funds at the moment to invest in energy equipment, or buy CO_2 credits at current prices. The market forecasts an increase in CO_2 credit prices in forthcoming months, and Alphatech does not wish to consider futures contracts in case prices should fall resulting in heavy financial losses. The company has a small pocket of funds which it could use on buying call options.

Market scenario, in February 2003

Say, a day in February 2003. Market data:

February 2003
Futures market prices, US\$/tonne CO_2 (second row)
Call option market prices, US\$/tonne CO_2 (third and fourth rows)

Mar 2003	June 2003	Sept 2003	Dec 2003	
30.00	30.50	30.75	31.25	
30.00	30.50	30.75	31.25	(Strike)
0.60	0.80	1.00	1.60	(Premium)

Spot CO_2 Index at the exchange: US\$ 30.00/tonne CO_2

Alphatech's view: Fears that prices in May 2003 will rise above current price of US\$ 30/t, as reflected by futures prices for June 2003, and perhaps by an additional amount determined by market conditions in the future. Alphatech buys June 2003 call options, hence receiving the right to buy credits at the strike price US\$ 30.50/t, call options purchased at premium of US\$ 0.80/t.

Market scenario in early May 2003

Early May 2003
Futures market prices, US\$/tonne CO_2 (second row)
Call option market prices, US\$/tonne CO_2 (third and fourth rows)

June 2003	Sept 2003	Dec 2003	Jan 2004	
31.50	31.75	32.00	32.50	
31.50	31.75	32.00	32.50	(Strike)
0.70	0.90	1.30	1.90	(Premium)

Spot CO_2 Index at the exchange: US\$ 31.25/tonne CO_2

Action by Alphatech Plc in early May 2003

CO_2 market prices has increased as anticipated. Alphatech decides to exercise option (anytime at or before June 2003) at the strike price of US$ 30.50/t, which were bought at the premium of US$ 0.80/t.

- Physical market:

Buys CO_2 at US$ 30.50/t = 10,000 t × US$ 30.50/t = US$ 305,000

- Options market:

Premium cost	= 10,000 t × US$ 0.80/t = US$ 8,000
Net result	= US$ 31,300
Equivalent result	= US$ 31.30 per tonne

- Do nothing scenario:

Buy at market price	= 10,000 t × US$ 31.25/t
Net result	= US$ 31,250
Equivalent result	= US$ 31.25 per tonne

Overall hedge position

Alphatech, which could have bought CO_2 credits in February 2003 at US$ 30.00 per tonne, but could not because of lack of funds, secured a lower price in May 2003 at the rate of US$ 29.70 per tonne by using call options.

If Alphatech had done nothing, it would be buying the CO_2 credits in May 2003 at a price of US$ 31.25 per tonne, an increase of US$ 1.25 per tonne of CO_2 credits. (Of course, market prices could have fallen over the period – if so – Alphatech would abandon or expire the option and purchased CO_2 credits at the lower price).

9.8.6 Options Hedging Example, Firm with Net CO_2 Emissions Reductions

Case: Hedging with an option strategy for a firm with net CO_2 emissions reductions – case example on Megapaper and Ultrapulp is re-visited.

Market scenario, early March 2003
In March 2003, Megapaper Plc, a UK manufacturing company has a surplus of 20,000 tonnes of CO_2 permits which it intends to sell to the market by December 2003.

The market value of CO_2 permits in March 2003 is US$ 29.50 per tonne, which is about half the current cost of control technology for CO_2 reduction in the United Kingdom. This market value hence reflects about half Megapaper's investment cost in CO_2 control. Any fall in the market price of CO_2 permits will reduce this asset value in Megapaper's balance sheet. The firm is naturally concerned that market prices of CO_2 permits has shown a downward trend as reflected by futures prices:

Early March 2003
Futures market prices, US$/tonne CO_2 (second row)
Put option market prices, US$/tonne CO_2 (third and fourth rows)

Mar 2003	June 2003	Sept 2003	Dec 2003	
29.50	29.50	29.25	29.00	
29.50	29.50	29.25	29.00	(Strike)
0.40	0.45	0.55	0.60	(Premium)

Spot CO_2 Index at the exchange: US$ 29.50/tonne

The futures market indicates a gradual fall of permits prices over the next nine months. Megapaper locates a buyer in the same industry, Ultrapulp, to which it will sell all its CO_2 permits at the Spot CO_2 Index price (quoted at the exchange) in December 2003.

As Megapaper will incur a loss in the physical market if the futures market trend turns out as predicted over the next nine months, it decides to adopt a hedge using a futures contract. If futures prices do fall, then it would realise a gain by selling December futures quoted in March 2003 and then closing out its position in say, November or early December.

However, Megapaper has learnt from certain sources in the market that there may exist a shortage of CO_2 permits over the last quarter of the year due to massive purchases by a sector of industry – which may push futures prices up. It would incur losses on futures contracts if this information turned out to be reliable.

After ample consideration, Megapaper decides to use options, and *not* futures contracts, to make profits if the market moves in its favour. Megapaper *buys a put option*, hence it acquires the right to sell CO_2 permits at a certain strike price at or before the option expiry. If the market price of CO_2 permits in say, November, is lower than the strike price, then it will exercise the option and realise a revenue equal to the price arbitrage. If the converse occurs, it will abandon or expire the option, and sell its permits at the higher market rate, and realise a revenue according to the price arbitrage in this event. However, the premium cost should be subtracted from any revenue to provide a net income figure for Megapaper.

Market scenario, November 2003
There will be two outcomes: A or B, assuming prices do not remain static.

- Outcome A: futures prices fall

Futures market prices, US$/tonne CO_2, Nov 2003

Dec 2003	Mar 2004	June 2004	Sept 2004
28.00	27.75	27.50	27.25

Put option market prices, US$/tonne CO_2, Nov 2003

Dec 2003	Mar 2004	June 2004	Sept 2004	
28.00	27.75	27.50	27.25	(Strike)
0.25	0.30	0.35	0.40	(Premium)

Spot CO_2 Index at the exchange: US$ 28.00/tonne CO_2, Nov 2003

- Outcome B: futures price rise

Futures market prices, US$/tonne CO_2, Nov 2003

Dec 2003	Mar 2004	June 2004	Sept 2004
31.00	31.25	31.50	31.75

Put option market prices, US$/tonne CO_2, Nov 2003

Dec 2003	Mar 2004	June 2004	Sept 2004	
31.00	31.25	31.50	31.75	(Strike)
0.50	0.60	0.75	0.95	(Premium)

Spot CO_2 Index at the exchange: US$ 31.00/tonne CO_2, Nov 2003

Action by Megapaper in November 2003

- *Outcome A*

CO_2 market prices fall as anticipated. Megapaper exercises option at strike of US$ 29.00/t, at option premium of US$ 0.60/t. Sells CO_2 permits to Ultrapulp.

Physical market:
Premium cost = 20,000 t × US$ 29.00/t = US$ 580,000

Options market:
Premium cost = 20,000 t × US$ 0.60/t = US$ 12,000
Net result = US$ 568,000
Equivalent result = US$ 28.40 per tonne

- *Outcome B*

CO_2 market prices rise unexpectedly.

Megapaper adandons or expires option, as the market price in November 2003 (US$ 31.00/t) is higher than the strike price (US$ 29.00/t) (the right it acquired in March 2003). The firm however, incurs a loss equal to the price of the premium, although not utilised. By expiring the put option, Megapulp has no obligation to sell CO_2 permits to Ultrapulp, but sells it to other willing buyers, which may include Ultrapulp.

Physical market:
Sales to willing buyers = 20,000 t × US$ 31.00/t = US$ 620,000

Options market:
Premium cost = 20,000 t × US$ 0.60/t = US$ 12,000
Net result = US$ 608,000
Equivalent result = US$ 30.40 per tonne

Do nothing scenario, lower market prices in Nov 2003:
A loss from sales of US$ 29.50 – US$ 28.00 = US$ 1.50 per tonne

Overall hedge position

If market prices of CO_2 permits fell as feared, Megapaper would have achieved a reasonable hedge – its revenue of US$ 28.40/t resulted in a loss of only US$ 1.10/t less than the price (US$ 29.50) it would had hoped for if prices remained stable from March to December 2003.

If market prices had risen, Megapulp would gain a profit of US$ 30.40 – US$ 29.50; of US$ 0.90 per tonne of CO_2 permits sold.

If the firm had done nothing, and market prices fell, it would have lost US$ 1.50/t – a loss greater than that incurred if it had bought the put option.

Comments on the case example

1. Depending on the outcome of future market prices, Megapaper achieved a sufficiently effective hedge. If market prices fell or rose, its overall financial position was better off than it had done nothing at all.
2. In practice, buying options on energy products may not be always 100% effective, due to to the cost of the premium, but opportunities of profit making may exist if markets are volatile and move in favour of the hedger.
3. Delta hedging

If market prices change, option premiums may change at different rates, or remain static, relative to the change in market prices. This relative change, known as delta, can produce gains for a hedger using options. Most ATM calls and puts have delta values of 0.5, that is, the value of premiums will move by half (half of the original premium value) for a change in price of the underlying asset. Deep ITM calls and puts have deltas close to 1.

Using the Megapaper/Ultrapulp example:

In March 2003, the December 2003 put option premium was US$ 0.60/t, a strike price for December 2003 of US$ 29/t, and market price of CO_2 permits US$ 29.50/t.

If the market price of CO_2 permits had actually *fallen* from US$ 29.50/t to US$ 28/t in November 2003, and the put option premium decreased from US$ 0.60/t to US$ 0.25/t, then the delta value would be slightly in excess of 1.0 for the movement in the premium.

Megapaper, if it had anticipated this trend in the market, could have, *at the start of the hedge*, bought an *excess* of put options at US$ 0.60/t – with the right to sell CO_2 permits at US$ 29/t. These put options, bought in March 2003 at US$ 0.60/t would substantially gain in value by November 2003, as a holder of these put options have the right the sell CO_2 permits at a level US$ 1.50 higher than the market price (US$ 29.50/t – US$ 28/t).

Although Megapulp's position in the physical market would be a financial loss from the sale of its CO_2 permits to Ultrapulp in November 2003 (equal to the difference between the market price of permits in March 2003 and November 2003), by using the delta hedging technique, a gain would be realised with the re-sale of the higher premium put option in November 2003. The net hedge result is a larger profit for Megapulp compared to that by just exercising purchased options.

Based on the previous case examples of risk hedging of future carbon prices, risk-return profiles of futures and options hedging are presented on pages 219 and 220.

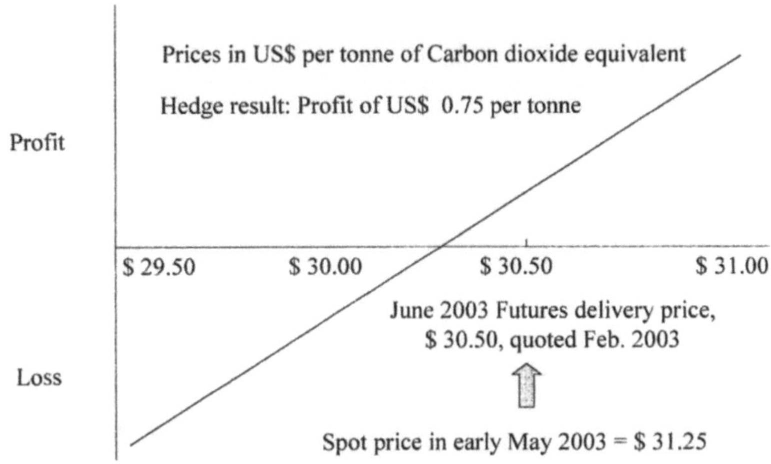

Fig. 9.2. Futures hedging for a buyer of carbon credits

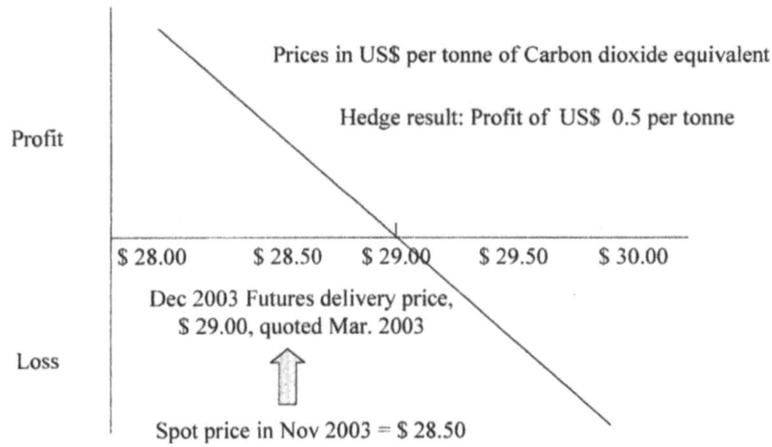

Fig. 9.3. Futures hedging for a seller of carbon credits

9.8.7
Swaps – "Physicals" Swaps and "Credit Risk" Swaps

In the commodities market, a "physicals swap" enables a futures contract counterpart to buy or sell the physical at a fixed price in the future while bearing the risk of fluctuating market prices. The buyer is obligated to buy at a fixed price in the future over a specified period of market exposure – any falls in the market price relative to the agreed fixed price at any time over the period is compensated to the seller, and vice-versa.

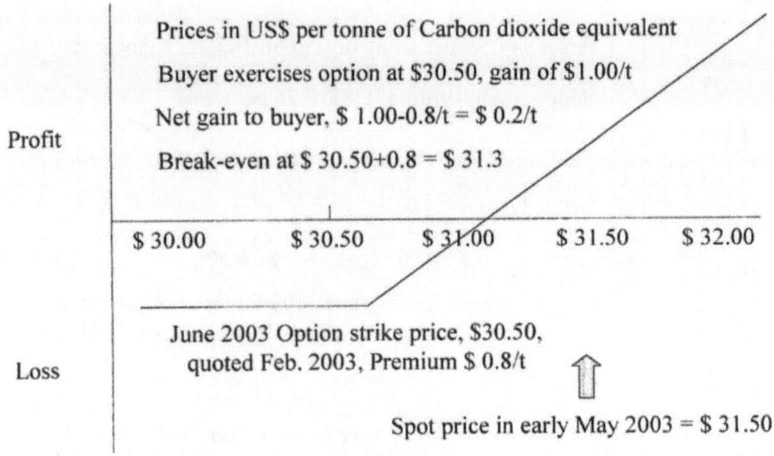

Fig. 9.4. Option hedging for a buyer of carbon credits

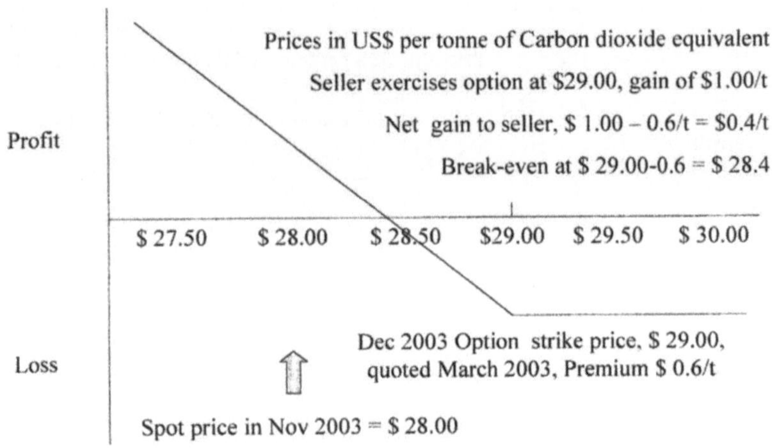

Fig. 9.5. Option hedging for a seller of carbon credits

Total profits and losses are possible; for example, if the market price should fall below the agreed fixed price for over, say, three months, then the buyer will incur mounting losses. The buyer pays the seller a "swap" rate to hedge itself from any increase in the market price over the period. Both traders are fully exposed to market risk.

In a "credit risk" swap, an exchange of mutually agreed percentages of the relative credit strengths of two parties allows each party to conduct business at a lower expense.

9.8.8
Swaps – Case Example

CO_2 emissions credits or permits, based on two buyers; fixed and floating prices

Two buyers with different financial risk profiles could participate in a CO_2 price swap, based on different dual pricing – one fixed and one floating in the same currency – on sales of CO_2 credits or permits to these parties.

Case example:
July 2003
Company X may be quoted a fixed price of US$ 30.00/tonne of CO_2 and a floating price of the Spot CO_2 Index rate plus 50 basis points (minimum price movement) for 10,000 tonnes of CO_2 emissions permits, for delivery in September 2003.
Company Y (with a slightly lower financial risk profile than company X) may be quoted a fixed price of US$ 31.00/t of CO_2 and a floating rate of the Spot CO_2 Index plus 25 basis points for an equivalent volume of CO_2 emissions permits and same delivery date.

The Spot CO_2 Index is US$ 29.00/t.

Company Y has a comparative advantage in the floating price offered it, while company X has advantage in the fixed rate. X borrows at Y's floating rate at an opportunity cost which is paid to Y, while Y borrows at X's fixed rate which is lower than that offered to Y. Both companies benefit in the end by borrowing at lower interest rates, but may suffer losses if the Spot CO_2 Index should fluctuate by large margins against their original positions.

A financial intermediary usually handles the swap transaction between the two parties, and earns a commission from the swap transaction.

9.8.9
Derivatives – Recommendations on Pricing and Applications

There is no historical precedent for the pricing of spot and futures financial instruments for the six greenhouse gas emissions as specified by the Kyoto Protocol. The closest equivalent is the US sulphur dioxide emissions trading system based primarily on spot transactions.

Recommended pricing strategies for greenhouse gas reductions

Spot trading
Cash transactions can be made on the present value of the average cost of GHG reduction based on annuitised capital cost or levelised capital cost. The calculation of spot prices would be straightforward for different technologies in different locations and conditions.

Futures trading
Traditional investment methods take into account the value of dividends or interest between the current period and a future point in time when calculating the futures

price. For example, the futures price of carbon dioxide permits or credits currently trading at US$ 25 per tonne spot would be calculated at 1.035 × US$ 25 per tonne in six months time if the risk-free rate of interest (government bond) is 7 percent per annum.

Options trading

The Black-Scholes options pricing model relies on historical data of stock prices and standard deviations per period of rate of return on stock. Its application to currency, bonds and futures contracts has been generally successful, but would not be suitable for greenhouse gas assets until sufficient market data on emissions trading prices and price volatilities is gathered over a reasonable period of time – perhaps at least twelve months after trading commencement.

Options prices could, however, be based on futures prices (simplistic method as proposed in futures trading section above) and a set of factors which could affect futures prices such as quality of technology, compliance to additionality criteria and other major risk elements which could affect the full value of emissions assets (Section 9.7 examines the range of factors that investors should consider).

Swaps trading

In commodity trading, both parties assume financial risks based on a fixed price in the future – in emissions trading, the parties could agree on a futures price of a greenhouse gas product and settle any cash differences at the date of maturity of the swap contract. The swap price would include an intermediary's cost if it was required to conduct the swap.

Recommended applications for national permits trading

Emissions trading within a country are likely to be conducted on a "cap and trade" basis – annual permits with progressively reduced quotas of emissions could be distributed on the "grandfathering" principle with a small percentage, say 5% to 10%, auctioned off.

To suit the financial requirements of various buyers and sellers, futures, options and swap contracts would offer the market the tools for managing their mandatory emission reductions at levels of risks that traders are willing to undertake.

Revisiting the Megapaper and Ultrapulp corporate case studies, which were based on the Clean Development Mechanism, the same input factors could be applied for permits trading if Megapaper and Ultrapulp were companies in an Annex I country. In a national scheme, there would be less sovereign, political and currency risk compared to a CDM project, but parity issues between CDM emission reductions acquired by Megapaper and/or Ultrapulp and national permits would have to be resolved or agreed upon if CDM credits were to traded internally.

9.9 International Emissions Trading

The estimated total amount of reductions for Annex B countries (the majority of which are Annex I countries except for Croatia and Slovenia) is about 600 million tonnes of carbon or 2,200 million tonnes of carbon dioxide equivalent, based on

Table 9.1. Size of the Clean Development Mechanism market

Proportion of total Annex I commitment offset by CDM		Clean Development Mechanism: Market value based on market prices Market price (US$)/t CO_2 equivalent				
%	m. tonnes	$ 5 $ m	$ 10 $ m	$ 15 $ m	$ 20 $ m	$ 25 $ m
10%	220	1,100	2,200	3,300	4,400	5,500
20%	440	2,200	4,400	6,600	8,800	11,000
30%	660	3,300	6,600	9,900	13,200	16,500
40%	880	4,400	8,800	13,200	17,600	22,000
50%	1,100	5,500	11,000	16,500	22,000	27,500

UNFCCC, International Energy Agency and Kyoto Protocol sources. The market size for the Clean Development Mechanism, assuming proportions of total Annex I reduction obligations can be acquired from non-Annex I countries, is shown in the Table 9.1.

If 10% of Annex I commitments were satisfied by emissions reductions in CDM projects, the value of contracts per year (say, 2005) assuming 10 transactions per day for 300 days at a price of US$ 10/t CO_2 would be:

$$220 \text{ million t } CO_2 \times \text{US\$ } 10/\text{t } CO_2 \times 300 \times 10 = \text{US\$ } 6{,}600{,}000 \text{ million} = \text{US\$ } 6.6 \text{ trillion.}$$

In contrast, non-CDM transactions can be conducted via inter-country Annex B trading. If Russia (Annex I and Annex B lists) were allowed to sell its surplus of about 140 million tonnes of carbon to Annex I country buyers, the trade would be worth US$ 700 million at US$ 18.35 per tonne of carbon (if traded at US$ 5 per tonne of CO_2 equivalent).

International emissions trading of GHGs – AIJ and corporate experiences

The Activities Implemented Jointly (AIJ) programme – precursor to Joint Implementation under Article 6 of the Protocol – initiated carbon-offset projects. It is estimated that between three-quarters and one billion dollars have been invested in AIJ projects at end 1999. The experience gained from AIJ projects should enable Annex I countries to implement JI projects more efficiently with partner countries in the millennium. Most of AIJ projects are registered with the UNFCCC in Bonn.

A. Brunello's thesis "The Cost Effectiveness of Joint Implementation Projects for Securing Reductions in Greenhouse Gases", University College London, provides a comprehensive survey of 78 AIJ projects, as shown in Table 9.2.

Other emissions reduction projects have been conducted internally within firms or subsidiaries of firms in one or more countries.

Table 9.2. Activities implemented jointly (AIJ) projects – financial data (carbon abatement costs are undiscounted)

Host country	No. projects	Project type	Average US$/t (preliminary figures)	Carbon abatement cost US$ /tonne High	Low
Belize	1	Carbon sequestration	1.72	–	–
Bhutan	1	Micro-hydro	53.51	–	–
Burkino Faso	1	Sustainable energy	6.13	–	–
Costa Rica	8	Carbon sequestration	150.43	861	2.0
Czech Rep.	2	Fuel switching	10.24	12	8.4
Estonia	16	n/a	120.70	406.1	120.7
Honduras	4	Biomass power gen. and renewable energy	121.04	418.0	121.0
Hungary	2	Fuel switching and energy efficiency	293.03	588.7	2.3
Indonesia	1	Renewable energy	12.07	–	–
Latvia	19	Wind energy	115.77	653.3	11.9
Lithuania	8	n/a	50.54	147.9	19.7
Mexico	3	Carbon sequestration and energy efficiency	10.02 (c seques)	n/a	n/a
Nicaragua	1	Geothermal	113.59	–	–
Panama	1	Carbon sequestration	242.44	–	–
Poland	1	Fuel switching	73.14	–	–
Romania	1	Energy efficiency	2.83	–	–
Russia	7	Carbon sequestration, energy efficiency, fugitive gas capture	60.96	384.62	0.01
Solomon Is.	1	n/a	19.34	–	–

Notes:
Carbon seq. = carbon sequestration, which includes forestry management.
Carbon abatement cost for Panama forestry project appears to be on high side.
Investor countries were the USA, Sweden, Netherlands, Norway, Germany, UK, Australia, Belgium, France and Denmark in descending number of projects per country.

Potential size of Annex B/emissions trading market

If 10% of quantified emission limitation and reduction commitments (QELRCs) of Annex B countries were traded at US$ 10 and US$ 20 per carbon dioxide equivalent tonne, the Joint Implementation initiative may generate a market of US$ 600 m and US$ 1,200 m respectively based on figures estimated for the CDM. It remains to be seen whether emissions trading of ERUs (credits created by Joint Implementation projects) or CERs (credits via CDM projects) will dominate global markets in the future. Table 9.3 shows the project details of five international emissions trading schemes undertaken by energy and energy-reliant companies in five continents. Shell's internal trading scheme includes its corporate units in Africa.

Table 9.3. Emission reduction projects – corporate experiences

Company	Trading regime/ project partner	Project type	Trade or project details/tonnes of carbon dioxide equivalents
BP Amoco, UK	Internal trading scheme	Energy efficiency and process improvements	About 300,000 tonnes of CO_2 eq. traded by March 2000 at average price \$ 10/t.
	Example of trading, Two BP units, North-Sea	Oil and gas development operations	10,000 t. CO_2 eq. @ \$ 17/t.
Niagara Mohawk, USA – seller	Suncor Energy, Canada – buyer	Energy efficiency and fuel switching	100,000 t.
Mitsubishi and Sumitomo	Unified Energy Systems, Russia	Energy efficiency and fuel switching	10 m. t. per year
Shell	Internal trading scheme Shell Tradable Emissions Permit System (STEPS)	Energy efficiency and fuel switching	About 500,000 tonnes of CO_2 traded by March 2000, price range \$ 7–14/t CO_2 eq.
Tokyo Electric Power (TEPCO)	New South Wales, Australia	Carbon sequestration	1,000 hectares of forests absorbing about 300 tonnes of CO_2 over lifetime. Possible increment to 40,000 hectares.

9.10 Trading Commissions

Transactions of GHG reductions over an exchange or Over-the-Counter (OTC) could incur commissions which vary between one to two percent of the contract price – perhaps higher if matching of buyers and sellers or negotiations between the parties is required.

At two percent commission, the value to all exchanges or intermediaries would be US\$ 44 million on the basis of 10% of Annex I commitments purchased through CDM projects at an average price of US\$ 10 per contract; US\$ 88 million at US\$ 20 per contract. Contracts are likely to be traded several times as markets become more efficient and prices achieve more transparency – hence if contracts were traded 10 times daily for 300 days in 2001 (10% of Annex I commitments at US\$ 10/t CO_2), the revenue to exchanges or intermediaries could rise to US\$ 132,000 million or 132 billion in that year.

Comparison of estimated values of financial flows vs emissions contract commissions vs 1% contribution to proposed LDC/IRISK equity fund

The value of foreign direct investments in CDM projects, based on, say, average plant capital costs of US\$ 50 and US\$ 100/tonne of CO_2 reduction (various non-renewable and renewable energy technologies, 10% discount rate, 10% Annex I commitments) are estimated at:

220 million tonnes (US$ 50, US$ 100 per tonne CO_2 reduction)
= US$ 11 billion and US$ 22 billion, respectively.

Of these projected amounts, China, India, South Africa and Brazil are expected to attract the majority share of capital investments - while LDCs, non-energy intensive countries and islands-at-risks will be largely ignored. The author's earlier proposal of a global LDC/IRISK (islands-at-risk) equity fund stands justified in the light of these figures.

Trading commissions to exchanges, intermediaries and financial institutions are expected to outstrip total capital investment flows - by a factor of about 5 to 10 times (US$ 132 billion compared to US$ 11 and 22 billion respectively, comparison with estimates above, assuming 10% Annex I commitments met by CDM). This explains the strong growth of emissions trading agencies and intermediaries in the private sector in Europe and USA in the first six months of 2000.

By comparison, as worked out earlier in Chapter 7 on equitable treatment and baselines, the allocation of 1% of the value of all traded emissions contracts at US$ 10/t CO_2 to the author's proposed LDC/IRISK fund would amount to US$ 66 billion, which is half of the total trading commission revenue (2% of contract value) to financial intermediaries.

9.11 Role of National, Regional and Global Exchanges and OTC Market

Financial intermediaries are necessary for international emissions trading - it is envisaged that OM Environmental Exchange (part of OM Gruppen, Stockholm) and International Petroleum Exchange, both based in London; New York Mercantile Exchange (NYMEX), Chicago Board of Trade, Sydney Futures Exchange and the Singapore International Monetary Exchange (SIMEX) will be amongst the leading exchanges to facilitate carbon transactions post 2000. As CDM rules are not expected to be finalised until end 2000, options and futures contracts appear to be the most likely instruments for emissions trading.

Contracts should include as much information on asset quality and project certifiability. The proposed set of terms and conditions for GHG trading contracts as described in the offer document (Section 9.7) is recommended. The terms also cover the key risks that investors may have to undertake, and seller's liabilities on non-performance of contract.

The setting up of Over-The-Counter (OTC) markets could be delayed as investors would initially prefer higher transparency of prices and issuers. OTC trades may follow once prices become less volatile and market players become more established.

In executing CDM transactions, Figure 9.6 (page 236) shows the exchange of documents of entitlement, cash/deferred financial payments, obligations of buyers/sellers, and the role of national exchanges and a CDM certification board appointed by Conference of the Parties. Verification and certification may become the most time-consuming process in the trading of carbon credits. The scheme is likely to be simplified with installation of dedicated IT exchange networks and user-friendly trading software. For Joint Implementation projects, a similar structure could be adopted, perhaps at less cost.

Price volatility

Prices of emission reduction assets traded globally are expected to experience large variations in the first three to six months due to highly speculative buying by investors, different types of technology costs and different values of risk-assessed contracts.

Several kinds of technologies will be applied to CDM and JI projects in countries with wide differences in labour costs and imported equipment discount prices. Contract prices will also be subjected to additional costs if asset evaluation by experts is required. Price stability is expected after a full year to eighteen months following the first cross-border trades through exchanges, OTC and direct bilateral or multi-party agreements.

Market price stability – hedging with global benchmarks

The prices of traded emissions contracts for each of the six targeted gases can be collated by stock exchanges operating within a time zone and relayed to central administration offices in New York, London and Tokyo to establish regional or a global price index (e.g. for carbon dioxide, an average of closing prices on the previous day) so that investors and speculators can manage basis risk, that is, the risk between current prices and a benchmark. In the futures market, analysts may be able to forecast futures prices and futures price indices for each respective greenhouse gas based on current and past price volatilities.

9.12 Packaged Commodity and Emissions Trading (PACOMET)

A dual commodity-emissions packaged trading (PACOMET) contract could be designed to complement and possibly maximise the efficiency of internal and cross-border emissions trading. This hybrid financial instrument would be particularly useful to companies producing bulk quantities of energy commodities (e.g. gas, oil, electricity) or companies which consume large amounts of energy in the process and manufacturing sectors (e.g. steel, cement, paper, chemicals, motor vehicles) – if these companies are required to achieve emissions reduction targets under the Kyoto Protocol. An example of how a PACOMET contract could work for the electricity generation industry is described next.

Packaged electricity and carbon dioxide trading (PACOMET-ELCO$_2$)

A power generation company X in an Annex I country which exports electricity to a power generator Y in another Annex I country, perhaps due to an upsurge of demand by company Y's customers may have the opportunity to sell emissions reductions required of Y. A dual trading of both electricty and emissions could hence be conducted between the parties.

The transaction could be acronymed PACOMET-ELCO$_2$ (Packaged Electricity and Carbon Dioxide Emissions Trading – seller's industry and greenhouse gas follows PACOMET) if electricity and carbon dioxide reductions are sold on spot, futures or options contracts. The pricing of the contract could be made more competitive than

Table 9.4. Potential PACOMET sellers and buyers

Primary industry	Primary product sold	Emissions traded (e.g.)	PACOMET sub-title	Likely clients other than from primary industries (first column)
Cement	Building products	CO_2	$CECO_2$	Large commercial buildings.
Chemicals	Chemicals, petchemicals	CO_2	$CHCO_2$	Plastics, polymer manufacture.
CHP	Steam, electricity	CO_2, CH_4	$CHCO_2/CH_4$	Large commercial buildings.
Gas	Natural gas, oil	CO_2, CH_4	$NGCO_2/CH_4$	Iron/steel, cement, chemicals.
Landfill	Electricity, steam	CO_2, CH_4	$LACO_2/CH_4$	Power generators, gas companies.
Motor veh.	Cars, trucks	CO_2	$MOCO_2$	Firms hiring/owning transportion.
Paper	Paper, pulp	CO_2	$PACO_2$	Supermarkets, stores.
Power	Electricity, steam	CO_2, CH_4	PCO_2/CH_4	Large commercial buildings
Steel	Steel products	CO_2, CH_4	$STCO_2/CH_4$	Car, machinery manufacturers.

singular electricity or emission transactions to attract a larger market in this particular sector. More complex permutations of primary product and emissions contracts, e.g. electricity futures and CO_2 options or vice-versa, and contracts based on fixed and floating prices could be developed once a liquid market is established.

Other PACOMET contracts

Similarly, a gas production company G could sell natural gas to another company H in the same industry and combine the primary sale of gas with secondary sale of emission reductions (achieved by G) if H has emission reductions requirements. The contract could be acronymed PACOMET-$NGCH_4$ (Packaged Natural Gas and Methane Emissions Trading).

9.13 Impact of Kyoto Protocol on Financial Markets and Issued Securities

The Kyoto Protocol will affect the structure of capital and equity markets in Annex I and non-Annex I countries, as well as investors of securities in both these markets.

In terms of project finance, 18 to 42 billion dollars may be required for new or retrofit CDM and JI projects based on average costs of reducing about 600 million tonnes of carbon equivalent (US$ 30/t C, author's projection of World Bank estimate

Table 9.5. Sources of financing for CDM and JI projects

Instrument	Financiers Annex I, non-Annex I	Expected returns
Equity	Direct investors	Emissions credits, plus project revenue.
	Indirect investors	Emissions credits, possibly project revenue.
Debt – senior debt	Banks, bilateral country debt	Interest payments, plus (possibly) part of emissions credits to Annex I government lenders.
Bonds	Indirect investors	Interest payments
Bonds + warrant	Indirect investors	Interest payments, plus emissions credits

Notes:
Direct investors are principally, the sponsoring consortium, who borrow capital and obtain equity from financial markets, e.g. foreign power utility company and the local power generation company. Indirect investors are "retail" investors, e.g. financial institutions, speculators, Annex I governments and firms, non-Annex I local investors.
Emissions trading between non-Annex I and Annex I companies are expected from the year 2000, in addition to intra-Annex B emissions trading.
Annex I and Annex B companies and possibly, Annex I and Annex B governments, are expected to be the final purchasers of emissions credits, as these credits do not represent any intrinsic or financial value to other entities or intermediaries.

which include carbon sequestration techniques and emissions trading quotas; US$ 70/t C, author's independent estimate based on average discounted capital costs of thermal power technology, non-carbon sequestration projects). The World Bank's estimate are mainly based on AIJ projects, some of which do not include discounted costs. If renewable energy projects are to be considered then the average cost of carbon reduction may be only 5% to 10% higher as the percentage of these projects of the total portfolio is expected to be marginal.

Traditional market instruments would be utilised to raise project finance. Emission credits, which represent tangible financial assets, would offer additional security for financing. Table 9.5 shows the utility of market instruments in Kyoto Protocol projects.

9.14 Kyoto Protocol's Financial Impact on Credit Ratings and Net Asset Values

The Kyoto Protocol's Clean Development Mechanis, Joint Implementation and Emissions Trading initiatives will have major impacts on all companies which operate in the industry sectors identified in Annex A (see Annex section of the book) in terms of net liability and credit ratings of securities.

Credit rating impact on obligated companies

An Annex I "obligated" company (a company has identified by the government for reduction of emissions) may have to examine its greenhouse gas inventory before it seeks to raise finance in local or international markets. Capital funding sources may require details of a company's GHG inventory, which could be available from corporate environmental reports, commercial sources, government reports or "green" credit rating agencies. The most likely questions potential capital providers would ask are:

1. What are the reduction levels, and when must the company meet its obligation?
2. If it is borrowing and/or issuing equity, what are the expected returns? What are the risks of these projects?

If the company intends to raise finance from the year 2001 to the first commitment period (2008–2012), this will represent a long-term liability in its balance sheet – which should be reflected in credit ratings of all its issued debt.

The credit rating of an obligated company should be lower if its capital structure is increased as a result of future borrowings and/or equity issues to finance the technology to meet its emission targets. However, the company should benefit from fuel savings, hence improve overall price competitivess in the intermediate to long-term and realise higher sales. A full cost benefit analysis should be performed.

Factors to consider for a Non-Annex I company

A non-Annex I company has no obligatory emission reduction targets, e.g. Albania, Bosnia and Herzegovina, China, India. If it chooses to raise finance via debt and equity, then its credit ratings should increase as its future projected revenue is augmented by GHG reduction sales to service its lenders and investors.

Credit rating impact on obligated countries

The capital market size of GHG obligatory reductions in Annex I countries will depend on the average cost of abatement technology for each country – which is likely to vary according to cost of materials and equipment, availability of energy resources, skilled labour and the relative costs of primary production (e.g. electricity, process steam, chemicals, steel). Table 9.6 compares the relative sovereign borrowings by Annex I countries if obligated companies decide not to invest in technology from its current balance sheets.

Net asset value impact on M&A

Due diligence should be conducted by evaluating the emissions performance of the target company before a merger or acquisition. The net asset values should be discounted by the appropriate market rate. For an Annex I target: the net asset value of the company would be less by that amount of capital investment required for meeting its emissions obligations, or higher by that amount which increases its future revenue if it has surplus reductions to trade. For a non-Annex I target: the net asset value of the company would be increased by that amount which can traded on the emissions market. Table 9.7 lists the impacts of technology investment and emissions trading on net asset values of firms.

Table 9.6. Potential borrowings by Annex I countries to meet obligations (energy industries only)

Western Europe: key countries	Reduction (+) requirements or excess (–) inventory	$ 10/t C $ m	Market value, US$ $ 20/t C $ m	$ 30/t C $ m
UK	+36 mt C	360	720	1,080
Germany	+32 mt C	320	640	960
Italy	+27 mt C	270	540	810
France	+7 mt C	70	140	210
Total EU	+145 mt C	1,450	2,900	4,350
Eastern Europe/CIS: key countries				
Russia, Ukraine	–174 mt C (surplus)	1,740	3,480	5,220
Total EE/CIS	–134 mt C (surplus)	1,340	2,680	4,020
Comparison				
USA	+480 mt C	4,800	9,600	14,400
Canada	+41 mt C	410	820	1,230
Japan	+38 mt C	380	760	1,140

Notes:
Russia and Ukraine has a surplus of emissions reductions (projected 2010 vs. 1990, 2010 taken as mid-point of the commitment period 2008–2012) due to economic inactivity. There is much debate on whether these countries can actually sell "hot-air".

Table 9.7. Corporate choices and impact on net asset value

GHG reduction method	Advantages	Disadvantages/Risks
"Permanent technology" solutions	More efficient process. More competitive pricing. Attracts "green" investors. Entitlement to credits. High prices of credits. New markets expansion.	Early cash call. Interest rates may fall in future. Additionality factors. Certification of credits. Project risks.
Emissions trading	Postponed cash call	Price exposure on credits. Plant/entity becomes uncompetitive at lower fuel efficiencies. Increased cost of technology if interest rates rise in future.

Cost-Benefit Test: The method which delivers a higher net present value, which implies a higher net asset value and internal rate of return, is chosen.

The assumption of emission reduction obligations by parties in a merger or by the acquiror should be clearly stated in M&A documentation. Acquirors of GHG emitting assets should hence be well advised of outstanding emission reduction obligations unless all parties agree on sharing the costs of abatement.

Technology investments improve net asset value

Annex I companies would incur capital expenditures equal to the amount of capital needed to (i) invest in long-term GHG reduction technologies or (ii) purchase emission credits at costs perhaps half or quarter of "permanent" technological solutions.

Fuel saving benefits over longer periods are realised as a consequence of investments in modern technology, compared to buying emission reductions indefinitely. The net present values of high technology projects should be positive in the long-term. Meeting targets by purchase of emissions credits result in negative NPVs and possibly higher negative NPVs if emissions credits become more expensive at unchanged or increased technology costs.

Generic flow diagram of an emissions trading network

A likely configuration of an emissions trading network is shown in Figure 9.6. The minimum period to execute a transaction may ultimately depend on the certification and verification process.

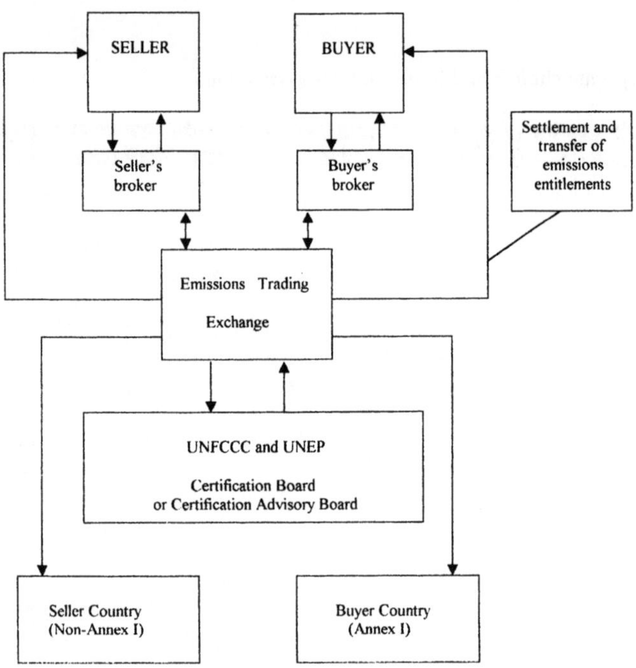

Fig. 9.6. Emissions trading – operations in an exchange

10
Conclusions

The Kyoto Protocol is an unprecedented initiative forged with the best intentions by countries which participated in the United Nations' third annual conference on climate change in 1997. It could well be the last straw that could keep the world afloat in a seemingly self-destructing environment as evidenced by a series of recurrent natural disasters since the 1980s.

The book describes the three principal instruments that holds the Protocol in cohesion – the Clean Development Mechanism (CDM), Joint Implementation (JI), and Emissions Trading. As Asia and South American countries replace OECD nations as the industrial powerhouses for global production and economic growth by the year 2050, early action must be taken to secure a sustainable environment against this quantum demand for resources. The emissions "additionality" requirements are imperative for both CDM and JI projects, but should be interpreted with common sense so that investments are not deterred.

Emissions Trading – impact on effective functioning of Clean Development Mechanism

Although various enabling technologies and risk management aspects of CDM and JI projects are introduced and reviewed in the book, the "project investment exit" pillar of the Kyoto Protocol – that of emissions trading – is likely to become the most controversial issue facing the Conference of the Parties, and COP VI in the Hague in November 2000. The concluding part of this publication examines the political factors which will influence emissions trading.

Developing countries may not play the emissions trading game at their expense

Huge inequities emerge with regard to the scale of financial benefits to emissions traders and Annex I investors, compared to the benefits to developing countries. The results from a credible number-crunching exercise performed by the author in Chapter 7 suggest that CDM countries will benefit less than Annex I investors in terms of net financial flows, but benefits to CDM and Annex I countries will be dwarfed by estimated revenues to emissions traders – the biggest winners of the Kyoto Protocol. It is a politically significant revelation, upon which the proverbial geese that lay the golden eggs – CDM and low GNP Annex I countries – must act to ensure equitable treatment of benefits.

Experts in developing countries and least developed countries are likely to arrive at the same inequitable ratios of benefits as calculated by the author – it is better to lay on a level playing ground for the international community than to invite hostile repercussions or high levels of "North-South" suspicion. As more countries participate, markets will become more efficient with higher liquidity and price transparency.

If emissions were traded at standard trading units of 100 tonnes at US$ 10/t, then trades would be similar to Eurobonds at US$ 1000 nominal. Net commissions after paying out costs could be less than 2%, but companies that own trading operations may not settle for less than 1% in a new market that needs to recover costs on technology investments and cross-border risk exposures.

The author has proposed to UNFCCC a one percent mandatory contribution of the value of every CER contract traded to the country of origin of the traded CER; 0.7% and 0.5% to low and medium GNP Annex I host countries; or multiples of these percentages; i.e. 4%, 3% and 2%, rounded to the nearest percentage unit, respectively. In addition, 1% to 2% of every CER and ERU contract traded could be allocated to a new charity, the LDC/IRISK Climate Change Fund – which could function as a precursor to the UNFCCC's proposed "Adaptation Fund". Small and large international relief agencies could be invited to advise and participate in the LDC/IRISK Fund, but foundations with broader responsibilities could act as fund trustees. If CDM and JI governments do not allow local and international relief agencies to receive a share of proceeds from the secondary of contracts, then the Conference of the Parties should vote for inclusion of this allocation – a minimum 1% of contract value is suggested – as a standard emissions trading requirement. The basis and rationale for these proposals are provided in Chapter 7.

Why mandatory contributions make sense

If Conference of the Parties approve of mandatory contributions to the CER or ERU country of origin, the total cost of a contract to a seller of emission credits will be the sum of the trader's commission (a flat fee or percentage of settled contract price) and the mandatory percentage of settled contract price. Secondary sellers (not project company which reduced emission levels) would hence raise prices to realise higher net profits. The effects are twofold – Annex I companies might decide to implement control technologies at home, which will reinforce domestic action on emissions reductions; and project companies would be able to sell at higher market prices. Buyers should pay the opportunity cost of securing emissions credits at prices lower than permits in the home market – the cost of additional mandatory contributions are expected to be marginal to buyers' savings. Emissions trading is a zero sum game in which all parties must benefit equitably.

Annex I cross-border investments at negative costs?

If Annex I investors insisted on a share of mandatory contributions, then their cash flows may end up with a largesse of capital higher than their investments in developing countries or low GNP Annex I countries.

If non-Annex I countries of CER origin oppose any sharing of mandatory contributions from traded contracts with Annex I countries, then *one possible solution* would be to apply an experimental percentage in the first quarter of trading, perhaps 1% of the value of traded contracts to CER host countries, and another 1% to Least Developed Countries – and then review the results before considering any contributions to those Annex I companies that make actual investments outside their countries.

Another equitable solution would be to assess the financial benefits accruing to "South" and "North" countries through all the mechanisms agreed by Conference of the Parties – which may include unilateral, bilateral or multilateral arrangements for the recycling of percentages of traded contract values to the host project country. Any marked imbalances of South/North benefits could be negotiated at the next COP meeting.

Renewable energy

To stimulate foreign investments in the renewable energy sectors of developing countries, the author recommends the allocation of a heavier weighting of the mandatory percentage of contributions from every contract traded to local stakeholders of renewable energy projects, compared to non-renewable energy projects. For example, if the Conference of the Parties agreed on 3% mandatory contribution of every CER contract to the CER country of origin, the government could receive, say, 2%; domestic stakeholders of renewable energy projects and non-renewable energy projects, 0.4% and 0.2%, respectively; non-energy projects 0.2%, and the humanitarian sector 0.2%. Foreign investors, hence, would be enticed to invest in non-renewables to benefit from any arrangements they may negotiate with local partners.

Progress in Asian markets

By end of 2000, Euro-Asia Pac Corp Ltd, a "boutique" energy, environmental and emissions trading consultancy based in London, has developed a multi-sector CDM projects portfolio for eleven countries in Asia and Africa. A total of about 2 million tonnes of carbon dioxide equivalents at a price range of US$ 20–30 per tonne is offered to investors and buyers. A prospectus is currently being prepared for institutional investors.

Global markets, technologies and the way forwards

The global survey of CDM host countries in Chapter 5, cost-benefit analysis of emissions reduction technologies in Chapter 8, sample calculation guides and case studies throughout the book, and derivative instruments for emissions trading in Part 9 provide most of the data investors, financiers and technology companies need to plan their business strategies. With relevant information on Annex I countries, investors should be able to apply the information provided in this book to JI projects with equal effectiveness.

In the absence or non-ratification of the Protocol, the author encourages the European Union to take up the gauntlet – at least to protect the regions within its borders as well as to future EU member states in Eastern Europe and CIS. Other regional economic groupings such as ASEAN and CARICOM could implement their own climate change initiatives. It is hoped that national efforts by China and India may lead to talks on bilateral cooperation.

Non-ratification of the Protocol by the USA could be counteracted by sufficient votes from the EU and other Annex I countries, but it remains to be seen if the world's largest economic power should choose to support the Protocol or renegotiate the goal posts.

Nevertheless, the author hopes that the contents of this book offers corporates, governments and academic institutions the tools to repair an earth in crisis.

Appendix 1

Contact Addresses of Environmental Authorities/Relevant Government Authorities in Developing Countries

Basis of listing: Developing countries which require needy investment in the environmental sector, countries experiencing transformation to free market economies, and countries whose GDP per capita are below the average of the top twenty economies in the world. The list hence includes certain countries under the Annex I grouping (UNFCCC) which meet one or more of these criteria. For countries which are members of the Caribbean Community, the CARICOM Secretariat is the coordination centre for climate change projects.

For some countries, the government ministry of foreign investments, economic planning, energy or international cooperation is the appropriate contact point for environmental matters and projects relating to environmental improvement.

For Argentina and Chile, the Ministry of Social Development in Santa Martini, Buenos Aires and Commission National del Medio Ambiente (CONAMA) in Santiago have recently been appointed to coordinate climate change policies and Kyoto Protocol projects.

CARICOM countries - Caribbean Community and Common Market: Anguilla, Antigua & Barbuda, Bahamas, Bardados, Belize, British Virgin Islands, Dominica, Dominican Republic, Grenada, Grenada, Guyuna, Haiti, Jamaica, Monserrat, St Kitts & Nevis, St Vincent & the Grenadines, St Lucia, Suriname, Turks & Caicos Islands (island states include territories and dependencies under foreign influence or jurisdiction).

CARICOM Secretariat
Bank of Guyana Building
PO Box 10827
Georgetown
Guyana
T: (592) (2) 692-281-9
F: (592) (2) 592-2-67816/57341/58039

Afghanistan
Ministry of Energy
Shah Rahi Sedarat
Kabul
T: (93) 251-09

Albania
Ministry of Trade & Economic Cooperation
Tirana
T: (355) (42) 346-68
F: (355) (42) 346-58

Algeria
Ministry of Energy & Mines
80 Ave Ahmed Ghermoul
Algiers
T: (213) (2) 67-33-00
F: (213) (2) 65-27-83

Angola
Ministry of Fisheries & Environment
Avda 4 de Fevereiro 25
Predio Atlantico
Luanda
T: (244) (2) 392-782

Argentina
Ministry of the Economy
Hipolito Yrigoyen 250
1310 Buenos Aires
T: (54) (11) 4342-6411
F: (54) (11) 4331-7426

Armenia
Ministry of Environmental Protection
35 Moskovian Street
375012 Yerevan
T: (374) (2) 530-741
F: (374) (2) 534-902

Azerbaijan
Ministry of Trade
40 Uzevir Hajibeyov St
370016 Baku
T: (994) (12) 985-074
F: (994) (12) 987-431

Bangladesh
Ministry of Housing, Municipalities & Environment
Diplomatic Area, Manama
Dhaka
T: (880) (2) 404-255
F: (880) (2) 865-741

Belize
Ministry of Natural Resoources & Environment
Market Square
Belmopan
T: (501) (8) 222-49
F: (501) (8) 223-33

Benin
Ministry of Environment
01 BP 3621
Cotonu
T: (229) 315-596
F: (229) 315-081

Bhutan
Ministry of Trade & Industry
Tashichodzong, PO Box 141
Thimphu
T: (975) 221-59
F: (975) 223-507

Bolivia
Ministry of Sustainable Development & Environment
Casilla 3116
La Paz
T: (591) (2) 359-820
F: (591) (2) 392-892

Bosnia & Herzegovina
Sector for Multilateral Relationship
Ministry of Foreign Affairs
Musala, Omladinska 2
Sarajevo
T: (387) (71) 667-977/663-813
F: (387) (71) 472-188

Botswana
Ministry of Mineral Resources & Energy
Private Bag 0018
Gaborone
T: (267) 352-454
F: (267) 372-738

Brazil
Ministry of Environment
SAIN, Av L4 Norte, Edif. Sede Terreo
70800 Brasilia
T: (55) (61) 226-8221
F: (55) (61) 322-1058

Bulgaria
Ministry of Environment
67 William Gladstone St
1000 Sofia
T: (359) (2) 847-220
F: (359) (2) 810-509

Burkino Faso
Ministry of Environment
565 rue Agostino Neto
Secteur 4, Koulouba
03 Ouagadougou
T: (226) 324-074
F: (226) 307-039

Burundi
Ministry of Energy & Mines
BP 745
Bujumbura
T: (257) (2) 259-09
F: (257) (2) 233-37

Cambodia
Ministry of Environment
48 blvd Sihanouk
Phnom-Penh
T: (855) (23) 427-894
F: (855) (23) 427-844

Cameroon
Ministry of External Relations
Yaounde
T: (237) 234-040
F: (237) 222-704

Cape Verde
Ministry of Agriculture, Food & Environment
Ponta Belem, Praia
Santiago
T: (238) 615-713
F: (238) 614-054

Central African Republic
Ministry of Mines, Energy & Petroleum
N/Djamena
T: (236) 515-603
F: (236) 512-565

Chad
Ministry of Mines, Energy & Petroleum
N'Djamena
T: (235) 515-603
F: (235) 515-565

Chile
Ministry of Economy
Teatinos 120
10 Santiago
T: (56) (2) 672-5164
F: (56) (2) 696-6305

China
Ministry of Foreign Trade & Economic Cooperation
Dongchangan Jie, Dongcheng Qu
100731 Beijing
T: (86) (10) 6519-8328
F: (86) (10) 6512-9568

China - Taiwan
Environmental Protection Administration
41 Chung Hua Rd, Sec 1
Taipei
T: (886) (2) 2322-8751
F: (886) (2) 2351-6227

Columbia
Ministry of Environment
Calle 38, No. 8-61
Santafe de Bogota
T: (57) (1) 288-6010
F: (57) (1) 243-3004

Comoros
Ministry of Environment
c/o Ministry of Economy
BP 324, Moroni
T: (269) (73) 2767
Telex: 245

Congo, Democratic Rep. of
Ministry of Environment
15 ave des Cliniques
BP 12348, Kinshasha-Gombe
T: (243) (12) 312-52

Congo, Republic of
Ministry of Foreign Affairs & Cooperation
BP 2070, Brazzaville
T: (242) 832-028
Telex: 5210

Costa Rica
Ministry of Natural Resources
Avdas 8 y 10, Calle 25
Apdo 100104, 1000 San José
T: (506) 257-1417
F: (506) 257-0697

Coté D'Ivoire
Ministry of Foreign Trade Promotion
BP 2302
04 Abidjan
T: (225) 213-707
F: (225) 219-172

Croatia
Croatian Agency for Promotion of Foreign Investments
Avenue of Dubrovnik
Zagreb
T: (385) (1) 655-4560
F: (385) (1) 655-4563

Cuba
Ministry of Foreign Investment & Economic Cooperation
Calle 1, No 201, Vedado
Havana
T: (53) (7) 366-61
F: (53) (7) 511-297

Czech Republic
Ministry of Environment
Vrsovicka 65
10010 Prague 10
T: (420) (2) 6712-1111
F: (420) (2) 6731-0308

Djibouti
Ministry of Environment
BP 2086, Djibouti
T: (253) 351-280
F: (253) 351-812

Ecuador
Ministry of Environment
Avda 10 de Agosta, 3560 y Mariana de Jesus
Quito
T: (593) (2) 393-5147
F: (593) (2) 390-8159

El Salvador
Ministry of Environment, c/o Ministry of Economy
1A Calle, Poniente 2310
Col. Escalon, San Salvador
T: (503) 224-2159
F: (503) 998-1965

Equatorial Guinea
Ministry of Environment
c/o Ministry of Trade & Industry
PO Box 1515, Asmara
T: (240) (1) 117-944
F: (240) (1) 120-586

Eritrea
Ministry of Finance & Development
PO Box 896
Asmara
T: (291) (1) 113-633
F: (291) (1) 117-947

Estonia
Ministry of Environment
Toompuiestee 24
0100 Tallinn
T: (372) (2) 626-2802
F: (372) (2) 626-2801

Ethiopia
Ministry of Environment
c/o Ministry of Commerce & Industry
PO Box 704, Addis Ababa
T: (251) (1) 518-025
F: (251) (1) 515-411

Fiji
Ministry of Environment
Government Bldgs, PO Box 2131
Suva
T: (679) 211-310
F: (679) 303-515

Gabon
Ministry of Environment
BP 178
Libreville
T: (241) 763-462
F: (241) 5711

Gambia
Department of State for Economic Affairs
Quadrangle
Banjul
T: (220) 228-291
F: (220) 227-954

Georgia
Ministry of Environmental Protection
68A Kostava
380015 Tbilisi
T: (995) (32) 230-664
F: (995) (32) 983-420

Ghana
Ministry of Environment
PO Box M232
Accra
T: (233) (21) 662-264
F: (233) (21) 666-828

Guatemala
Ministry of Energy & Mines
Diagonal 17, No. 29-78, Zona 11
Guatemala City
T: (502) (2) 760-679
F: (502) (2) 763-175

Guinea
Ministry of Natural Resources & Energy
BP 295, Conakry
T: (224) 441-186
F: (224) 414-913

Guinea-Bissau
Ministry of Foreign Affairs & Int'l Cooperation
Rua General Omar Torrijo
Bissau
T: (245) 202-752
F: (245) 202-378

Honduras
Ministry of Environment
c/o Ministry of Industry & Commerce
Edif. Salame, 5a Avda, 4a Calle
Tegucigalpa
T: (504) 382-025
F: (504) 372-836

Hungary
Ministry of Environment
PO Box 351
1011 Budapest
T: (36) (1) 201-3843
F: (36) (1) 201-2846

India
Ministry of Environment & Forest
Paryavaran Bhavan, CGO Complex
Phase II, Lodi Road
110 003 New Delhi
T: (91) (11) 436-0721
F: (91) (11) 436-0678

Indonesia
State Ministry for Environmental Pollution Control
Jalan DI. Panjaitan, Kav 42
134110 Jakarta Timur
T: (62) (21) 858-0103
F: (62) (21) 858-0101

Iran
Ministry of Cooperatives
76, Bozorgmehr St
Vali-e-Asr Avenue
14169 Tehran
T: (98) (21) 640-0938
F: (98) (21) 641-7041

Israel
Ministry of Environment
PO Box 34033, 5 Kanfei Nesharim St
Givat Shaul
95464 Jerusalem
T: (972) (2) 655-3777
F: (972) (2) 653-5934

Jordan
Ministry of Environment
PO Box 1799
Amman
T: (962) (6) 464-1393
F: (962) (6) 464-9341

Kazakhstan
Ministry of Energy, Industry & Trade
Mira 37
473002 Astana
T: (7) (3172) 337-133
F: (7) (3172) 337-164

Kenya
Ministry of Environment
Kencom House, Moi Avenue
30126 Nairobi
T: (254) (2) 229-261

Kiribati
Ministry of Environment
PO Box 234, Bikenibau
Tarawa
T: (686) 282-11
F: (686) 283-34

Korea, Dem. People's Republic
Ministry of Land & Environmental Protection
c/o Committee of External Economic Cooperation
Pyongyang
T: (850) (2) 333-974
F: (850) (2) 381-4498

Kyrgyz Republic
State Committee on Foreign Investments
58a blvd Erkindlk
720874 Bishkek
T: (996) (3312) 223-292
F: (996) (3312) 661-075

Laos
Ministry of Agriculture & Forestry
Ventiane
T: (856) (21) 412-359
F: (856) (21) 412-344

Latvia
Ministry of Environmental Protection
25 Peldu St
1494 Riga
T: (371) 702-6470
F: (371) 782-0442

Lebanon
Ministry of Environment
c/o Ministry of Economy
Rue Artois, Beirut
T: (961) (1) 340-503

Lesotho
Ministry of Environment
c/o Ministry of Development Planning
PO Box 395
100 Maseru
T: (266) 311-100
F: (266) 310-157

Liberia
Ministry of Energy
Capitol Hill, PO Box 10-9024
1000 Monrovia 10

Lithuania
Ministry of Environment
A. Tuozapaviciaus 9
Vilnius 2600
T: (370) (2) 725-868
F: (370) (2) 722-029

Macedonia FYR
Ministry of Environment
91000 Skopje
T: (389) (91) 366-595
F: (389) (91) 366-931

Madagascar
Ministry of Environment
Ampandrianomby
101 Antananarivo
T: (261) (20) 224-0908

Malawi
Ministry of Environment
c/o Ministry of Commerce & Industry
PO Box 30366, Capitol City, 3 Lilongwe
T: (265) 732-711
F: (265) 780-680

Malaysia
Ministry of Science, Technology & Environment
Wisma Sime Darby, Jalan Raja Laut
50662 Kuala Lumpur
T: (60) (3) 293-8955
F: (60) (3) 293-6006

Maldives
Ministry of Environment
Huravee Bldg, Ameer Ahmed Magu
Malé
T: (960) 323-820
F: (960) 324-739

Mali
Ministry of Trade & Finance
Quartier du Fleuve, Kolouba
Bamako
T: (223) 225-687
F: (223) 228-853

Marshall Islands
Ministry of Environment
PO Box 2
MH 969960 Majuro
T: (692) (625) 3445
F: (692) (625) 3649

Mauritania
Ministry of Environment
BP 366, Nouakchott
T: (222) 251-500
F: (222) 257-475

Mauritius
Ministry of Environment
Ken Lee Tower, Barracks St
Port Louis
T: (230) 212-8332
F: (230) 212-9407

Mexico
Secretariat of State for the Environment
Anillo Periferico, Sur 4209, No. 3
Col. Jardines en la Montana
14210 Mexico City
T: (52) (5) 628-0604
F: (52) (5) 628-0644

Micronesia
Department of Economic Affairs
Palikir, Pohnpei
FM 96941 Eastern Caroline Islands
T: (691) 320-2646
F: (691) 320-5854

Moldova
Ministry of Environmental Proetction
73 Blvd Stefan cel Mare
2001 Chisinau
T: (373) (2) 225-144
F: (373) (2) 233-806

Mongolia (Outer)
Ministry of Environment
Government Bldg 3, Ulan Bator
T: (976) 322-011

Morocco
Ministry of Economy
Ave Muhammad V
Quartier des Ministeries, Rabat
T: (212) (7) 763-171
F: (212) (7) 761-575

Mozambique
Centre for Promotion of Foreign Investments
Rua de Impressa 332 r/c, CP 4635
Maputo
T: (258) (1) 422-525
F: (258) (1) 422-604

Myanmar
Ministry of National Planning & Economic Dev.
Theinbyu St, Botahtaung Township
Yangon
T: (95) (1) 280-816
F: (95) (1) 282-101

Namibia
Ministry of Environment
Swabou Bldg, Post St Mall
Windhoek
T: (264) (61) 284-111
F: (264) (61) 221-930

Nauru
Ministry of Environment
c/o Ministry of Finance
Aiwo
T: (674) 444-3140
F: (674) 555-4477

Nepal
Ministry of Foreign Affairs
Shital Niwas, Maharajganj
Kathmandu
T: (977) 416-011
F: (977) 416-016

Nicaragua
Ministry of Environment & Natural Resources
km 12$^1/_2$, Carretera Norte, Apdo 5123
Managua
T: (505) (2) 63-1271
F: (505) (2) 63-1274

Niger
Ministry of Foreign Affirs & Cooperation
BP 396, Niamey
T: (227) 722-907
F: (227) 735-231

Nigeria
Ministry of Industry
Gwagwalada Area, PMB 24
Abuja
T: (234) (9) 243-1250

Pakistan
Ministry of Works & Environment
UBL Bldg, Jinnah Avenue
Islamabad
T: (92) (51) 922-4578
F: (92) (51) 824-122

Palau
Ministry of Interior
PO Box 6031
PW 96940, Koror
T: (680) 488-2655
F: (680) 488-2649

Palestine
Ministry of Economy & Commerce
Gaza
T: (972) (7) 998-2370
F: (972) (7) 544-9032

Panama
Ministry of Planning & Economic Policy
Edif. Ooawa, Via Espana, Apdo 2694
Panama 3
T: (507) 269-2169
F: (507) 269-6822

Papua New Guinea
Department of Environment
Central Government Offices, PO Box 6601
Boroko
T: (675) 301-1606
F: (675) 301-1691

Paraguay
Ministry of Industry & Commerce
323 Avda Espana
Asuncion
T: (595) (21) 204-638
F: (595) (21) 213-529

Peru
Ministry of Economy
Jiron Junin 339-4
Lima
T: (51) 427-3930
F: (51) 431-7836

Philippines
Department of Energy
(administers CDM/JI projects)
Merritt Rd, Fort Bonifacio
Taguig, Metro Manila
T: (63) (2) 844-1021
F: (63) (2) 817-8603

Poland
Ministry of Environmental Protection
ul. Wawelska 52154
00-922 Warsaw
T: (48) (22) 825-0001
F: (48) (22) 825-3972

Romania
Ministry of Environmental Protection
12 Blvd Liertatii
76106 Bucharest
T: (40) (1) 410-0482
F: (40) (1) 312-4227

Russia
Ministry of Fuel & Power Industry
1 Kitaigorodskii proyedz
103074 Moscow
T: (7) (095) 220-5500
F: (7) (095) 220-5656

Rwanda
Ministry of Finance & Economic Planning
BP 158, Kigali
T: (250) 754-10
Telex: 502

Samoa (Western)
Ministry of Environment
Apia
T: (685) 224-81
F: (685) 231-76

Sao Tomé & Principe
Ministry of Economy
Largo Alfandega
Sao Tomé
T: (239) 223-72
F: (239) 221-82

Senegal
Ministry of Environment
Bloc Calmette
Dakar
T: (221) 821-1126

Seychelles
Ministry of Industry
International Conference Centre
PO Box 648, Victoria
T: (248) 225-060
F: (248) 225-086

Sierra Leone
Ministry of Environment
c/o Ministry of Economic Planning
Secretariat Bldg, George St
Freetown
T: (232) (22) 226-911
Telex: 3363

Slovak Republic
Ministry of Environment
nam L'Stura
81235 Bratislava
T: (421) (7) 515-2459
F: (421) (7) 516-2451

Slovenia
Ministry of Environment
c 48, Dunasjska
1000 Ljubljana
T: (386) (61) 178-3600
F: (386) (61) 178-7422

Solomon Islands
Ministry of Environment
PO Box G24
Honiara
T: (677) 258-48
F: (677) 212-45

Somalia
Ministry of Finance & Economy
Mogadishu
T: (252) (1) 330-90

South Africa
Ministry of Environmental Affairs
240 Vermeulen St
0002 Pretoria
T: (27) (12) 321-9587
F: (27) (12) 323-5181

Sri Lanka
Ministry of Forestry & Environment
Unity Plaza Bldg
Colombo 4
T: (94) (1) 588-274
F: (94) (1) 502-566

Sudan
Ministry of Environment
c/o Ministry of Finance & National Economy
PO Box 700, Khartoum
T: (249) (11) 77-003
Telex: 22-324

Swaziland
Ministry of Economic Planning & Development
PO Box 602, Mbabane
T: (268) 437-65
F: (268) 421-57

Syria
Ministry of Economy
rue Maysaloun, Damascus
T: (963) (11) 221-3513
Telex: 411-982

Tajikstan
Ministry of Environment
Kuchai Bokhtar 12
734025 Dunshabe
T: (7) (3772) 232-878
F: (7) (3772) 275-581

Tanzania
Ministry of Energy & Mineral Resources
PO Box 2000
Dar es Salaam
T: (255) (2) 110-414
F: (255) (2) 116-719

Thailand
Ministry of Science, Technology & Environment
Thanon Phra Ram VI, Ratchathewi
10400 Bangkok
T: (66) (2) 246-0064
F: (66) (2) 246-8106

Togo
Ministry of Environemnt
c/o Ministry of Economic Affairs
Ancien Palais, Ave de la Marina
BP 387, Lomé
T: (228) (2) 213-554
F: (228) (2) 210-905

Tonga
Ministry of Commerce
PO Box 110, Nuku'alofa
T: (676) 236-88
F: (676) 238-87

Tunisia
Ministry of Environment
Centre Urbain Nord
2080 Ariana, Tunis
T: (216) (1) 704-000
F: (216) (1) 702-431

Turkey
Ministry of Environment
Gevre Bakanligi
Istanbul Cad. 88
Istkitler, Ankara
T: (90) (312) 342-3900

Turkmenistan
Ministry of Environmental Protection
c/o Ministry of Economy & Finance
ul Nurberdy Pomma 4
744000 Ashgabat
T: (993) (1) 251-0563
F: (993) (1) 251-1823

Tuvalu
Ministry of Works & Energy
PMB, Vaiaku, Funafuti
T: (688) 200-55
F: (688) 207-22

Uganda
Ministry of Environment
c/o Ministry of Finance & Economic Development
PO Box 8147, Kampala
T: (256) (41) 234-700
F: (256) (41) 230-163

Ukraine
Ministry of Environmental Protection
vul. Kreshchatik 5
252601 Kiev
T: (380) (44) 228-0644
F: (380) (44) 229-8383

Uruguay
Department of Environment
Zabala 1427
Montevideo
T: (598) (2) 915-0211
F: (598) (2) 916-2914

Uzbekistan
Ministry of Foreign Economic Relations
ul Bujuk Ipak Yuli 75
700077 Tashkent
T: (7) (3712) 687-567
F: (7) (3712) 687-477

Vanuata
Ministry of Industry & Trade
PMB 030, Port Vila
T: (678) 227-70
F: (678) 256-40

Venezuela
Ministry of Environment
Edif. Centro Simon Bolivar
Torre Sur, Piso 25
El Selancio, Caracas
T: (58) (2) 408-7008/1001
F: (58) (2) 483-2445/2209

Vietnam
Ministry of Science, Technology & Environment
39 Tran Hung Dao
Hanoi
T: (84) (4) 825-2731
F: (84) (4) 825-1518

Yemen
Ministry of Planing & Development
PO Box 175, San'a
T: (967) (1) 250-101
F: (967) (1) 251-503

Yugoslavia
Ministry of Environment
22-26 Nemanjina
11000 Belgrade
T: (381) (11) 657-143
F: (381) (11) 642-242

Zaire – see Dem. People's Rep. of Congo

Zambia
Ministry of Environment
c/o Ministry of Economic Development
Finance Bldg, PO Box RW 50062
Lusaka
T: (260) (1) 213-822
F: (260) (1) 422-21

Zimbabwe
Ministry of Environment
Karigamombe Centre
Private Bag 7753, Causeway, Harare
T: (263) (4) 751-720
F: (263) (4) 734-075

Appendix 2

Contact addresses of United Nations and Multilateral Organisations Concerned with Environmental Policies and Financing of Projects

African Development Bank
PO Box 1387
01 Abidjan
Cote D'Ivoire
T: (225) 204-168
F: (225) 204-964

Asian Development Bank
6 ADB Avenue
Madaluyong City
0401 Metro Manila
Philippines
T: (63) (632) 632-4444
F: (63) (632) 636-2444

European Bank for Reconstruction & Development
One Exchange Square
London EC2A 2EH
United Kingdom
T: (44) (207) 338-6000
F: (44) (207) 338-6100

Global Environmental Facility
GEF Department, The World Bank
1818 H Street, NW
Washington, DC 20433
USA
T: (1) (202) 473-1053
F: (1) (202) 522-3240

Global Green CDM Fund, LDC/IRISK Climate Change Fund
c/o Euro-Asia Pac Ltd
St Alphage House
Fore St, London EC2Y 5DA
United Kingdom
T: (44) (207) 638-2019
F: (44) (207) 256-6930

Inter-American Development Bank
1300 New York Avenue, NW
Washington, DC 20577
USA
T: (1) (202) 623-1000
F: (1) (202) 623-3096

International Finance Corporation
Environmental Division
1850 I St, NW
Washington, DC 20433
USA
T: (1) (202) 477-1234
F: (1) (202) 477-6391

United Nations Development Programme
1775 K St, NW – Suite 420
Washington, DC 20006
USA
T: (1) (202) 331-9130
F: (1) (202) 331-9636

United Nations Environmental Programme
PO Box 30552
Nairobi, Kenya
T: (254) (2) 624-001
F: (254) (2) 226-895

World Bank
1818 H St, NW
Washington, DC 20433
USA
T: (1) (202) 477-1234
F: (1) (202) 477-6391

Clean Development Mechanism Fund and LDC/IRISK Climate Change Fund (Least Developed Countries/Islands-at-Risk)

1. Euro-Asia Environmental
   Euro-Asia Pac Corp Ltd
   St Alphage House Fore Street
   London EC2Y 5DA Great Britain

   Tel: 44-(0)20-7638-2019, (0)20-7628-2019
   Fax: 44-(0)20-7256-6930, (0)20-7628-1018
   E: euro_asiapac@email.com
   michael@greenpac.freeserve.co.uk

Africa, Middle East - contact Michael See, Abike Daniyan
China, Greater China - contact Qin He-Qing, Xing Su-Ping, Zheng Yu-Wei
South-East Asia, South Pacific, South America, Europe - contact Michael See

2. Euro-Asia Environmental/Glonet Technology & Trade
12 Castel Way
Folksworth
Peterborough PE7 3TX
Great Britain

Tel: 44-(0)1733-244-066
Fax: 44-(0)1733-240-477
E: Shantanu@glonet.freeserve.co.uk

India, South Asia - contact Shantanu Banerjee

3. Euro-Asia Environmental/Glonet Baroda
International Trade Links
FF-30 Avishkar Complex
Old Padra Road
Baroda 390015, India

Tel: 91-265-301-600/341-663
Fax: 91-265-301-474
E: itl.brd@smn.sprintprg.ems.vsnl.net.in

South Asia - contact Mr Jayesh Tanna

4. Euro-Asia Environmental/Glonet Calcutta
8/5 Citizen's Estate
103 Maniktala Main Road
Calcutta 700 054

Tel: 91-33-355-7289
Fax: 91-33-355-1620
E: glonetindia@hotmail.com

India, Bangladesh - Contact Kalyan Chattopadhyaya

Appendix 3

Emissions Measurement and Monitoring

A 3.1
Technical Methods

The quantified emission limitation and reduction commitments (QELRCs) of Parties of the Kyoto Protocol are to be achieved by domestic action, cross-border transfers of emission credits generated by investments in JI/CDM host countries, and emissions trading of credits from JI/CDM projects. Articles 3, 5, 7, 8 and 12 of the Kyoto Protocol refer to procedures and issues on GHG reporting, monitoring, certification and assessment.

The baseline years for emission reduction measurements and methodologies for establishing emission targets were elaborated in Chapter 7. This section deals with the technical aspects of physical measurements of the Kyoto Protocol greenhouse gases. Physical methods may be required by UNFCCC to verify any numerical calculations of greenhouse gas reductions. Table A3.1 provides a summary of current measurement techniques.

A 3.2
Maintaining Sustainability in Kyoto Protocol Projects

The environmental, economic and social conditions of the location in which the project is to be implemented must meet sustainability requirements. Measurements of sustainability should be conducted in spatial and temporal terms.

Areas of measurement

Three areas of priority should be established: (i) primary level – the host community; (ii) secondary level – the host region which could be defined as the area with a radius of fifty kilometres from the centre of the host community; and (iii) tertiary level – the surrounding environment defined by that area with a radius of, say, up to one hundred kilometres from centre of the host community.

Periodic surveys

Surveys should be conducted at least one month before; one year, two years, three and five years after; construction or work commencement.

Table A3.1. Technical methods for emissions measurement and monitoring

Greenhouse gas	Method/technique of measurement
Carbon dioxide	Gas chromotography Non-dispersive ultra-violet spectrum Non-dispersive infra-red spectrum Gas filter analyser, Luft analyser, microflow sensor Supercritical fluid (liquid) chromotography
Hydrofluorocarbons and Perfluorocarbons	Absorption of infra-red radiation (infra-red spectrum) Chemical analysis (coulometry, titration, colorimetry) Flame ionisation (of gas) detector Heat evolved during combustion Photo ionisation (of gas) detector
Methane	Combustible gas detector Gas chromotography Non-dispersive infra-red spectrum Gas filter correlation, Luft analyser, microflow sensor, single beam analyser Non-dispersive ultra violet fluorescence Rotating filter analyser Semi-conductor sensor
Nitrous oxide	Gas chromotography Gas filter correlation infra-red Non-dispersive infra-red spectrum Gas filter correlation, Luft analyser, single beam analyser
Sulphur hexafluoride	Supercritical fluid chromotography

Notes:
The author is grateful to Dr Andrew G. Clarke of Leeds University for the preparation of this section. Dr Clarke's book *Industrial Air Pollution Monitoring*, published by Chapman & Hall/International Thomson, is an authoritative reference guidebook on source emissions monitoring.

Economic impact and social welfare

The economic output of the community or region where the project is to be implemented can be measured in terms of income per inhabitant. Government contributions to local inhabitants should not be reduced as a consequence of capital inflows into the community. Social welfare inputs to the community – usually financed through direct government benefits to individuals or the local authority, and through upkeep or construction of clinics, schools and care centres – should neither be compromised.

Non-Kyoto Protocol environmental emissions and effluents

The six greenhouse gases targeted by the Protocol were elaborated in Chapter 2. Emissions additionality of non-Protocol gases such as oxides of sulphur and nitrogen is not mentioned anywhere in the Protocol, but could be applied to the sustainability requirement.

To meet the sustainability criteria, non-Kyoto Protocol emissions or pollutants should not exceed their current levels. Current pollutant levels which do not adhere to the minimum safety requirements set by international organisations such as the World Health Organisation and World Bank should comply with these requirements.

A 3.3
World Bank Guidelines on Environmental Standards

Pollution control targets for coal-fired thermal power plants

Table A3.2. Maximum emission limits set by the World Bank for coal-fired power stations and WHO recommendations for ambient air quality

Pollutant	Removal efficiency required (World Bank) mg/Nm^3	Emission concentration required (World Bank) t/day/MW$_e$	Specific emission levels (World Bank) mg/Nm^3	Max. emission increment, 24 hour mean value (WHO) mg/Nm^3	Max. emission increment, annual average (WHO) mg/Nm^3
All particles	99%	–	50	100–150	–
PM_{10}	98%	–	50	100–150	–
NO_x	40%	–	750 (6% excess O_2)	500	100
SO_x	–	0.20 (0.1 incremental for plants >1,000 MW$_e$)	2,000	100–500	10–50

Notes:
Nm^3 = Normal cubic metre, measured at normal temperature and pressure of 0 degree Centigrade, 101.325 kN/m^3
PM_{10} = particles of 10 microns or less in size

Maximum emission limits set by the World Bank on oil and gas-fired power plants

Limits for the power plants using oil and/or gas as fuels are:
Hydrogen sulphide: 5 mg/Nm^3
NOx emissions: oil fired – 360 mg/Nm^3 or 100 ng/J (liquid fuels)
gas fired – 240 mg/Nm^3 or 65 ng/J (gaseous fuels)
Particulates and SOx: as for coal-fired power plants
Volatile organic compounds (VOCs) including benzene: 20 mg/Nm^3

Table A3.2. *Continued*

Pollutant	Removal efficiency required (World Bank)	Emission concentration required (World Bank)	Specific emission levels (World Bank)	Max. emission increment, 24 hour mean value (WHO)	Max. emission increment, annual average (WHO)
	mg/Nm^3	t/day/MW_e	mg/Nm^3	mg/Nm^3	mg/Nm^3

Emission limit values, pH and temperature specifications for effluents from thermal power plants

World Bank standards for selected effluents are shown in the next table.

Emission limit, specification	Maximum limit
Chlorine, total residual	0.2 mg/l
Chromium, total	0.5 mg/l
Chromium, hexavalent	0.1 mg/l
Copper	0.5 mg/l
Iron	1.0 mg/l
Nickel	0.5 mg/l
Oil and grease	10 mg/l
Suspended solids	50 mg/l
Zinc	1.0 mg/l
pH	6–9
Temperature increase	Less or equal to 3 degrees Centigrade*

* The temperature increase should be measured at the edge of the zone where initial mixing or dilution occurs; if the zone is not defined, 100 metres from the point of discharge should be regarded as the temperature boundary limit.

Waste incineration – emission limits

The Environmental Protection Agency of the USA has set the following emission limits for incineration of waste containing hazardous materials; limits are set for boilers/furnaces but not for the plant as a whole:

Hazardous material	Incinerator limits	Boiler/industrial furnace
Antimony	nr	14–31,000 g/h
Arsenic	nr	0.11–240 g/h
Barium	nr	2,400–5,000,000 g/h
Beryllium	nr	0.26–580 /h
Cadmium	nr	0.04–86 g/h
Carbon monoxide, CO*	Continuous monitoring	Continuous monitoring and <100 p.p.m. or; >100 p.p.m. and <20 p.p.m. hydrocarbons or; >100 p.p.m. and acceptable risk
Chromium	nr	0.2–430 g/h
Chlorine	nr	19–41,000 g/h
Dioxin and furan	30 ng/Nm^3 (new MSW) 60–250 ng/Nm^3 (old MSW)	Incinerators with dry particulate removers operated at 230–400 degrees Centigrade must show acceptable risk

Table A3.2. *Continued*

Pollutant	Removal efficiency required (World Bank) mg/Nm³	Emission concentration required (World Bank) t/day/MW$_e$	Specific emission levels (World Bank) mg/Nm³	Max. emission increment, 24 hour mean value (WHO) mg/Nm³	Max. emission increment, annual average (WHO) mg/Nm³
Lead		nr		4.3–9,200 g/h	
Mercury		nr		14–31,000 g/h	
Nickel		nr		950–2,000,000 g/h	
Selenium		nr		190–410,000 g/h	
Silver		nr		140–310,000 g/h	
Thallium		nr		14–31,000 g/h	
Hydrocarbon		nr		Continuous monitoring if CO > 100 p.p.m. 330–720,000 g/h	
Hydrochloric acid		1.8 kg/h or 99%		180 mg/Nm³	
Particle		180 mg/Normal m³**			

Notes
* measurement adjusted to 7% oxygen content in flue gas.
** volume measured at normal temperature and pressure, i.e. 0 degree Centigrade and 101.325 kN/m^{-2}.
nr not regulated by EPA – this is dependent on the number of boilers/furnaces per facility. Emission limits vary for boiler/furnaces according to the lowest and highest levels, and conditions of stack emission and atmospheric dispersion.
Limits from boilers/furnaces burning medical waste are expected to more stringent than MSW sources.

IGCC projects – gasification of heavy oil feeds

The gasification of heavy oil products from refineries produces off-take residues in the hopper which contain high levels of vanadium, nickel, lead, mercury, and cyanides. International standards for vanadium and nickel are concentrations in drinking water not exceeding 0.1 mg/l in most developed countries although the World Health Organisation advocates a lower level of 0.02 mg/l; for lead, not more than 0.01 mg/l (WHO); for mercury not more than 0.001 mg/l (WHO); and for cyanide, 50 microgrammes per litre (USA, UK).

The gasification process also produces oxides from these elements which are far more toxic if inhaled; hence the design and construction of the reactor vessel as well as gas treatment units must ensure effective containment and removal of the pollutants.

Appendix 4

The Green 500 Index Project – a National Green Index to Rank Firms

Case Study: United Kingdom

Project:	The project creates the Green 500 (UK): league listings of quoted and unquoted companies based on greenhouse gas emissions reductions.
League listings:	Section A: Publicly quoted firms on the London Stock Exchange Section B: Private limited firms
League classification by industry:	Industry sectors:

Bricks manufacturing
Cement manufacturing
Ceramic potteries
Electrical engineering
Food, drinks and tobacco production
Glass manufacturing
Industrial chemicals, petrochemicals
Iron and steel
Mechanical engineering
Non-ferrous metal processing, smelting
Non-metallic metal processing, smelting
Packaging
Pharmaceutical production
Plastics and rubber
Pulp production and paper manufacturing
Textiles, leather and clothing

Vehicle manufacturing
Agriculture
Coke ovens
Gas production
Oil refineries
Power generation
Waste processing and utilisation

Other industries

League classification by firm size:	Small: sales £ 600,000 – 6 m. (US$ 0.1 – 10 m.), or employee strength below 100. Medium: sales £ 6 – 60 m. (US$ 10 – 100 m.), or employee strength below 1,000. Large: sales £ 60 – 600 m. (US$ 100 – 1,000 m.), or employee strength below 5,000. Very Large: sales above £ 600 m. (above US$ 1 bn.), or employee strength above 5,000. Corporate sales and corporate employee criteria based on activities in the United Kingdom only.
Emissions:	Carbon dioxide, methane, nitrous oxide, hydrofluoro-carbons, perfluorocarbons, sulphur hexafluoride.
Emissions reduction measurement:	Carbon dioxide equivalent units from source based on Global Warming Potential (GWP).
Unit of emissions:	Metric tonnes of gas.
Emissions sources:	Stationery and mobile sources including transport.
Country market:	United Kingdom and possibly Ireland.
Time frame:	Test launch 6 June 2000 (World Environment Day).
Index publication dates:	First Monday of March, June, September and December. Reports must be made no later than the last three working days of February, May, August and November.
Reporting:	Reports can either be submitted voluntarily or verified by independent or statutory auditors authorised by, or registered with, the Department of the Environment or equivalent government branch or agency.
Basis of reporting:	Voluntary (vol) or independently verified (inv).
Green Index:	Net reduction of Greenhouse Gas Emissions ×100 since 1990 or operational year post-1990 before next quarterly review Greenhouse Gas Emissions in 1990 year or operational year post-1990

	Net reduction = emissions abatement in excess of the firm's target, with reference to the UK government's obligations to UNFCCC or revised national policies.
Green Index calculation:	As proposed in Chapter 9. Other methodologies developed could also be used.
Relevant factors:	Kyoto Protocol and the UN Framework Convention on Climate Change

Proposed tax incentive scheme for investors in Green 500 Index companies

Modus Operandi:	Taxable investors receive tax exemption on profits (on sale of share) or dividends (if share not traded) of the listed company; at the Green Index rate.
Government involvement:	Department of Trade & Industry (regulator). Environmental Agency (compliance, enforcement). Treasury, Stock Exchange. Inland Revenue (taxation issues) – the project will require a change in taxation legislation. DTI and Parliament will have to approve project first.
Benefit to listed company:	More equity purchased by investors; expands capital base
Benefit to investors:	Higher return on investment to private investors as well as institutional investors/investment funds. Identification of companies which may have binding commitments to reduce greenhouse gas emissions per national obligations to the Kyoto Protocol. Firms which meet or overperform their emission targets have lower environmental liabilities and financially healthier balance sheets.
Benefit to environment:	Reduction of greenhouse gas emissions.
Market effects:	May improve sales of quoted and unquoted firms, and market capitalisation of quoted firms listed in the Index. Imminent market comparisons with non-Indexed firms.

Case Example 1

Emissions reductions from different corporate plants commissioned in separate years

The case example is shown as a set of data and calculations in Table A4.1. The Green Index is determined in the last column.

Table A4.1. Green 500 Index project case study: plants built in different years (Carbon dioxide emissions reduction of 20% below 1990 levels for large power generators)

Carbon dioxide emissions and emissions permits, units of 10,000 tonnes						Pro-rata operational year to 2010		
Firm	1990 (i)	1998 (ii)	1999 (iii)	Emissions reduction (iv)	Permits allocation, end 1999 (v)	Theoretical abatement obligations, end 2010 (vi)	Net reductions or emissions, = net tradable permits, end 1999 (vii) = (iv) – (vi)	Green Index = net reductions or emissions, from first operational yr, end 1999 +reduction –emission
A	2.0 PFC	0.9 CCGT	0.8	1.2	2.0	0.40	0.8	+40% (0.5*)
B	–	1.1 CCGT	0.9	0.2	1.1	0.13	0.7	+64%
C	–	1.0 CHP	0.9	0.1	1.0	0.12	–0.02	–2%
D	–	1.5 PFC	1.4	0.1	1.5	0.18	–0.08	–5.3%
E	–	2.0 PFC	1.9	0.1	2.0	0.24	–0.14	–7%

Notes

A operated a pulverised fuel (coal-fired) combustion plant in 1990, converted to CCGT in 1997. A, B and C implemented energy efficiency measures from 1998–99.

A is only allowed to hold only 1.6 of the allocated 2.0 units of permits at end 2010, by which time it is obligated to return 0.4 units to the government. It can sell 0.8 units to any buyer if it does not increase its own emissions beyond 1.6 units.

* If A sells, say, 0.3 units of permits at end January 2000, its ranking by emissions reductions will not be affected, but a second ranking based on net tradable emissions could be indicated (0.8 – 0.3 = 0.5).

In terms of allocations of carbon dioxide permits – in theory – A should receive 2 units; B, 1.1 units; and C, 1 unit, based on the "grandfathering" principle.

D constructed an advanced pulverised fuel plant in 1998, about the same capacity as A.

E constructed a conventional pulverised fuel plant in 1998, about the same capacity as A.

Case Example 2

Company X – manufacturing and distribution activities

Revenue origins:	Manufacturing (core activity). Trading. Distribution of goods.
Bulk of greenhouse gases emitted by:	Manufacturing plants.
Other sources:	Administration and office buildings. Transport – commercial and passenger vehicles.
Corporate strategies:	Promote corporate image to increase the sales of products and services, share value in the stock market, and possible trading of credits through initiatives of the Kyoto Protocol. Green Index Project is publicised.

Data to evaluate Green Index

Basis of evaluation

- UK's carbon dioxide reduction target of 20% below 1990 levels by 2010.
- Measurement of emissions at "normalised baseload condition" for specified year.

"Baseload condition" is defined as 80% of the full operating capacity of the plant or facility. Emissions at the original baseload condition (first year of plant operation) provides the best benchmark for emissions comparisons in future years.

Table A4.2. Green 500 Index project case study: manufacturing plants

	Carbon dioxide data (million metric tonnes)				
	Emissions 1 yr post construc.	1 Feb 99 –1 Feb 00	Emissions reduction	Pro-rata reduction obligations	Net reduction or emissions*
20 built in 1980s	2.5	2.8	–0.3	n/a	n/a
10 built in early 1993	1.5	1.1	+0.4	0.25	+0.15
5 built in early 1997	1.0	0.8	+0.2	0.13	+0.07
3 in 2001	0.5 (pred.)	-			Green Index
6 in 2006	0.5 (pred.)	-			=8.8%

Notes

+indicates reductions, – emissions; in column 4, with reference to the first operational year. Reductions calculated on annual basis in this example.

+indicates net reductions, – net emissions; in column 6, with reference to national target by 2010.

Green Index calculated by adding total net reductions, dividing the result by total emissions in the first operational years, multiplied by 100 percent.

The firm has about three weeks to finalise reports to the Green Index review committee, for publication on the first Monday of March 2000.

For manufacturing and industrial plants

Measurement is taken at quarterly intervals (four times a year) at baseload condition. The method is a fair means of comparison for different firms in the country which may experience different cycles of economic activity or business output. For example, one firm may emit increased carbon dioxide levels than the previous year due to higher demand for its products – such an increase should not reflect inefficient energy practice. Conversely, lower economic activity may result in reduced emissions, but should not reflect improvements in energy efficiency. Testing emissions at the baseload level would represent a fair basis of emissions comparison between different firms in the country. Theoretical adjustments to stimulate the baseload condition could be made from a higher or lower emissions level each quarter from observed data to minimise fuel and business costs – such self-corrective procedures if approved could be described as measurements at "normalised baseload condition" to provide equitable comparisons with other plants.

For buildings

Carbon dioxide emissions calculated on the basis of carbon burnt per unit of electricity and heating imported; and measured/calculated emissions from boilers based on amount of fuel used. The emissions baseline for buildings is that level measured or calculated in the first year of operation, although adjustments may have to be made if the firm experiences changes in business activity or energy demand.

- Method of calculation: UNEP/NPI's Creating a Standard for Carbon Dioxide/Greenhouse Gas Indicator is recommended, although other methods of calculation may be applicable.

Environmental reports and reporting

Ranking of companies in the Green 500 Index relies on detailed reporting of GHG emissions from large plants, factories and semi-industrial establishments – they would include industries listed in Annex A of the Kyoto Protocol. Small and medium-size enterprises (SMEs) which may not be identified for emissions reductions could be encouraged to participate in the Index – which could expanded to include these entities. SMEs would be able to market their green image on equal footing with large enterprises, hence attract more business and investments.

Environmental reports and specific emissions reductions should be independently audited by environmental consultancies and agencies in developed countries – preferably authorised by the relevant government department.

Extension of the national Green 500 Index ranking to global firms

A global Index could be created for companies in Annex I countries. This would be useful for investment and market analysts in appraising the environmental credit worthiness and reduction requirements of companies which consume large amounts of fuel or electricity – particularly on the impact of environmental liabilities or assets on issued securities.

Appendix 5

Carbon Accounting and Carbon Taxes

The Association of Certified Chartered Accountants has proposed a series of measures in a consultation process between the UK government and British business on Climate Change policy. Carbon permits under the Joint Implementation initiative may be "legal tender" for a one-year period – hence necessitating the holder to adopt positive measures or purchase more permits before permit expiry. The Clean Development Mechanism provides greater flexibility; GHG reduction units acquired could be banked or carried over a number of years – hence CDM credits would appear more attractive to emitters.

The European Union currently leads the world in environmental reporting – the European Environmental Reporting Award (EERA) scheme was launched in 1997 by founder countries the Netherlands, Belgium, Denmark and the United Kingdom. By 1999, it had expanded to include Germany, France and Portugal.

A 5.1
Accounting for Carbon Assets

The treatment of carbon permits in corporate accounts, assuming an efficient and liquid market, would be dependent on whether such permits are:

1. Issued for tweleve months or for a considerable period of time, and if these permits could be fixed or current assets.
2. Granted to existing operators free or would have to be purchased from the government.

Permits would be treated as assets – current if they have "annual lives", or fixed if they are valid and hold value for more than one year.

Fixed assets

As intangible fixed assets, permits would be shown at cost and amortised against earnings over their useful life, or as used in line with the holder's emissions. If the market value of permits fell below cost, there would be no automatic write-down in the company's profits.

Permits issued free or unused purchased permits could be revalued to a higher market value, but any gain would be be reflected in the company's reserves and not benefit its profits. Any permits issued free, but held a value to the market, could be sold and repurchased to convert nil cost into a purchase price – by potentially taking the cost and related profit in different periods, one as an operating cost and the other as an exceptional profit. An arbitrageur would be only too willing to oblige. He would simply deposit "x" dollars with the permit holder and is reimbursed 1.01 "x" dollars a few weeks later.

Current assets

Permits with one year life would be treated as current assets. A write-down might have to be made if the market value at year's end fell below cost, even though if such permits were issued free. If market values of permits rose above cost, the gain may not be recognised, unless a sale and repurchase transaction were to be effected by the holder. A charge against operating profits would be made as permits or that proportion of permits are consumed.

Off-balance sheet Considerations

Derivative instruments may be used for trading of permits, e.g. futures and options contracts, which may be settled before exercise or expiry dates within or longer than twelve months. In most developed countries, derivatives trading are accounted for as off-balance sheet items.

A 5.2
Effect of Taxation on Carbon Dioxide Emissions

Government taxes based on energy and/or carbon consumption are fiscal instruments imposed on both industry and domestic sectors to reduce environmental emissions and promote energy conservation. Western Europe – in particular the Netherlands, Sweden, Finland, Denmark, Austria and Germany – leads the world in energy and carbon taxes.

The US, as evidenced by Congressional consensus, favours emissions trading and is strongly opposed to mandatory taxation or command-and-control (CAC) measures in meeting its Kyoto Protocol targets. A number of US think-tanks have estimated that the country would incur a third or quarter of actual CAC costs in its climate change obligations through emissions trading, JI and CDM. However, as shown earlier in Chapter 4, external solutions rather than domestic action in the US may exacerbate global warming.

Energy vs. carbon tax

An energy tax is based on the calorific value (net favoured to gross) of a fossil fuel, and carbon tax on its carbon content. Natural gas and gasoline have higher calorific values than coal per unit mass, and already command high prices in the market. Consumers would hence pay more for natural gas and gasoline on the basis of an energy tax. The price of coal would be more competitive than these "premium" fuels under an energy tax, but could be relatively expensive under a carbon tax. The carbon content of coal (depending on its origin) per unit mass could be as high as the carbon content of gasoline or natural gas, but offers lower heat rates per unit mass – hence more coal would have to be burnt (purchased) at higher prices for a particular thermal requirement, e.g. electricity or steam generation.

How energy/carbon taxes may be calculated and collected

A number of EU countries apply a self-assessment tax accounting procedure – electricity and energy consumers are responsible for calculating how much electricity or gas they have used in a previous quarter of the year or *ad valorem* tax period, and submit such payments as necessary to the tax authorities. The scale of taxation may be based on a national "flat" rate of carbon dioxide emissions per kWh_e of electricity or increase with higher carbon dioxide emissions from the source. Electricity generators would not be taxed for fuel consumption, as business or private users are responsible for implementing fuel efficiency measures if they are to avoid paying higher energy or carbon taxes.

A national "flat" tax rate could be calculated based on the average of all carbon dioxide emissions in the country divided by the annual total electricity output for a particular year. The flat rate may hence change as a consequence of conversions of coal-fired to gas-fired power plants from one year to another.

The tax rate for consumers of natural gas (e.g. steam generation for paper/pulp industry or district heating), diesel, fuel oil, gasoline (transport) or coal are calculated on straightforward heat and mass balance equations in terms of carbon dioxide emissions per unit mass of fuel purchased. Some industries emit greenhouse gases other than carbon dioxide, e.g. methane, nitrous oxide – these emissions may have to be measured individually and converted to carbon dioxide equivalents.

Energy/carbon taxes vs permits – who pays more, or less?

A uniform tax rate (energy/carbon) across all businesses in a country would a represent a fair policy to encourage energy conservation. Most countries would avoid domestic taxation as the majority of elderly or disadvantaged citizens may not be able to meet the burden.

Certain countries may impose both energy/carbon taxes and emission permits; a double-barrelled strategy to abate emissions. Larger enterprises, which include electricity generators, would be entitled more permits under the "grandfathering" principle of allocation. These entities may want to negotiate for lower rates of tax as they have been singled out to meet additional costs. Certain medium-size and small and medium-size enterprises (SMEs), on the other hand, which have reached critical size and which may not be entitled to permits under the grandfathering scheme would oppose this as they might have to pay higher net taxes.

Reduced taxes could be considered for energy-intensive industries like iron and steel works and cement manufacturing plants where modern technology like combined heat and power do not provide practical solutions. In the UK, the CHP sector and plants which use CHP technology (e.g. paper, chemicals, food processing) have lobbied for lower taxes or tax rebates because of very high fuel efficiencies – about 75% to 80% – of these plants.

Appendix 6

National Pilot Schemes for Emissions Permits Trading Outside Annex I Countries

Unilateral schemes for non-Annex I countries

More than a hundred developing countries could experiment with internal emissions trading schemes. The key rationale is to avert external costs to the national economy as a result of pollution by local industries – particularly environmental damage costs and expenditures on welfare, health and rehabilitation programmes. The models described in Chapter 7 and Appendix 4 are suitable blueprints to implement trading schemes, although strict "additionality" and sustainability criteria imposed on JI and CDM projects may have to be revised to national capabilities.

Unilateral and bilateral schemes for India and China

The author and his CDM-JI team has produced a framework document for unilateral and bilateral emissions trading schemes for India and China. These two countries are expected to contribute to about 40% of the world's emissions of greenhouse gases in the year 2050.

A 6.1
National Permits Trading Scheme for India and China: Bilateral Cooperation Between the World's Two Most Populous Countries

An introduction to permits trading – an open letter to China and India

Permits trading, in the context of environmental pollution control, is a means of reducing atmospheric pollutants by the efficient sharing of technological resources between companies of different sizes and capabilities. It is the preferred option for companies which are looking to achieve financial savings compared to meeting the full costs of emissions control which they may not be able to afford. Permits trading provide more flexibility to legislative energy or carbon taxes – particularly in North America and EU countries – in their current efforts to meet their Kyoto Protocol targets.

Smaller manufacturers and producers – which may be less efficient in controlling emissions – may benefit from the economies of scale of emissions reductions of larger companies or more efficient emissions reducers. A minimum emissions control target could be imposed on companies to ensure that they meet, say, at least 50% of its obligations – with the exception of old plants to be decommissioned.

A simplified illustrative example: analysis performed on 7 April 1999

In 1998, the BP permits scheme which involved international subsidiaries resulted in one trade quoted at US$ 17 per tonne of carbon dioxide (US$ 62 per tonne of carbon). One BP company had to pay this amount to another (which had achieved the reduction) in order for it to comply to its emissions obligations under the scheme, or risk being fined by an amount much higher than the permits price it paid for.

If we assume that both companies involved in the trade, which may not be of similar sizes, had invested in technology, and that

- one company, X, was more efficient in controlling emissions than the other, Y, and had achieved a surplus in emissions reductions against its target level;
- Y could only afford partial investment in technology to reduce emissions to say, half its target level;

Y, by purchasing permits from X, had met its target by paying a total amount which is less than the full cost of control. This is the basis of "resource sharing in meeting each party's environmental obligations" – which enables large companies and SMEs to help each other out in meeting their emissions obligations – in the absence or presence of energy or carbon taxes.

The concept of permits trading is already well established in US's national programme to reduce sulphur dioxide emissions. The US government is not in favour of carbon taxes. In Europe, developed countries such as Austria, Denmark, Finland, Sweden, Netherlands, and Norway impose energy and/or carbon taxes with some form of tax rebate for industries which are energy intensive or which demonstrate reductions in fuel consumption. In the UK, an energy tax will be applied to business consumers of electricity and fuels from April 2001. Much interest in trading permits using carbon or carbon dioxide units has been expressed by Finland, France, the Netherlands, Denmark and the United Kingdom.

How the scheme could work for India and China

India and China, in 1995, were responsible for 4% and 14% of the world's emissions of carbon dioxide, a major constituent of the global warming effect.

A experimental permits trading scheme on an private enterprise or regional level, leading to a national scheme would be beneficial to each country's objective of fuel efficiency as well as mitigating the long-term effects of climate change. Improving energy efficiency reduces fuel consumption as well as emissions of sulphur and nitrogen oxides – which will greatly alleviate economic and social costs of environmental pollution for both governments.

Allocation of permits

95% of permits could be allocated free-of-charge to those industries and businesses which consume large amounts of energy including electricity, based on a more "lenient" emissions target of 1990 levels in the year 2012, compared to the Kyoto Protocol.

5% of permits would be auctioned off to new entrants in the market after the launch of the scheme, or to existing (usually smaller) enterprises which may require more permits because they are less efficient in controlling emissions.

Industries and businesses which consume energy or electricity, including power generators would be allocated permits – each permit specifying target emission levels and expiry date – based on the historical record of the entity's emissions. Entities formed after 1990 would be allocated permits based on obligatory targets of the entity between its first year of operation and the year 2012 – such entities may be entititled to "excess" permits equal to improvements in emissions reductions in the first year compared to say, the average of 80% of similar size units in the country operating at base load capacity.

Verification of emissions reductions

Measurements, monitoring and verification of emissions reductions should be conducted by independent environmental auditors or experts authorised by the national agencies, with guidance on verification procedures by UNFCCC.

Trading of permits

Trading could be conducted on a bilateral arrangement basis or through a trading exchange. Relevant expertise, training, hard and software technology will need to be provided.

Impact of Kyoto Protocol's Clean Development Mechanism on permits trading schemes in India and China: local co-investors are not excluded from emissions reductions obligations

Certified Emissions Reductions or CERs, gained from emissions reductions in CDM host countries are shared between co-investors of the projects, and some proportion is allocated to the host government and economies which require urgent adaptation to climate change.

Domestic investor companies in CDM projects in India and China are rewarded with CERs which they will (most likely) sell to Annex I co-investor companies or other companies from the Annex I investor country, under the terms of CDM. Sales of CERs for hard currency would command higher prices than the domestic market, however, such a domestic company may be required to retain part of its CERs to meet mandatory emissions requirements or may have to purchase additional rights through the local permits trading market.

The Kyoto Protocol's Clean Development Mechanism initiative may only access a fraction of the GHG market in India and China in the first years of implementation. The amounts of capital are expected to be small, and dedicated to high profile experimental projects. The CDM is also the subject of intense political debate by governments as to the eligibility of projects in terms of "additionality" – that is, emissions reductions should be additional to those projects which would have taken place in the absence of CDM. In the absence of CDM, one would argue, such projects may not have taken place at all – particularly in a recessive economic environment.

India and China could implement internal permits trading schemes to achieve lower energy costs for industries and higher manufacturing price competitiveness, with or without CDM.

A 6.2
Model Questionnaire to Enterprises in Developing Countries

VOLUNTARY PARTICIPATION
IN A PILOT EMISSIONS PERMIT TRADING SCHEME:
PLANNING THE FIRST STEP

Questionnaire to interested firms

Organised by
ABC Group Limited
(the "agency")
on behalf of

XYX environmental authority

Please tick or respond as appropriate

A. Confidentiality status

1. Do you wish the agency to maintain full confidentiality of information? [Y] [N]
   Name of your firm (optional)

   ....

   Industry sector (optional)

   ....

   For purposes of the survey, you may like to leave these blank if you wish to remain anonymous. Your responses, however, would be useful in the planning of a pilot trading scheme).

B: Role of exchanges

1. Would you prefer more than one exchange to handle trading of permits? [Y] [N]
2. Would you prefer Over-The-Counter (OTC) transactions? [Y] [N]

C. Permits – Types of gases

1. Would carbon dioxide be a suitable greenhouse gas to start off with? [Y] [N]
2. Would you also consider methane and nitrous oxide at the same time? [Y] [N]
3. Would you also consider hydrofluorocarbons, perfluorocarbons and sulphur hexafluoride? [Y] [N]

D. UNEP/NPI has produced a Greenhouse Gas Indicator, which enables estimates of carbon dioxide to be calculated in terms of electricity consumption and burnt hydrocarbons for the year 2000. Average emission levels of carbon dioxide emissions can also be calculated from carbon-electricity data provided for earlier years. Would you consider using the Indicator to calculate your emission levels? [Y] [N]

E. Reporting
1. Do you prefer voluntary reporting of emissions? [Y] [N]
2. Continuous monitoring is a fair means of measuring emissions allowed. [Y] [N]
   Is your firm able to install this technology?
3. SMEs may not have capital to install equipment for continuous monitoring. [Y] [N]
   Would quarterly monitoring of emissions, at baseload capacity, represent an alternative means to continuous monitoring?

F. Prices of permits
1. What would you think is a fair price for carbon dioxide permits ($/tonne) [ ]
2. For methane ($/tonne) [ ]

G. System of allocating permits
1. Do you prefer the grandfathering system? [Y] [N]
2. Do you prefer auction? [Y] [N]

H. Launch year of permits allocation and trading
1. Would you participate in a pilot scheme in the year 2002? [Y] [N]
2. Permits could be allocated on a grandfathering and auction system, without monetary value. Administration costs, however, have to be met. Would you consider participating in a such pilot scheme? [Y] [N]

I. Annual permits or period (more than one year) permits
1. Do you prefer one year permits to be allocated for the next five years [Y] [N]
2. Do you prefer period permits of say, 2, 3, 4 year durations? [Y] [N]
3. Do you think that permits should be discounted in future years? [Y] [N]

J. Measuring past and current emissions
1. Are you confident in monitoring your emissions? [Y] [N]
2. If not, would you engage an independent firm to monitor your emissions? [Y] [N]

K. The next step

1. Would you be interested in the results of this survey? [Y] [N]
2. Methods for allocating permits and company ranking by a Green Index have been developed. Would you like to receive information on these methods? [Y] [N]
3. Are there questions that you like to raise with regard to a pilot scheme? [Y] [N]
4. Have you any suggestions on implementing a pilot scheme? [Y] [N]

If your answer is yes for questions K3 and/or K4
please write or type your query, comments or suggestions on a separate page

Please return this questionnaire to

The Environmental Officer
Department of the Environment

Appendix 7

Calculating the Cost of CO_2 Control

A 7.1
Annuitised and Levelised Costs of Carbon Abatement

The "annuitised" cost method and the "levelised cost" method – which are the most popular techniques employed by energy project managers to calculate the cost of utility output – are used in this section to calculate the cost of GHG reductions. Both methods are considered equivalent and hence should provide the same results – however, the timing and magnitude of project revenues may result in slight differences in these results.

Annuitised cost method

The method provides an approximation of cost of GHG reduction using three key inputs; annual interest on loan, annual operational and maintenance (O&M) costs, and annual fuel cost (not applicable in non-carbonaceous renewable energy projects), over the project life or period of repayment as agreed by financiers. Annual loan repayments can be sourced from annuity tables in most accounting and corporate finance publications. If equity is provided, the annual repayment may also include dividends to investors.

Levelised cost method

The method is a more detailed technique of calculating the cost of GHG reduction based on elements of project costs which occur at irregular intervals over the life of the project. It is normally used for large projects which involve two to five years of construction and perhaps decommissioning – power plants, hydro dams and gas transportation networks are exemplary.

Annual costs which are incurred and annual emissions reductions to be achieved over the project life are discounted to provide net present values so that the current cost of reducing a unit quantity of the GHG can be calculated. The method should take into account different forms of financing and their expected returns, performance of GHG reduction equipment and possible replacement of plant inventories.

Return on project financing: project IRR and economic IRR

1. Project IRR

Project internal rates of return (IRR) are calculated using discounted cash flow techniques by applying a suitable hurdle rate of return on the project. The IRR is that discount rate at which the net present value of all future cash flows equals zero.

For projects which receive debt and equity financing, standard textbook techniques like weighted cost of capital (WACC), Miles-Ezzell and Modigliani and Miller's (MM) formulae could be used; although for the latter two methods, certain assumptions apply.

2. Economic IRR

The economic IRR differs from project IRR in that the expected returns on the national or regional economy is also taken into account. For example, a project may increase export sales or regional productivity, or reduce fiscal expenditure. These gains are evaluated in monetary terms and discounted to provide a value of the "economic" contribution from the project. Hence, economic IRRs are normally higher than project IRRs.

A 7.2
Pricing Credits or Permits: Calculating "Overnight" and "Discounted" Cost

Market pricing issues

The pricing of CO_2 permits and credits may not necessarily based on any single methodology or pricing model. Pricing policies are likely to be influenced by the private sector and could be different for developed and developing countries.

Permits

In the USA, SO_2 allowance prices range from a third to half of control costs. As the same allowances can be used to offset emissions targets or allow excess emissions above targets, permits and credits are interchangeble at par value.

Credits

Costs for CO_2 reductions in developing countries could vary widely depending on the cost of technology transfer, plant and capital equipment, subsidies and/or incentives. For Clean Development Mechanism projects, prices of CO_2 credits are expected to differ from one host country to another in the first few years of implementation. A crucial pricing factor for CDM countries would be different emissions baselines for calculating emissions reductions.

Price convergence and price consistency

If the prices were different, then the spread or gap differential will narrow if these assets were traded in a transparent market. However, convergence of prices based on a major hard currency or hard currencies may be complicated by quotations in local currency if markets in developing countries experienced volatility.

Emissions permits and credits should (generally) be based on control costs of technology. As permits are traded at fractions of the cost of control to promote market trading, control costs should represent a sensible benchmark for pricing permits as well as credits. The following technical example could serve as a generic pricing model.

A 7.3
Technical Example I for Pricing CO_2 Emissions Reductions

Gas Turbine to Combined Cycle Gas Turbine

How should a firm which intends to invest in a energy saving technology to meet its assigned target price emissions?

Case study – Pharmacon Plc.

Pharmacon Plc owns a number of manufacturing plants in the UK. It has been assigned its CO_2 emissions target by the Environmental Agency – a reduction of 5.2% of its 1990 levels by the year 2012 (UK's obligations under the EU "bubble" is higher over this period, but Pharmacon has conducted an energy conservation programme since the early 1900s).

Three of its factories in the north of England is supplied with electricity from a single gas turbine plant. Pharmacon decides to install a Combined-Cycle Gas Turbine (CCGT) plant to reduce CO_2 emissions and save on fuel costs.

Note: US$ is used for the example for convenience of comparing investment costs with other developed or developing countries.

Cost data:

Capital cost of Gas Turbine plant	: US$ 448/kW_e
Capital cost of CCGT plant	: US$ 720/kW_e
Capital investment needed	: US$ 272/kW_e

Assumption: Gas turbine in the original plant is used, as it is fully operational. The additional costs are the purchase of a waste heat boiler and a steam turbine.

Scenario, July 1999 – July 2000

Data on capital investment, fuel cost and CO_2 emissions:

Gas turbine, rated at 25 MW_e.
The capital investment needed is US$ 272/kW_e for a 25 MW_e plant = US$ 6.8 m.
Electrical output a year; say, 100 GWh_e a year based on operating 12 hours a day, 360 days.
Plant thermal efficiency = 32%

100 GWh_e/year from July 1999 to July 2000, requires 312.50 GWh_t of thermal energy (based on 32% thermal efficiency).
12×10^9 kWh_t = 397×10^6 therms (from energy data sources)
Therefore, 312.50 GWh_t or 312,500,000 kWh_t = 10.34×10^6 therms

Fuel cost: 10.34×10^6 therms of natural gas @ say, 23 cents per therm = US$ 2.38 m.

1 tonne of carbon will produce 3.67 tonnes of CO_2.
1 tonne of coal, which contains say, about 60% of carbon, will produce 2.2 tonnes of CO_2.
1 tonne of coal, when combusted, generates 2.2 tonnes of CO_2 and 397 therms of heat.

$$\text{Hence, } 10.34 \times 10^6 \text{ therms} = \frac{10.34 \times 10^6}{397} \times 2.4 \text{ tonnes of } CO_2$$
$$= 10.34 \times 10^6 \text{ tonnes of } CO_2$$

$$\text{For the plant} = \frac{57.29 \times 10^3}{100} \text{ tonnes of } CO_2/GWh_e$$
$$= 572.9 \text{ tonnes of } CO_2/GWh_e$$

Comment: the volume of CO_2 emitted/GWh will depend on what is assumed for the thermal efficiency of the gas turbine; heat output from natural gas of an assumed calorific value; and the percentage of carbon in the coal in the calculation.

Alternative method of calculating CO_2 GWh_e

100 GWh_e = 312.50 GWh_t @ 32% thermal efficiency
312.50 GWh_t = $312.50 \times 10^9 \times 3{,}600$ Joules

As the gross calorific value of natural gas is 48 MJ/kg or 35 MJ/m^3, assuming the density of natural gas (North Sea) at 15 degrees Centigrade and 101.325 kN/m^2 to be 0.72 kg/m^3,

$312.50 \times 10^9 \times 3{,}600$ Joules is supplied by 23,437 tonnes or 32,143 m^3 of natural gas.

1 tonne of methane (CH_4) produces 2.75 tonnes of CO_2 when combusted, but natural gas from the North Sea is a mixture of 95% methane, 3% ethane and other lighter hydrocarbon gases (the composition may vary from different fields). The combustion of a unit mass or volume of natural gas, hence, produces different amounts of

CO_2, based on the different chemical equations for combustion of the constituent gases.

If we assume that the North Sea natural gas used in the case example contained 95% methane and 5% ethane for simplifying calculations, 1 tonne of natural gas of this composition would produce less than 2.75 tonnes of CO_2, as there is less carbon per tonne of the fuel gas when methane and ethane are combusted.

If we assign a value of 2.50 tonnes of CO_2 produced by the natural gas in the case example,

23,437 tonnes of natural gas = 58,592 tonnes of CO_2
For the plant = 585.92 tonnes of CO_2/GWh_e

The result is close to that obtained previously using the "therms" method.

Scenario, July 2000 – July 2001

Data on fuel cost and CO_2 emissions:

CCGT plant operational from August 2000, assuming the waste heat boiler and steam turbine are installed over the second quarter of the year.

Plant thermal efficiency = 50%
100 GWh_e/year from July 2000 to July 2001 requires 2007 GWh_t.
200 GWh_t = 6.61×10^6 therms.

Fuel cost: 6.61×10^6 therms of natural gas @ say, same rate a year ago, 23 cents per therm = US$ 1.52 m.

Fuel savings compared to previous year

= US$ 2.38 – 1.52 m.
= US$ 0.86 m.

$$6.61\times10^6 \text{ therms} = \frac{6.61\times10^6}{397}\times 2.2 \text{ tonnes of } CO_2$$
$$= 36.63\times10^3 \text{ tonnes of } CO_2$$

$$\text{For the plant} = \frac{36.63\times10^3}{100} \text{ tonnes of } CO_2/GWh_e$$
$$= 366.3 \text{ tonnes of } CO_2/GWh_e$$

CO_2 reduction compared to previous year = 572.9 – 366.3 tonnes of CO_2/GWh_e
= 206.6 tonnes of CO_2/GWh_e.

Calculating "Overnight" cost of CO_2 reduction

Analysis of capital investment and CO_2 reduction:

A capital investment of US\$ 6.8 m. was required to reduce CO_2 emissions in the original gas turbine plant by 206.6 tonnes/GWh_e.

For the 100 GWh plant, a total of 20,660 tonnes of CO_2 was reduced (July 2000 to July 2001 compared to July 1999 to July 2000).

$$\text{The cost per tonne of } CO_2 \text{ reduced} = \frac{\text{US\$6.8m}}{20{,}660} = \text{US\$329.14.}$$

If fuel savings were taken into account for the cost per tonne of CO_2 reduced: (July 2000 to July 2001 compared to July 1999 to July 2000)

$$\text{then the cost per tonne of } CO_2 \text{ reduced} = \frac{\text{US\$6.8} - 0.86\text{m}}{20{,}660} = \text{US\$288.35.}$$

The results so far only consider capital and fuel costs, but neglects annual operation and maintenance (O&M) costs – which normally vary between 1%–5% of capital costs depending on plant complexity.

Calculating "Discounted" cost of CO_2 reduction

Analysis of capital investment and CO_2 reduction:
Using annuity tables,
Annuitised cost of capital investment @ 10% over 30 years (plant useful life) =

$$\frac{\$6.8\text{m.} \times 10\%}{1 - \frac{1}{(1+10\%)^{30}}} = \frac{\$0.68\text{m}}{0.9427} = \$0.72\text{m. per annum.}$$

$$\text{Discounted cost of } CO_2 \text{ reduction} = \frac{\$0.72\text{m.}}{20{,}600 \text{ tonnes}} = \$\,3.49/\text{tonne of } CO_2 \text{ reduction.}$$

The impact of fuel savings on the cost of CO_2 reduction:
If savings in fuel costs of US\$ 0.86 m. per year over thirty years of the project described in the case study were taken into account, and the cash flow of these savings were discounted at 10%,

$$\text{Net cost per tonne of } CO_2 \text{ reduction} = \frac{\$0.72\text{m.} - \$0.86\text{m}}{20{,}600} = \text{negative cost, i.e. pure profit generated.}$$

Analysis of case example's results with the USA's SO_2 reduction programme and the World Bank's recommended price for CO_2 credits

The "overnight" cost of reduction for CO_2 based on the case example is almost the same as for marginal cost for SO_2 reduction in the USA from 1995 to 2000. According a recent World Bank study, discounted CO_2 credits are expected to be priced between US$ 5 to US$ 10 per tonne, but this is an average value based on a portfolio of carbon sequestration and high technology projects.

The discounted cost of CO_2 reduction without and with fuel savings for the Pharmacon project – a fuel switching project – are US$ 31 and US$ 29 per tonne, respectively – about three to six times the World Bank figures based on a portfolio of technologies which include (low cost) carbon sequestration projects.

A 7.4
Technical Example II for Pricing CO_2 Emission Reductions

Combined Heat and Power (CHP) plant

Case study: Polychem Plc.

Polychem Plc is an oil refinery. It currently buys electricity from the national grid and uses process steam from its boiler. It decides to invest in a CHP plant, so that it can generate its own electricity and produce the process steam using the same boiler. The company may earn extra revenue by exporting electricity back to the grid. The current plant was built in 1995.

Note: US$ is used for the example for convenience of comparing investment costs with other developed or developing countries.

Basics of Combined Heat and Power technology

To simplify the principle of CHP, if, say, 100 Joules of heat energy enters a gas turbine, 30 Joules of electrical energy is produced, and 70 Joules of heat energy leaves the turbine. A large proportion of the input heat energy is not utilised by the gas turbine due to the high temperature of combustion gases at the turbine exit. These 70 Joules of heat energy can be used to provide, say, process heat via a heat exchanger. The overall thermal efficiency of a combined heat and power plant can hence be as high as 80%, depending on system losses.

Plant data:

The refinery uses 100 GWh_e per year. The gas turbine-generator unit should be rated about 25 MW_e, assuming the refinery operates an equivalent 12 hours a day, 360 days a year.

Amount of power consumed from the grid is 100 GWh_e.
Heat input to boiler is 170 GWh_t, boiler efficiency is 90%.

Assumptions:

1. Boiler from existing plant is retained to generate steam (reconversion undertaken)
2. CHP plant operates at 1.7 : 1 heat to power ratio.

Scenario, July 1999–July 2000

Electricity purchased from the grid, separate boiler.

Data on electricity and plant costs, CO_2 emissions and fuel cost:

Electricity cost

Cost of electricity from grid	: US$ 0.06 per kWh_e (tariff band)
Cost for 100 GWh_e per year	: US$ 6 m.

Plant cost:

Capital (historic) cost of boiler system	: US$ 8 m. (unchanged for CHP investment)
Capital cost of gas turbine and generator	: US$ 20 m.
Capital investment needed for CHP plant	: US$ 20 m.

Emissions data, boiler emissions:

At a boiler efficiency rating of 90%, 170 GWh_t would require a thermal input of

$$= \frac{170}{0.9} = 187\ GWh_t \text{ (rounded figure)}$$

187 GWh of heat $= 1.87 \times 10^9 \times 3600$ Joules

A typical gross calorific value of North Sea natural gas = 48 MJ/kg or 35 MJ/m^3

$$187\text{ GWh of heat} = \frac{187 \times 10^9 \times 3600}{48 \times 10^6 \times 1000} \text{ tonnes}$$

$$= 14{,}025 \text{ tonnes of natural gas}$$

$$= 19.234 \times 10^3 m^3 \text{ of natural gas.}$$

One tonne of natural gas produces 2.50 tonnes of CO_2 (calculation based on Gas Turbine/CCGT case example)

187 GWh of heat = 35,062 tonnes of CO_2
Boiler emissions = 35,062 tonnes of CO_2.

Emissions data, electricity generated for UK national grid:
CO_2 emissions from UK grid generators * as a result of refinery's demand

= 525 tonnes per GWh_e

Refinery demand

100 GWh_e = 52,500 tonnes of CO_2

* based on average CO_2 emitted by generators (in kg per kWh_e) using different fuels in the United Kingdom in the year 2000. A value of 0.525 kg per kWh_e (525 tonnes per GWh_e) was predicted by the Department of Environment, Transport and the Regions in 1999 (refer asterisk to previous page).

Total CO_2 emissions = 87,562 tonnes

Fuel cost: 740 m^3 of North Sea natural gas cost, say, US$ 79 (from energy cost data)
34,251 × $10^3 m^3$ of natural gas will cost US$ 3,656,000 = US$ 3.65 m.
Total cost of electricity and natural gas = US$ 3.65 + 6 m = US$ 9.65 m.

Scenario, July 2000–July 2001

CHP plant is operational. Grid electricity is imported to support start-up and operation of the plant until it produces sufficient electricity to make net exports to the grid. In practice, the plant imports grid electricity as a steady source and exports the surplus throughout the day.

Data on electricity costs, CO_2 emissions and fuel cost:

Cost of electricity from grid:	nil
Emissions data, boiler emissions:	35,062 tonnes of CO_2
Emissions from power generated for grid	nil
Fuel (natural gas) cost:	US$ 3.65 m. (assume unchanged). The quantity of natural gas for the boiler is replaced by the same quantity of the same fuel entering the gas turbine after combustion in a separate chamber.

Calculating "Overnight" cost of CO_2 reduction

Analysis of capital investment and CO_2 reduction:

A capital investment of US$ 20 m. was required to reduce CO_2 emissions in the refinery by 52,500 tonnes (87,562 minus 35,062) – which is equal to the amount of emissions produced by power generators for the national grid to meet the refinery's original demand. This works out to a cost of US$ 381 per tonne of CO_2.

The overnight cost of CO_2 reduction for the plant is US$ 381 per tonne – based on capital and fuel costs but neglecting annual O&M costs, which represent a relatively small percentage of the capital cost.

If electricity savings were taken into account for the cost per tonne of CO_2 reduced

$$\text{then the cost per tonne of } CO_2 \text{ reduced } = \frac{\text{US\$20} - \text{US\$6m}}{52{,}500}$$

"Overnight" cost, neglecting O& M cost = US$ 266.66/tonne of CO_2 reduction.

Calculating "Discounted" cost of CO_2 reduction

Analysis of capital investment and CO_2 reduction:
Using annuity tables,

Annuitised cost of capital investment @ 10% over 30 years (plant useful life) =

$$\frac{\$20\text{m.} \times 10\%}{1 - \frac{1}{(1+10\%)^{30}}} = \frac{\$2.0\text{m}}{0.9427} = \$2.12\text{m. per annum.}$$

$$\text{Discounted cost of } CO_2 \text{ reduction} = \frac{\$2.12\text{m.}}{52{,}500} = \$40.38/\text{tonne of } CO_2 \text{ reduction}$$

The impact of electricity savings on the cost of CO_2 reduction:

If savings in electricity of US$ 6m. per year over thirty years of the project described in the case study were taken into account, and the cash flow of these savings were discounted at 10%,

$$\text{Net cost per tonne of } CO_2 \text{ reduction} = \frac{\$2.12\text{m} - \$6\text{m}}{52{,}500}$$

$$= \text{negative cost, i.e. pure profit generated}$$

If electricity was exported to the grid, revenue earned would reduce this net cost per tonne.

Analysis

1. If a higher heat: power ratio is required for our case example, then additional heat must be supplied to the system as the gas turbine and reconverted boiler arrangement limits the the amount of heat leaving the gas turbine to 1.7 times the power generated. A combination of *a gas turbine plus a waste heat boiler package, plus original boiler* (left intact) would offer the best solution to provide heat: power ratios of 3 to 5.
2. Cost calculations do not anticipate rises in natural gas prices, inflation and O&M costs, and any applicable energy or carbon taxation.

3. CCGT vs CHP:

Discounted costs of CO_2 reduction: Polychem's CHP cost of CO_2 abatement is higher than Pharmacon's CCGT, but if fuel and electricity cost savings were taken into account for both plants, respectively, then net costs for Polychem are lower than Pharmacon.

A 7.5
Economics of Energy Conversion – Coal to Natural Gas for Power Stations

Note: A similar financial calculation could be done for change from coal to fuel oil.

Data available to perform the calculations in this section

Fuels, fuel prices and heat energy equivalents:
(Spot prices of fuels quoted – price sources: FT and McCloskey, September 1998)

Coal	14 cents per Therm
Crude oil	33 cents per Therm
Natural gas	23 cents per Therm

1 million tonnes of oil = 1.5 million tonnes of coal = 397 million Therms = 40 million million (trillion) BTUs = 1.11 billion cubic metres of natural gas

1 tonne of coal = 0.67 tonne of oil = 740 cubic metres of natural gas = 397 Therms (same heat value)

To produce the same quantity of heat, the following is required:
1 tonne of coal = US$ 30 (marker price)
0.67 tonne of crude oil (f.o.b., Brent blend) = US$ 63
0.67 tonne of gas oil (c.i.f., UK) = US$ 84

740 cubic metres of natural gas (Bacton, Oct) = US$ 79 (397 × 20 cents per Therm)

Rationale for coal plant conversion

An industrial user of coal should pay for the cost of environmental impact of carbon dioxide (global warming or localised atmospheric warming), damage to environment due to sulphur dioxide (acid rain), and nitrogen dioxide/nitric oxide (photochemical smog and acid rain). These costs, if not met by the polluter, will be met by existing public taxes, or an increase on existing taxes, to counter effects on the human and natural environment, and to avert GDP erosion. Governments should not tax civilians but the polluter.

Hence, if market values of carbon tax and sulphur tax were to be applied to the owners or operators of coal-fired power stations without sulphur dioxide control equipment, the price of coal with carbon and sulphur taxes can be calculated using the information provided in the next page.

Coal and sulphur data

1 tonne of coal with 2 percent of sulphur: 980 kg of carbon, 20 kg of sulphur. 1 tonne of coal = US$ 30 (spot, September 1998).

Carbon tax, per tonne of carbon (indicative, EU, USA) = US$ 50–100 (assume an average value for 980 kg of carbon).

Sulphur tax per tonne of sulphur dioxide = US$ 2,000 (USA).
20 kg S produces 40 kg of sulphur dioxide on reaction with oxygen.
US$ 2,000 per tonne of sulphur dioxide works out to US$ 1,000 per tonne of sulphur on an elemental basis; hence the sulphur tax on 20 kg S is US$ 20.

Calculations

The total price of coal, with C and S tax imposed = US$ 100 to US$ 150 per tonne (average of US$ 125 per tonne). However, we must include the carbon tax on natural gas to make a sensible comparison.

To produce the same amount of heat as 1 tonne of coal, only 0.53 tonne of natural gas is needed (based on North Sea gas, and density of the gas under normal conditions). The amount of carbon dioxide emitted by this amount of natural gas is about 0.6 of that emitted by 1 tonne of coal.

The carbon tax on natural gas, mostly methane, based on North Sea gas in this exercise, would be about US$ 30–60 based on the same amount of heat released by 1 tonne of coal. Hence, the price of (North Sea) gas with carbon tax would be about US$ 110–140; which works out at an average of US$ 125 for the same amount of heat released as 1 tonne of coal.

This is a prime reason why industrial sources using coal as fuel (power stations, factories) in most OECD countries are rapidly switching from coal to natural gas – the switch of fuels does not entail any significant change in costs in view of environmental taxation. To produce the same output, the size of natural gas-fired power stations could be about a third less than conventional coal-fired plants.

Analysis

A (UK) power generator would incur no significant change in costs if he changed fuel from coal to natural gas (two and a half to three times as expensive as coal based on market price) as a result of legislative environmental costs of carbon dioxide and sulphur emissions levied on the two fuels. These comparative figures, however, are based on North Sea gas containing negligible sulphur gases.

There would be other costs if coal was used: NOx removal from fuel nitrogen, dust and particulate removal, turbine blade clean-up (depending on power technology and turbine usage), coal handling equipment, sulphur removal technology and disposal, and ash disposal.

When conversion to natural gas may not be commercially viable

The capital cost of a coal-fired plant may exceed a natural gas-fired plant, but the cost would depend on labour, quality of materials and technology. In China and India where coal reserves are large and natural gas is scarce, the price of coal is estimated at about US$ 5–10 per tonne – hence the cost of electricity from coal plants is much lower than gas plants.

In the absence of environmental legislation or penalties, there is little incentive to change from coal-fired to gas-fired power plants for carbon intensive developing countries, unless carbon credits offset total costs.

Appendix 8

Cost Data – Indicative Costs of Key Equipment for Thermal Power Projects

Average 1998 capital costs are provided for turbines and boilers manufactured in North-West Europe. Manufacturer's prices may vary between as much as 20% on the figures quoted.

Turbines

Gross output MW_e	Steam turbines US$ m/MW_e	Gas turbines US$ m/MW_e
100	0.300	0.210
200	0.210	0.190
300	0.180	0.165
400	0.165	0.150
500	0.160	0.145
600	0.145	0.125
700	0.142	0.120
800	0.140	0.118
900	0.138	0.115
1,000	0.136	0.113

Industrial boilers

Gross output MW_t	Conventional coal US$ m/MW_t	Oil-fired US$ m/MW_t	Gas-fired US$ m/MW_t
250	0.180	0.090	0.070
500	0.150	0.060	0.040
750	0.145	0.050	0.030
1,000	0.140	0.045	0.028
1,250	0.139	0.040	0.025
1,500	0.138	0.039	0.023

Note: Biomass boiler costs can be estimated from the average of oil-fired and conventional coal boilers quoted.

Environmental equipment

Basis of 200 MW_e plant	Flue-gas desulphur. (Wet FGD)	NOx reduction LNB	LNB/OFA	Particle removal ESP	Pulse jet filter
Capital cost, US$/$kW_e$	180	33	35	55	50
O&M cost		neg	neg		
fixed cost/yr, US$/$kW_e$	13			neg	neg
variable cost/yr, US$/$kWh_e$	0.002			0.0015	0.0019

Notes: ESP Electrostatic precipitator
LNB Low NOx burner
OFA Overhead-fire air
neg negligible cost (but may amount to 7% of capital cost for NO_X reduction equipment)

Appendix 9

The Clean Development Mechanism: How to Implement the CDM* – A *Modus Operandi* Framework Document Prepared by Euro-Asia Pacific Corporation Ltd, UK

*This section is based on a *revised version* of the original framework document "Proposal to the Conference of the Parties and Intergovernmental Panel on Climate Change to discuss Practical Implementation of the Clean Development Mechanism", presented to the UNFCCC, Bonn by the CDM team of Euro-Asia Pacific Corporation Limited (UK) on 22 February 1999.

THE CLEAN DEVELOPMENT MECHANISM AS PROPOSED IN THE KYOTO PROTOCOL INITIATIVE OF THE UNITED NATIONS FRAMEWORK CONVENTION ON CLIMATE CHANGE

A PROPOSAL TO THE CONFERENCE OF THE PARTIES AND THE INTERGOVERNMENTAL PANEL ON CLIMATE CHANGE TO DISCUSS PRACTICAL IMPLEMENTATION OF THE CLEAN DEVELOPMENT MECHANISM

Prepared with consultation with
the CDM Project Team
of Euro-Asia Pacific Corp Ltd (UK)

London, The United Kingdom.

Originally submitted on 22 February 1999
to UNFCCC Secretariat in Bonn, Germany:
updated on 1 August 2000.

A 9.1
The Clean Development Mechanism

Summary

The Clean Development Mechanism enables countries listed in Annex I of the Convention to invest in projects in countries not included in Annex 1 of the Convention – any reductions in greenhouse gas emissions in the host country can be credited to the investor country or company to meet its own emissions targets, or sold to any country or company which is listed in Annex B of the Kyoto Protocol.

The Kyoto Protocol aims to reduce the volumes of six greenhouse gases by the year 2012 in thirty six countries, as described in Annex I of the Convention, to about five percent below 1990 levels in these countries. A scheme for cross-border emissions trading between countries listed in Annex B of the Kyoto Protocol was proposed to facilitate the reduction of the gases.

Under Article 12 of the Protocol, which specifically relates to the Clean Development Mechanism, there is no obligation for the COP or governments to provide any financing for the Mechanism. Any Annex I financing, hence, could be voluntary; but may be limited in the case of OECD countries which have quotas on grants/aid for overseas projects.

Private sector financing represents a potententially huge market for investors and lenders. In the case of large projects, an investor company may not have sufficient capital to meet total project costs, hence supplementary financing from the private sector and development banks could make up the shortfall.

The Benefits of the Clean Development Mechanism (CDM)

There are two main types of benefits – technical and commercial.

1. Technical benefits via reduction of greenhouse gas emissions and global warming impact.

Prevention of:

ecological damage to agricultural land – hence reducing crop yields;
hydrological disruption – reducing water supplies, desertification of semi-arid land;
increased human mortality rates from water-borne diseases in absence of clean water;
rise in sea-levels, and sinking of lowlands and small island states;
spread of diseases in tropical regions as mosquitoes become more prevalent

CDM may effectively preserve human lives and their standards of living, maintain economic growth in GDP terms, and safeguard the environment for future generations.

2. Commercial benefits

The investor company can achieve new market penetration, market expansion, and diversify domestic risk in overseas markets. It can gain higher revenues through follow-up exports of goods, services and ancillary products to the host country.

Greenhouse gas emissions reductions in the host country can be used to off-set the investor company's own target emissions, or traded for hard currency. The cost of emission reductions in a developing country (non-Annex I) could be substantially smaller compared to an Annex I developed country under the Clean Development Mechanism – by perhaps a straightforward application of an advanced pollution reduction technology.

Developing countries would benefit from lower fiscal expenditure, inward investments and technology transfer. Any threats to its environment could be reduced. Governments which invest in CDM projects in the form of equity stakes could utilise emission reductions to meet their own targets – perhaps at lower costs – if their companies are not successful in bidding for CDM projects or are unable to afford carbon, energy taxes or permits.

How the Clean Development Mechanism Could Start

Free-for-all or tenders or large company/SME investor quotas for CDM projects?

The Clean Development Mechanism commenced on 1 January 2000. An extraordinary meeting of the COP or the executive board of the CDM is necessary to jump - start projects. Key factors for consideration:

- Free-for-all scenario

The 1-to-1 project development; one country or company negotiating directly with a CDM host country or company on a potential project could result in early project implementation - perhaps the quickest way, until other avenues are explored.

Disadvantages: some companies and countries may complain of irregularities, which often include anti-competitive behaviour, if they do not succeed in project procurement.

Relevant factors: in some CDM host countries, direct contacts for project procurement may not constitute a breach of anti-competitive behaviour, if this block exemption exists in its sovereign or regional community statute.

- Tender scenario

Tenders and the framework for organising tenders needs to be agreed upon by COP or the executive board of CDM.

Disadvantages: an agreed framework will most likely delay the launch date.

Advantages: an agreed framework may solve any disputes or complaints by companies and countries if they do not succeed in securing projects.

- Large company/SME quota scenario

To ensure fairness in project procurement amongst investor companies, a large company/SME ratio of 60/40 or 70/30 is proposed. Hence, for a particular CDM project, regardless of the project value, a CDM beneficiary country could adopt such a criteria to maintain the balance of investor interests. The ratio system would not apply in absence of SME interest.

In a preliminary meeting between Parkin, Osman and See (see Discussions CDM/EAPCO 1, References) in December 1998, a first grouping of five potential projects needs to be identified. A CDM project requires a company from one of the 36 developed countries in Annex I of the Convention to invest in any developing country not listed in the Annex.

CDM projects could involve, say, a company from the USA investing in Brazil or Bosnia and Herzegovina; Australia in Thailand or Indonesia; USA, Germany, United Kingdom or France in China or India; or other possible investor-host combinations.

The first grouping of projects should be closely monitored and results used to improve on future projects which could start six months later. Projects should be screened to identify recent, current and future environmental liabilities before they can be considered for funding.

Financial returns to the project company are expected to match or exceed rates of similar projects without environmental screening, i.e. projects with unknown environmental risks. A case example of how the Clean Development Mechanism could operate for a European company intending to invest in a developing Asian country is presented in page 313.

Euro-Asia Pac's proposal to COP/UNFCCC

The current document is an initiative of the Euro-Asia Pacific Corporation Limited to assist global international efforts in protecting the environment, and has been compiled in good faith. The author assumes full responsibility for the text – the information contained herein bears no claim to complete correctness or accuracy.

1 CDM planners

a) COP, UNFCCC, IPCC
in close working relationship with

b) Private, semi-private sector and NGOs

Say, representatives
10 from Asia-Pacific region origin or linkages
10 from European region origin or linkages
10 from North, Central and South American origin or linkages
10 from African origin or linkages

Environmental research organisations, consultancies
Experts, individuals
Companies

2 Financing

2.1 Sources and size — To be determined by Annex I investor companies.

Co-financing by:

Equity investors
Export credits

Project bonds
Venture capital
Syndicated bank loans
Loans secured on futures and options contracts
Possibly governmental participation
Other sources.

2.2 Currency of financing	Choice by Annex I investor company, co-financiers.
2.3 Principal objective	Emission reductions of greenhouse gases as specified in the Kyoto Protocol in CDM host country. Such reductions would improve fuel conservation, hence promote economic growth.

The Clean Development Mechanism shall assist the financing of cross-border projects which are responsible for greenhouse gases emissions in the CDM beneficiary country. 20% of the financing could be ear-marked for Combined Heat and Power projects in "optimal" locations where high levels of electricity and heat are demanded. Renewable energy pojects should be prioritised in less developed countries where natural resources are scarce or in areas where the cost of grid connection is prohibitive.

2.4 Environmental agenda for projects

Projects shall have a broad environmental agenda as follows:

Reduction in emissions of the six greenhouse gases (Annex A, Kyoto Protocol) through energy-saving or appropriate technologies, compared to existing levels of emissions without such technologies in CDM beneficiary countries.

Target sectors: Power generation, manufacturing/production, transport vehicles, agriculture, fuel handling, forestration, others. According to the International Energy Agency, transport vehicles produce more carbon dioxide than any other source in the world – however, this would be true for a large number of OECD countries, and not developing countries like China and India.

A proposed methodology for choosing a suitable datum for measuring emissions reductions in CDM host countries is described in Appendix 9.11 of this document.

The reference datum for emissions reductions and the datum year from which reductions are measured may be determined by CDM beneficiary governments. The conditions of emissions additionality, financial additionality and sustainability must be satisfied in all projects.

2.5 Countries covered by financing: basis of selection

The country listing (page 307) is based on classifications of developed and developing countries by the World Resources Institute, World Bank, United Nations Framework

Convention on Climate Change, Annex I of the Covention, United Nations Kyoto Protocol (1997) and Annex B of the Protocol. At the time of compiling the listing, countries which are not party to UNFCCC may be included.

Notes: Existing trade sanctions by United Nations on Iraq and Libya are enforced at the time of preparation of this document. Chinese Taipei/Taiwan is not included. It would be classified as a developing country by the World Resources Institute and World Bank. Palestine could considered for financing if it decides to sign up to UNFCCC.

There are three points worth noting in the list – parameters that investors and financiers of the Clean Development Mechanism and/or Joint Implementation initiatives of the Kyoto Protocol – may like to consider in terms of environmental priority:

1. Definition of "developed countries"

a) per Annex I of the United Nations Framework Convention on Climate Change – basis: OECD countries, incumbent EU/OECD countries, industrialised countries;

b) according to GNP per capita criteria, World Bank Atlas 1999

Note: Annex I countries from sub-section (a) above – Belarus, Bulgaria, Latvia, Lithuania, Slovak Republic, Romania, Russia and Ukraine may not be considered as developed under this criteria. However, they qualify for JI financing;

c) according to the World Resources Institute: basis: state of the environment for different countries – under this classification,

non-Annex I countries Albania, Bosnia and Herzegovina, Croatia, Macedonia, Moldova, Slovak Republic, Slovenia and Yugoslavia are considered "developed" in terms of potential risk of exposure to polluted outdoor air, and air polluted with leaded gasoline. However, the WRI's classification system for developed countries does not take into account water and nutrition risk exposure – as applied rigorously to developing countries – which may exist for these Balkan nations;

d) according to the United Nations Conference on Trade and Development: basis: 48 LDCs as identified by UNCTAD – average GDP/capita of US$ 230. Sao Tome and Principe is not included in WRI's list of developing countries.

2. Developing countries: environmental and economic criteria

For some countries, e.g. Kuwait, Singapore, South Korea and United Arab Emirates, the World Resources Institute has assigned "developing country" status, but these countries are classified as high-income economies by the World Bank.

A "needs" prioritisation approach in selecting country participants to benefit from cross-border investments is advocated: firstly on environmental protection, secondly on economic insufficiency.

The countries Antigua and Barbuda, Bahamas, Barbados, Cape Verde, Comoros, the Cook Islands, Djibouti, Grenada, Nauru, Niue, Saint Kitts and Nevis, Saint Lucia, Saint Vincent and the Grenadines, Samoa, San Marino, Seychelles, Tuvalu: which do not appear in the World Resources Institute list of developing countries (possibly because of the lack of environmental data), but which are parties to the Kyoto Protocol, are included.

Developing countries: World Resources Institute classification

Afghanistan (LDC/UNCTAD)
Algeria
Angola (LDC/UNCTAD)
Antigua and Barbuda
Argentina
Armenia
Azerbaijan
Bahamas (World Bank Atlas 1999 classification: high income economy)
Bahrain
Bangladesh (LDC/UNCTAD)
Barbados
Belize
Benin (LDC/UNCTAD)
Bhutan (LDC/UNCTAD)
Bolivia
Botswana
Brazil
Burkina Faso (LDC/UNCTAD)
Burundi (LDC/UNCTAD)
Cambodia (LDC/UNCTAD)
Cameroon
Cape Verde (LDC/UNCTAD)
Central African Rep (LDC/UNCTAD)
Chad (LDC/UNCTAD)
Chile
China
Colombia
Comoros (LDC/UNCTAD)
Congo, Dem. Rep. (LDC/UNCTAD)
Congo, Rep.
Cook Islands
Costa Rica
Cote d'Ivoire
Cuba
Djibouti (LDC/UNCTAD)
Dominican Republic
Ecuador
Egypt

El Salvador
Equatorial Guinea (LDC/UNCTAD)
Eritrea (LDC/UNCTAD)
Ethiopia (LDC/UNCTAD)
Fiji
Gabon
Gambia (LDC/UNCTAD)
Georgia
Ghana
Grenada
Guatemala
Guinea (LDC/UNCTAD)
Guinea-Bissau (LDC/UNCTAD)
Guyana
Haiti (LDC/UNCTAD)
Honduras
India
Indonesia
Iran
Iraq
Israel
Jamaica
Jordan
Kazakhstan
Kenya
Kiribati (LDC/UNCTAD): high risk from sea-level rise/Commonwealth Secretariat
Korea, Dem. People's Rep. (North)
Korea, Rep. (South)
Kuwait (World Bank Atlas 1999 classification: high income economy)
Kyrgyz Republic
Laos, People's Dem. Rep. (LDC/UNCTAD)
Lebanon
Lesotho (LDC/UNCTAD)
Liberia (LDC/UNCTAD)
Libya
Madagascar (LDC/UNCTAD)
Malawi (LDC/UNCTAD)
Malaysia
Maldives (LDC/UNCTAD): extreme risk from sea-level rise
Mali (LDC/UNCTAD)
Marshall Islands
Mauritania (LDC/UNCTAD)
Mauritius
Mexico
Micronesia
Mongolia
Morocco

Mozambique (LDC/UNCTAD)
Myanmar (LDC/UNCTAD)
Namibia
Nauru
Nepal (LDC/UNCTAD)
Nicaragua
Niger (LDC/UNCTAD)
Nigeria
Niue
Oman
Pakistan
Palau
Palestine (West Bank and Gaza)
Panama
Papua New Guinea
Paraguay
Peru
Philippines
Rwanda (LDC/UNCTAD)
Saint Kitts and Nevis
Saint Lucia
Saint Vincent and the Grenadines
Samoa (LDC/UNCTAD): high risk from sea-level rise/Commonwealth Secretariat
San Marino
Saudi Arabia
Senegal
Seychelles
Sierra Leone (LDC/UNCTAD)
Singapore (World Bank Atlas 1999 classification: high income economy)
Solomon Islands (LDC/UNCTAD)
Somalia (LDC/UNCTAD)
South Africa
Sri Lanka
Sudan (LDC/UNCTAD)
Suriname
Swaziland
Syrian Arab Rep.
Tajikstan
Tanzania, United Rep (LDC/UNCTAD)
Thailand
Togo (LDC/UNCTAD)
Trinidad and Tobago
Tunisia
Turkey
Turkmenistan
Tuvalu (LDC/UNCTAD): extreme risk from sea-level rise
Uganda (LDC/UNCTAD)

United Arab Emirates
Uruguay
Uzbekistan
Vanuatu (LDC/UNCTAD)
Venezuela
Vietnam
Yemen (LDC/UNCTAD)
Zambia (LDC/UNCTAD)
Zimbabwe

Developed countries excluding Europe: World Resources Institute classification

Australia	Annex I, Annex B
Brunei (World Bank Atlas 1999 classification, high income economy)	
Canada	Annex I, Annex B
Japan	Annex I, Annex B
New Zealand	Annex I, Annex B
Qatar (World Bank Atlas 1999 classification, high income economy)	
USA	Annex I, Annex B

Europe: developed countries: World Resources Institute classification list
Europe: developing countries: World Bank classification *

The listing shows differences between the World Resources Institute and the World Bank's classification systems: WRI classifies countries based on environmental factors, the World Bank on GNP per capita. Andorra, Cyprus, Liechtenstein, Luxembourg, Malta and Monaco, which are not featured in the WRI list (possibly to the lack of environmental data for these countries, or perhaps due to other criteria) but regarded as high income economies by the World Bank, are included.

Albania*	
Andorra	
Austria	Annex I, Annex B
Belarus*	Annex I
Belgium	Annex I, Annex B
Bosnia and Herzegovina*	
Bulgaria*	Annex I, Annex B
Croatia*	Annex B
Cyprus	
Czech Republic*	Annex I, Annex B
Denmark	Annex I, Annex B
Estonia*	Annex I, Annex B
Finland	Annex I, Annex B
France	Annex I, Annex B
Germany	Annex I, Annex B
Greece	Annex I, Annex B
Hungary*	Annex I, Annex B

Iceland	Annex I, Annex B
Ireland	Annex I, Annex B
Italy	Annex I, Annex B
Latvia*	Annex I, Annex B
Liechtenstein	Annex B
Lithuania*	Annexe I, Annex B
Luxembourg	Annexe I, Annex B
Macedonia*	
Malta	
Moldova*	
Monaco	Annex B
Netherlands	Annex I, Annex B
Norway	Annex I, Annex B
Poland*	Annex I, Annex B
Portugal	Annex I, Annex B
Romania*	Annex I, Annex B
Russian Federation*	Annex I, Annex B
Slovak Republic*	Annex I, Annex B
Slovenia*	Annex B
Spain	Annex I, Annex B
Sweden	Annex I, Annex B
Switzerland	Annex I, Annex B
Ukraine*	Annex I, Annex B
United Kingdom	Annex I, Annex B
Yugoslavia*	

2.6 Conditions of Application and Financing – Proposals

Environmental requirements: All applicant companies for financing would be required to submit their mandate or programme in meeting their own emissions reductions specified by their governments in accordance with the Kyoto Protocol.

The Clean Development Mechanism enables emissions credits to be awarded to investor companies from Annex I countries (investors may raise capital from international financial markets) and stakeholders in projects which reduce greenhouse gas emissions in non-Annex I developing countries. Stakeholders include the host government and possibly Annex I government-investors. These credits can be sold to any company belonging to countries listed in Annex I of UNFCCC or Annex B of Protocol (including governments in these annexes) or used by the investor company to meet its own emissions targets.

1. Atmospheric pollution abatement criteria

Applicant companies must submit a schedule of their greenhouse gas emissions target and indicate how they intend to meet these targets; this is primary pre-requisite for financing. Annex I governments of the Convention should be able to advise applicant companies of their emissions targets in quantitative terms.

It is proposed that emissions credits can only be awarded if the investor company is able to lower greenhouse gas emissions to below 5% or more, of levels from current sources in the host country.

2. Financing preferences for CDM projects

Atmospheric emissions reductions levels:

Financing preference could be given to those projects which aim at reducing emissions at least 5% or more, of targeted greenhouse gases in the project compared to existing technologies in the host country. In terms of process or combustion emissions from fossil fuels, deviations not more than 10% of carbon, sulphur and/or nitrogen content in such fuels may be accepted for calculating emissions reductions.

Non-combustion sources of greenhouse gases emissions:

Financing preference could also be given to projects which reduce greenhouse gas emissions by at least 5% or more, of current levels from non-combustion sources.

Exemplary projects – change of agricultural process, technique, equipment or materials (e.g. optimum use of nitrogenous fertilisers or substitutes); improvements in methane production by fermentation, livestock management, extraction process of fossil fuels; bunker storage of fossil fuels; and sealing leakages in fuel gas transmission systems.

Additionality factors – emissions and financial; sustainability; and potential gains to all stakeholders will also affect financing preference.

3. Environmental audit

Applicant companies must have an environmental audit, performed no more than three years prior to the date of application. Only environmental audits certified by the Department or Ministry of Environment or relevant government authority are acceptable. The ISO 14001 standard may be used as a guide for performimg environmental audits.

Environmental audits must establish a verifiable record of an applicant company's past and current pollution emissions in the home country and if applicable, overseas countries. These levels will be compared with international standards and threshold concentrations recommended by WHO, World Bank, medical research bodies and expert panels.

4. Environmental impact assessment

Environmental impact assessments must be performed for projects which receive funding, and measures must be taken to reduce impacts which may be harmful to the environment before project implementation.

5. Environmental limits

Environmental pollution limits in threshold concentration and/or volume terms of specific pollutants must be observed in the host country; if the host country does not legislate on pollutant levels then international standards and/or recommended limits by medical research bodies will apply.

6. Capital structure of applicant companies

Corporate investors in projects may secure secondary funding from other sources, but their capital structure, that is, debt divided by equity, may not exceed 2.

All applicant companies may be required to declare the effects of potential and actual liabilities of carbon/energy taxes and emissions permits, as well as impacts of environmental investments on their cash flow and balance sheets.

7. Co-financiers' right to terminate financing

Co-financiers may cease contributions to the project based on any non-compliance of Kyoto Protocol articles or UNFCCC by the investor or the project company; non-observation of international law, UN agreements and conventions by the host country; and any factors which are considered detrimental to the objectives of the UNFCCC.

8. Legal advice

Financiers may engage one or more law firms, preferably with a global presence and which specialise in environmental and commercial laws, to advise on its financing or investments in CDM projects.

If disputes should arise between financiers and stakeholders, such disputes shall be settled in the host country through domestic procedures, or referred to UNCITRAL or International Chamber of Commerce.

2.7 Financing less privileged entities and countries

Financiers and investors are encouraged to support small and medium-size enterprise (SME) projects or those projects which involve SMEs, as SMEs are often excluded from capital financing in most developing countries.

The discount rates applied for calculation of net present values of cash flows for projects (in the absence of CDM credits), could be 1% to 2% higher than applied for large company projects – this may affect the financial additionality criteria.

2.8 An example of a CDM investment project (Financing data are case estimates).

Case example: a European company investing in a developing Asian country

Malaysia	Collateral, taxation and customs incentives may be offered.

The government is entitled for a share of emissions reductions credits, on the basis of its administrative costs, incentives and any assets or any financial input extended to the project.

Foreign investor	Up to 33.33% project cost or balance if required * (equity, debt)
Malaysian jv partner	Up to 33.33% project cost or balance if required (equity, debt)
CDM co-financing	Up to 33.33% project cost (equity, debt):

CDM co-financing may be sourced from institutional investment funds, "green" funds (equity funds such as World CDM Green Fund being developed by Euro-Asia Ltd), host country investor market and other private investors.

*UK investor, an Annex I corporate investor, may enter joint-venture partnerships with other "active" Annex I corporate investors to share capital or technology input.

Financing to meet balance of project cost if capital raised is not sufficient.

Other sources: debt
- Export credits from export credit agency of Annex I country.
- Multilateral bank soft loans; e.g. World Bank, IFC.
- Commercial bank lending syndicate.
- Project bonds and securities.

Equity
- Other Annex I companies with emissions targets investing equity in emissions reductions to be achieved in the project.
- Other Annex I countries which anticipate thay may not be able to meet their targets.
- World Bank's Global Environment Facility and Prototype Carbon Fund – provided their investments in humanitarian and social sectors are not diluted.

IRR on project:

Commercial IRR of 10–15% in US dollar terms, based on sales of emission credits (e.g. contract price per tonne of carbon dioxide reduced) and main product stream (e.g. electricity in power projects) – these two revenue sources would service debt providers, and reward equity investors and stakeholders. IRRs may be "fixed" by UNFCCC for certain projects to satisfy the financial additionality criteria.

Economic IRR, which is different from commercial IRR, is calculated on the basis of returns from the project plus economic benefit in terms of income generation in the host region or country.

Foreign investor absence:

Global private equity funds could finance up to 50% of the project to meet technology transfer and associated costs.

"Ethical" returns on investment:

Some groups of society may not invest in CDM projects with purely financial motives, although they are entitled to emissions credits under the Mechanism. They may hence invest in projects in countries where business risks are high and expected rates of return low or which are unquantifiable.

Consolatory "Climate Change Certificates", officially certified by UNFCCC, would represent at least, testimony of an ethical investor's financial contribution to decelerate global warming. Ethical returns are elaborated in page 325.

2.9 Financing issues

Non-exclusivity of origin of CDM financing for Annex I investors and financiers

To promote CDM projects on a global basis, there should be no restrictions on the origins of financing, although Article 12 of the Kyoto Protocol refer investments from Annex I countries into non-Annex I countries. It is proposed that funding from any source can be utilised for CDM project investments to supplement financing from Annex I companies – these additional resources are necessary for large projects which are capital intensive.

Annex I investors will be the ultimate buyers or traders of CDM credits as these assets represent no real value to non-Annex I entities. Some Annex B countries which have adopted voluntary targets may also receive and transact emission credits which originate from CDM projects through secondary trading with Annex I/Annex B countries.

Financial returns: investors, lenders and stakeholders from both developed and developing countries should share the same status as project beneficiaries in terms of financial returns on the project product, e.g. electricity sales.

Credits acquired by host governments, non-Annex I investors and financiers – sales could be effected through exchanges, OTC or an inventory created by UNFCCC for auction.

2.10 Financial efficiency to Annex I investors

Monitoring the financial efficiency of CDM projects

The progress and financial efficiency of CDM projects could be measured by:

*End year measurements of emissions levels achieved/not achieved by the project.

*Possible return on investments by project company – by offsetting domestic targets with emissions reductions (vs. domestic cost of control), sales of net emissions to companies in the home country or third countries listed in Annex B.

*Amount of export sales from Annex I company or country to CDM host country.

3
References

Asian Tax Profiles (individual countries), KPMG, International Tax Centre, Amsterdam.

Asian Country Profiles; Standard Chartered Bank, Hong Kong & Shanghai Banking Corp.

Dun & Bradstreet's Guide to Doing Business Around the World, Prentice Hall, 1997.

Euro-Asia Green Fund, 1998
A second draft paper prepared on 8 December 1998 by Euro-Asia Pacific Corp Ltd.

European Business Sourcebook: Foreign Direct Investments in East and West Europe, Michael See and Stefan Zhurek, International Thomson Publishing, 492 pp, 1998.

European Tax Handbook 1998, KPMG International Tax Centre, Amsterdam.

Kyoto Protocol to the United Nations Convention Framework on Climate Change, 1997
http://www.unfccc.de

Discussions CDM/EAPCO 1: Parkin, Osman and See
Notes from a meeting between Mr Adam Parkin, Mr David Osman and Mr Michael See – London, 7 December 1998.

Discussions CDM/EAPCO 2: Banerjee, Chew, Osman, Parkin and See
Notes from a meeting between Mr Shantanu Banerjee, Mr Christopher Chew, Mr David Osman, Mr Adam Parkin and Mr Michael See – London, 3 February 1999.

Discussions CDM/EAPCO 3: Banerjee, Qin and See
Notes from a meeting between Mr Shantanu Banerjee, Dr He-Qing Qin and Mr Michael See – London, 1 April 1999.

Discussions CDM/EAPCO 4: Cockle, Parkin and See
Notes from a meeting between Mr Paul Cockle, Mr Adam Parkin and Mr Michael See – London, 9 December 1999.

Least Developed Countries 1998 Report
United Nations Conference on Trade and Development, Geneva.

Tax Handbooks for North America, South America and Africa/Middle East, 1998
KPMG International Tax Centre, Amsterdam.

Tax-related Issues in Financing Infrastructure Projects in Asia-Pacific, 1998
Naresh Makhijani, KPMG Asia-Pacific Tax Centre, Melbourne.

World CDM Green Fund, 1998, Euro-Asia Green Fund, 1998, Global Green Fund, 1999
Proposals prepared over 8 Dec 1998/end 1999 period by Euro-Asia Pacific Corp Ltd.

World Resources 1998–99: A Guide to the Global Environment
The World Resources Institute/Oxford University Press, 1998.

9.2
Role of the Conference of the Parties (Governments) in the Clean Development Mechanism

Governments are now instrumental in "jump-starting" the Clean Development Mechanism, as a framework to implement the CDM by all Parties has not been agreed in Buenos Aires.

The roles of the Conference of the Parties and stakeholders in the CDM are proposed in this section – which could be adopted by individual Annex I investor countries.

Regulatory, Verification, Co-ordination, Supervisory, Compliance

1. COP/governments need to establish a register list of companies which are obligated to reduce their GHG emission levels as required by the Kyoto Protocol and/or national targets. Companies with net emission reduction requirements or net reduction surpluses can then conduct their investments abroad with a knowledge of how much emissions reduction or additional reductions they could achieve compared to their domestic emission levels.

2. Greenhouse gases emissions reductions should be measured and certified by UNFCCC inspectors or UNFCCC agencies. Environmental audits, assessments and certificates provided by investor companies or project stakeholders need to be authenticated by UNFCCC or UNFCCC appointed agencies in conjunction with the host government.

An independent verification panel should be established to ensure that the standards adopted by individual countries are acceptable to UNFCCC – this could be funded by the cost of CDM projects.

3. Projects need to be properly co-ordinated and supervised – COP/governments shall need cooperation and assistance at the highest level, with minimum bureaucracy – so that project objectives can be achieved within agreed periods.

4. In all Annex I investor countries, companies, enterprises and facilities which have not complied with emissions reduction target levels by a certain deadline should be notified of impending penalties or fines. These entities should be allowed a further grace period to install appropriate technology to meet their targets or given the option to purchase emission permits in the home market or emission credits from CDM countries (receive emission credits if they are investing in CDM host country) at prevalent market prices.

Emissions reduction compliance by all Annex I countries is the most important factor for the successful implementation of the Protocol. Without compliance, there would be no interest in pursuing CDM or JI projects and earning emissions reduction credits in non-Annex I developing countries.

Offering a Stimulus for Clean Energy Technology Implementation – Fiscal Incentives

At present, electricity pricing of a number of clean energy technologies are not as competitive as conventional technologies – for example, prices based on a coal gasification steam plant or a coal gasification combined cycle plant are much higher than a pulverised coal steam plant (costings exclude SOx abatement equipment). Renewable energy technologies such as solar and wave power incur high costs for electricity generation compared to conventional fossil-fuel fired plants – transmission costs rise if the source is far away from consumer locations.

It is hence recommended that CDM host governments provide preferential corporate tax incentives to encourage clean energy technology investments, at rates which enable these technologies to compete with the average energy or electricity tariff in the country. Such tax rates could be acronymed CLENTAXA – Clean Energy Tax Advantage. For example, if a clean/renewable energy technology costs 10 US cents per kWh_e to a grid supplier, and the average generation cost is 4 US cents per kWh_e, the government could allow a 33% reduction in the standard tax rate for the clean/renewable energy company.

Tax incentives could be offered to expand the capital base for CDM projects. Reduced rates of taxation on repatriation of share dividends, gains on equity sales, and loan interest are recommended. Tax incentives should also be given to domestic investors.

Governments as Direct Investors

Although the Kyoto Protocol does not require Annex I governments to provide any finance to CDM projects, there may arise a situation whereby an Annex I country discovers that it is unable to meet its emissions targets in the first commitment period of 2008–2012.

Difficult economic conditions in some Annex I countries may prevent companies from investing in emissions reductions technology. The country may then decide to invest in CDM projects to acquire the emissions reductions it needs to meet its targets rather than to pay possible heavier penalties for non-compliance.

It is unclear at the time of compilation of this report as to whether government financing of CDM projects would entitle the government to utilise emissions credits for trading, or partake in profits of the project company.

Appendix 9.3
The Role of the Private Sector and Trading Exchanges

Private Sector and NGOs

The private and semi-private sectors, and non-governmental organisations are crucial for the success of the Clean Development Mechanism, but vested commercial interests should not take precedent to the principal objectives of the Kyoto Protocol. Although the private sector may provide unlimited funds to viable CDM projects, popular opinion is also crucial.

Semi-private and NGOs provide a "balanced" input to the political acceptability of CDM projects, so that these projects can proceed smoothly with consensus on all parties as a whole. Pressure groups' participation is important as they offer true facts and figures compared to information provided by perhaps, biased market research companies.

Emissions Trading Exchanges – Targeted Greenhouse Gases, SOx and NOx

The Chicago Board of Trade currently leads emissions trading of sulphur dioxide in the USA, while the International Petroleum Exchange in London and Swedish group OM are currently planning to establish environmental exchanges for the trading of emissions and waste products in Europe. In addition to the six targeted greenhouse gases, trading of SOx and NOx in key commodity exchanges would encourage more energy producers to take greater commercial interest in reducing atmospheric pollutants from fossil fuel utilisation. Such exchanges would also be instrumental for the trading of emissions credits during the "grey period" until the USA ratifies the Protocol.

Appendix 9.4
Kyoto Protocol and Non-Protocol Project Risks

The primary risks based on the Kyoto Protocol agreement have been elaborated in Chapter 9 (specific provisions in the embodiment of the exemplary carbon dioxide contract). This section deals with risks inherent in international projects.

Non-Kyoto Protocol Project Risks and Risk Mitigation

Investors and financiers of CDM projects are exposed to all the risks that are associated with cross-border projects. Financial risks and risk mitigation strategies are described in the following table:

Non-Protocol project risk	*Risk aversion strategy*
• project risk	Screen technology
• inflation	Include projections in budget
• fuel prices	Futures contracts
• market risk	Analyse future demand for main product (e.g. electricity)
• country risk	Analyse political risk
• environmental liability	Perform environmental impact assessment
• tax increase	Anticipate corporate rate, import rates, VAT
• import controls	Determine percentage of raw materials that can be imported
• exchange controls	Has country signed up to IMF's article VIII on capital flows?
• mandatory surrender of hard currency	Determine percentage of hard currency to be surrendered and the official exchange rate to be applied
• exchange rate risk	Purchase forwards or currency contracts
• force majuere	Insurance against "acts of God"
• time-frame of projects	Ensure construction stage do not exceed two years for most renewable projects and five years for thermal power projects

Importance of Good Projects

CDM projects will only be successful if careful planning and due diligence are conducted. Investors should not only target countries which emit large quantitites of carbon dioxide, e.g. China, India, Brazil and Mexico, but also examine the conditions of sustainability, emissions and financial additionality required of CDM projects.

Professionals Are Required

Projects need competent management – in most cases, companies and investors will require astute financial expertise. There would be opportunities for investment bankers, accountants, tax planners and strategists for CDM projects. Personnel in trading exchages will need to be trained on the fundamentals of emissions spot, futures and options contracts.

Insurance

Project investors and financiers are well advised to take out insurance on CDM projects from commercial and multilateral agency insurers where perceived risks, e.g. political risk, *force majuere*, are considered high. Multilateral agencies, MIGA/World Bank and commercial agencies – AIG, Willis Corroon and Allianz are amongst the largest insurers.

Probably the Biggest Single Risk: Non-Ratification by the USA

Carbon dioxide and other greenhouse emission credits may not hold any intrinsic value if the Kyoto Protocol is not ratified by the USA. However, based on 1995 data, USA's share of the targeted greenhouse gas emissions is estimated at 30% to 40% – which falls short of the 55% minimum that other countries may be able to muster under the terms of the Protocol.

It is rumoured that the EU may not ratify the Protocol without the USA's participation as additional investments in energy technology would undermine EU's industrial and export competitiveness. During this "grey period" of ratification uncertainty, emissions credits could be traded "over the counter" and exchanges to parties which are interested in taking a position on the future worth of these credits. Futures, options and swaps contracts, described in Chapter 9 could be used by investors who are prepared to assume the underlying risks.

Appendix 9.5
Case for Private Sector Financing of CDM projects

As removal of government incentives and subsidies to stimulate domestic and global competition has been proposed in the Protocol – it is proposed that CDM tap private sector equity and capital markets. However, the removal of state support is highly contentious with a number of countries which already offer incentives to foreign investments.

The World CDM Green Fund (see pages 320 and 335) could emerge as one of the largest sources of capital for CDM projects. The case for private sector financing to expand the capital base for financing CDM projects are:

1. Cross-border projects are normally capital intensive. CDM projects will need all the help it can get from the private sector.
2. Asia is expected to increase energy consumption and greenhouse gases emissions at a faster rate than any other region in the world between 2000–2050. However, most Asian governments and firms are reluctant to invest in emissions reduction technologies.

   Most Western governments are also reluctant to finance CDM projects in developing countries unless there is a commercial objective or an environmental catastrophe occurs.
3. Two contributory factors: CDM projects are new investment products and the private sector (in developed countries) operates in efficient markets. Hence, CDM project companies are under pressure to make the best returns compared to existing global benchmarks, which can only be achieved by the highest possible cuts in emissions.
4. CDM projects should be very attractive for "pro-green" investors; most of whom are ordinary individuals from all age groups, backgrounds and beliefs. Current surveys show that the environment is amongst the top concerns of the global population.

"Ethical Rates" of Return on CDM Projects, Climate Change Certificates

CDM projects, perhaps even with the benefit of carbon dioxide and other greenhouse gas emissions reductions credits, may not be able to provide high performance market rates of return in the short or medium term compared to, say, FTSE 500, Standard & Poor 500 or Morgan Stanley Capital Index, due to introduction of new technologies at high capital costs. This could be true for solar, tidal and some biomass thermal projects in regions with very low economic output or skilled human resources.

However, some investors or lenders to CDM projects may have a green agenda and are prepared to accept "ethical" returns on their financing – as attested by a Climate Change Certificate (CCC), which could be certified by UNFCCC – to show their commitment to combat global warming.

Appendix 9.6
Issues Regarding "Tradability" of Emissions Credits Earned By Non-Annex I Firms

Non-Annex I companies investing in Annex I countries, non-Annex I companies investing in non-Annex I, Annex B countries

Non-Annex I Company Investing in a Annex I Country

A non-Annex I company, say, from Singapore, investing in, say, a project in an Annex I country, say, the United Kingdom, may reduce greenhouse gas emissions compared to, say, a previous year's level or the datum year 1990.

The case may be better illustrated if a large re-converted thermal power generation plant in the UK, as a result of investment by a Singapore company, emitted 100,000 tonnes less carbon dioxide than a year before. If the same thermal power plant was identified by the UK government as an emissions entity responsible for part of the UK's quantitative emission limitation and reduction commitment (QELRC), then the Singapore investor should be able to claim the right to sell any emissions reductions below the assigned amount for the plant – without any prejudice to the identity of the investor. The Singapore investor can hence sell the emissions reductions through the UK permits trading scheme or to an Annex B company.

An eventual issue facing COP is whether financing could be made available for a company from a non-Annex I country investing in projects which result in greenhouse gas emissions in either an Annex I country or a non-Annex I country.

The Protocol could allow companies from developing countries which invest in projects in Annex I countries to arbitrage any emission reductions they are able to achieve.

On the basis of equity, emissions credits "earned" by investor companies from developing countries in Annex I countries should be able to (i) sell these credits to any Annex I or Annex B company which may require such credits to meet their emission targets; (ii) assist voluntary emissions targets of the investor (developing) country; or (iii) convey as a token of goodwill without fee to any party of the Protocol.

Non-Annex I Company Investing in a Non-Annex I, Annex B Country

If a Brazilian company were to invest in Croatia – a non-Annex I, Annex B country – then on the same basis of equitable and national treatment, the company could sell any emission reductions below the project's assigned amount (Croatia has a voluntary QELRC) to any other Annex B company.

Appendix 9.7
Case of Governments Which Have Not, or May Not, Sign Up to the Kyoto Protocol

The executive board of the Clean Development Mechanism shall have to decide on whether developing countries which do not sign up to the Kyoto Protocol should be excluded from CDM projects. If excluded, these countries would not receive capital from investors who prefer CDM projects with potentially high returns.

A clear risk emerges from developing countries who have not signed the Protocol – perhaps by indifference or lack of commitment to certain articles. At mid-2000, several African, Central and South American, and Middle Eastern countries were absent from the signatory list which is open for signing at the United Nations headquarters in New York.

Appendix 9.8
Countries Listed in Annex I to the Covention Compared to Countries Listed in Annex B to the Protocol

Annex B of the Protocol is a listing of 38 countries which describes their specific quantified emissions limitations and reduction commitments as proposed in the Protocol; expressed mathematically, it includes the countries listed in Annex I of the Convention minus Belarus, plus Liechtenstein, Monaco and Slovenia. The reason why Belarus is missing in Annex B may be because its specific quantitative emissions limitations and reductions commitments have not been agreed; and the reason why Liechtenstein, Monaco and Slovenia – not listed in Annex I of the Convention -appear in Annex B of the Protocol may be because these three countries are not bound by those sub-articles under Article 3 which apply to Annex I countries, and/or that Annex B countries are allowed to conduct emissions trading.

Appendix 9.9
Definition of Small and Medium-Size Enterprises, and Large Companies

Small and Medium-size Enterprises (SMEs) should be included in an Annex I company's and/or government's financing plans.
For the purposes of CDM projects,

1. Small enterprises could be defined as those companies with between US$ 100,000 and US$ 10 million; and medium-size enterprises as those companies with between US$ 10 and US$ 100 million in group annual revenue.
2. Large companies are those with group annual revenues above US$ 100 million.

Annual revenues could based on the average of the past five years, as the performance of several companies in developing countries have been affected by the Asian economic crisis.

Appendix 9.10
Project Selection Criteria for CDM: Industry and Country Targets

Project selection would be based on Article 12, paragraph 5 (b):

Paragraph 5 of the Kyoto Protocol specifically states that

"emissions reductions resulting from each project activity shall be certified by operational entities to be designated by the Conference of the Parties serving as the meeting of the Parties to this Protocol, on the basis of:

(b) real, measurable, and long-term benefits related to the mitigation of climate change.

The targeted gases are carbon dioxide, methane and nitrous oxide, which form about 90% of the six greenhouse gases; hydrofluorocarbons, perfluorocarbons and sulphur hexafluoride.

Project selection – regulation of CDM emissions credits to ensure compliance in home country

Firms belonging to Annex I of the Convention and Annex B of the Protocol should not receive more than 50% of emissions reductions credits from the CDM, AIJ/JI and emissions trading initiatives against their own domestic targets, to assist the countries of corporate origin to meet their obligations under the Kyoto Protocol.

Long-term benefits criteria: To ensure greenhouse gases emissions are effectively reduced over a reasonably long period, projects which involve taking-over of an existing facility are required to ascertain that the remaining operational life of the facility should not be less than half the normal working life it was designed to operate. Preference would be given to more robust projects in terms of its life span based on this criteria – including new plants replacing less efficient ones. The internal rate of return of projects with longer operational periods are much higher compared to projects which have to meet recurrent costs of maintenance/repair.

Global "hot-spots": regions of CDM host countries only

Regions, whose atmospheric conditions are considered "hot spots" from satellite surveillance:

Caspian Sea, east of Caspian Sea areas
North-east Asia
South-east China
West and south-east parts of S. America.
North and north-east India
Java and East Kalimantan
Southern part of Africa.

Some of the regions do not have intensive industries normally associated with large emissions of carbon dioxide, but produce sufficient amounts of other greenhouse gases – methane and nitrous oxide – together with moderate emissions of carbon dioxide, to create hot-spots.

Largest emitters of carbon dioxide

Industry:	Power stations burning coal, oil and gas (in descending order). Cement and steel manufacturing plants. Gas flaring. Other industries, e.g. automotive, chemicals, paper, plants.
Private /Commercial:	Transport vehicles – vehicles that use gasoline and diesel fuels. In China, coal-stoves are used by a large proportion of the population.
Land use change:	Burning or clearing of forests and forest cover for commercial use.

Countries:	Africa	Algeria, Egypt, South Africa, Nigeria.
	Asia	China, India, Indonesia, Iran, Iraq, Kazakhstan, Malaysia, Korea (North), Korea (South), Pakistan, Saudi Arabia, Thailand, Uzbekistan
	Central America	Cuba, Mexico
	South America	Argentina, Brazil, Colombia, Venezuela.

Note: region/country classification according to the World Resources Institute.

Transition economies: Albania, Armenia, Azerbaijan, Bosnia and Herzegovina, Georgia, Kyrgyz Republic, Macedonia, Moldova, Mongolia, Tajikstan, Turkmenistan and Yugoslavia would also be potential markets for greenhouse gases emissions reductions.

Largest emitters of methane

Industry:	Oil/gas/coal extraction processes. Leakages from production (e.g. refineries, coking plants) and distribution systems of these fuels. Gas flaring. Combustion (incomplete) of fossil fuels in power stations. Agricultural - enteric fermentation and animal wastes. Bunker fuel storage discharges.
Private /Commercial:	Transport vehicles, depending on engine technology. Private farming - release from soils, burning of agricultural wastes. Burning of biomass, fuel wood and cleared forests. Waste emissions from landfills.
Countries:	China, Bangladesh, Ethiopia, Indonesia, Gambia, Kazakhstan, Nepal, Oman, Peru, Philippines, Venezuela.

Largest emitters of nitrous oxide

Industry:	Nitrogenous fertilisers: agriculture - application and excess usage. Chemical industry (particularly nitric acid and nylon production) and refineries. Burning of biomass, fuel wood and cleared forests. Combustion of fossil fuels - power stations, transport vehicles, industries.
Private /Commercial:	Private farming, fertiliser use, animal wastes. Transport vehicles - emissions from catalytic converters.
Countries:	China, Indonesia, Venezuela.

Other main sources of greenhouse gases emissions

Hydrofluorocarbons:	HCFC-22 manufacture, refrigerants, vehicle air conditioning.
Perfluorocarbons:	Aluminium production.
Sulphur hexafluoride:	Electrical switching gear, magnesium industry.

What kinds of technologies for CDM projects?

Combustion-generated sources

Clean energy technologies – e.g. Combined Heat and Power (cogeneration), fluidised bed combustion with cogeneration, and overhaul of existing power plants to improve fuel efficiency and NOx control. Vehicle emissions control – more efficient engine design and exhaust controls.

Non-combustion generated sources: agricultural industry and croplands

Advanced technology and equipment to reduce methane and nitrous oxide emissions from agricultural sources; and halo-carbons and sulphur hexafluoride from industrial processes.

The issue of direct investment in pollution control equipment

The CDM executive board shall have to decide on whether direct investments in a singular piece of equipment or component, e.g. acidification of carbon dioxide, methane removal or conversion, would qualify as certifiable projects.

Transport sector (long-term planning)

Transport infrastructure planning – installation of subway, underground transport systems or regulated automotive vehicles entry (e.g. alternate number plates) into major cities – would encourage use of public transport, hence enabling massive cuts in greenhouse gases emissions. The admissibility of such projects will have to be approved by the CDM's executive board.

Renewable energy sources

Solar, wind, geothermal, tidal, wave, biomass regrowth and regenerable horticulture produce zero greenhouse gas emissions compared to fossil fuels – if GHG emissions from construction of the plant or facility are not taken into account. Biomass regrowth and horticulture regrowth fuels are considered zero-GHG emitters by most proponents of climate change think-tanks, but investors should consult UNFCCC as to whether this would be the case as some biomass products with higher carbon content may not be admissible as credible carbon sinks.

Nuclear and hydro power are the most established renewable energy technologies while geothermal plants are steadily on the increase in tectonic plate regions. Biomass technology is already commercially applied on a large scale, while some tidal projects may be too capital intensive to attract investor interest.

CDM projects should concentrate on sectors where investment costs or returns are relatively high compared to current market returns – provided carbon credit returns are equally attractive.

Following a number of discussions with UK experts in solar, wind, geothermal and wave energy technologies, and research conducted in these energy sources, the author recommends the following locations for CDM projects:

Energy type	Locations
Geothermal	LE: Algeria, Argentina, China, Indonesia, Mexico, Philippines, Thailand SE: Algeria, Costa Rica, El Salvador, Kenya, Nicaragua
Solar	LE: Chile, India, Indonesia, Iran, Israel, Malaysia, Mexico, South Korea, Thailand, Turkey SE: Bolivia, Colombia, Ethiopia, Jordan, Kyrgyzstan, Senegal, Sri Lanka, Tanzania, Zambia
Wave	LE: Argentina, Brazil, Chile, India, Indonesia, Philippines, South Africa SE: Barbados, Guam, Mauritius, Micronesia, Sri Lanka, Tonga
Wind	LE: Argentina, Brazil, Chile, China, Egypt, India, Indonesia, Israel, Mexico, South Africa, South Korea, Thailand, Venezuela SE: Costa Rica, Jordan, Kyrgyzstan, Morocco, Namibia, Nepal

Notes: LE = large economies, SE = small economies

Appendix 9.11
Reference Datum for Measuring Emissions Reductions for CDM: Emissions Additionality

The basis of counting emissions reductions is described in Article 12, paragraph 5(c) of the Kyoto Protocol:

"reduction in emissions that are additional to any that would occur in the absence of certified project activity (i.e. in the absence of CDM)".

For maximum financial and emissions reductions credits gains, most Annex I companies will invest in projects with the highest emissions levels of greenhouse gases.

However, there are two main factors which will affect this investor criteria:

1. Certified project activity – activity criteria.
2. Certified projects – project conditions.

Certified Project Activity – Activity Criteria

The author recommends that projects which meet the activity criteria listed in Appendix 9.10 should qualify as projects in the functioning of CDM, and that reduction levels of 5% or more of existing GHG emissions – which may be weighted on a composite basis – should also form a key requirement.

Certified Projects – Project Conditions

Three project scenarios exist as regards interpretation of Article 12, paragraph 5(c):

Generic case example of a factory in Thailand (assuming project meets certification criteria): case 1 and case 2.

Case 1: An existing project with emissions reductions technology

The factory has already installed emissions reduction equipment – perhaps manufactured by a local manufacturer – and is currently achieving a reduction of, say, 10 million tonnes of carbon dioxide a year, say, for 1998.

A pro-active UK firm, proceeds to invest in the factory in, say, April 1999; hoping that the project would be considered a CDM project and additional emissions reductions achieved in addition to the factory's current emissions reductions, would be retrospectively credited. The company achieves an additional reduction of, say, 5 million tonnes of carbon dioxide at end of 1999, over the nine month period between April and December 1999.

The basis of certifying emissions credits in the above example would be straightforward after 1 January 2000, but there may be many Annex I companies which may have invested in projects before this start date.

Question: Would the CDM executive board/UNFCCC allow the UK company to be credited?

In principle, the board/UNFCCC would probably agree on ethical grounds, but this would would be regarded as a precedent for any other project which acheived similar results in the past – which could number hundreds for Thailand. On hindsight – probably not.

Case 2: An existing project without emissions reductions technology

Article 12, paragraph 5(c) would be interpreted as:

"reduction in emissions in the presence of certified project activity" – as in the absence of certified project activity, there has been no emission reductions.

The counting of emissions reductions from the Thai factory, if the year of investment was changed to 2000 and emissions reductions obtained post-2000, would be straightforward – provided there are no other technologies that may be implemented in the absence of CDM because of commercial non-viability or adverse economic conditions.

Case 3: A new project

The executive board of the CDM shall have to decide whether or not new projects, if they meet certification requirements, can be entitled to emissions reductions credits.

There may be objections to new projects, as they contribute to overall GHG emissions. However, building a new project to replace another which is to be decommissioned may be acceptable as a CDM project. Countries like India and China are already implementing a number of new projects to meet demand, and investor companies in these countries would certainly raise the issue of their entitlement to emissions reduction credits.

The author recommends that any new project which meet the project selection criteria in Appendix 9.10 should qualify for emissions reductions credits under the CDM to promote foreign investments in new projects in developing countries. A reference datum level for emissions reductions would be difficult to ascertain in the case of new projects. However, an improvised model for new projects could be helpful:

A new factory, with similar capacity as the generic case example, is to be constructed in Thailand. A sensible reference datum level for counting emissions reductions could be the emissions levels of "similar capacity case examples" - which exists in, say, another province of the country - which produce less greenhouse gas emissions.

The author recommends that the reference datum level for emissions reductions for new projects which do not have case models, or more importantly fair comparative models, base the datum level on, say, average emission levels of 70% or 80% of facilities of similar nature and capacity, rather than selectively identifying case examples with the highest emissions.

Verification of greenhouse gas emissions reductions

It is proposed that CDM project companies reduce emissions by 5% or more, compared to emissions using existing technology in the host country - as a general rule - and that these reductions could be verified by UNFCCC/IPCC inspectors in conjunction with the Ministry of the Environment or equivalent government authority.

Impact on economic cycle on emissions reductions measurements

Emissions reductions can only be effectively measured if measurements were made between two periods, say, 2005 and 1990 (or post-1990 year when plant was commissioned) at the same operating capacity. If emissions reduction equipment was installed in a year of slower economic growth, then emission reductions comparisons can not be realistically made.

Non-Kyoto Protocol gas emissions – fuel content specifications for combustion or thermal decomposition related projects

To meet sustainability requirements, there should be no new or additional emissions of non-Kyoto Protocol gases. For combustion or thermal decomposition generated emissions, the contents of nitrogen, sulphur, halogen compounds (chlorine, fluorine) and mineral matter of fuels may not exceed by ten percent from those fuels previously used in the project - if indeed the same type of fuel was employed. These fuel content specifications ensure that halo-carbon emissions do not increase, while hydrogen content may be varied for different heat outputs.

Counting of greenhouse gases emissions reductions in CDM projects

A reduction of a greenhouse gas in volume terms can only be counted once against a former emission level - for example, a reduction in carbon dioxide emissions in 2003 compared to levels in 1990 or, say, 1997, when the plant was built. Reductions can not be counted again – in 2004 for the preceding example, unless further improvements in lowering such emissions were made – or else more than 100% removal of emissions would be possible!

Emissions reduction measurement of a plant built in the year 2002 in CDM host country

How would one interpret the emissions additionality requirement for an emissions source, say, a power station commissioned and operational in 2002, when the reference year is 1990 (plant non-existent, but the CDM host country has chosen 1990 as the reference year)?

On the basis of carbon dioxide, emission reductions should be additional if the emissions in 2002 are lower than average emission levels of similar power stations in the year 1990. If new or advanced technology is implemented in 2002, and there are no similar plants in the country in 1990 to compare with, then the national average of emissions levels could be adopted as the baseline.

Appendix 9.12
How Annex I Countries Can Move Ahead with the CDM

Department	*Recommended strategy*
The Ministry or Department of Foreign Affairs, or equivalent	Contact Commercial sections of embassies of CDM beneficiary countries. 1. Identify current and future projects. – Project nature, details, location. – Project cost, partners? – Local environmental conditions. – Project emissions levels (of existing facility). – Host country incentives, taxes. – Other essential details. – Create a global database of potential projects. 2. Liaise visits by Annex I companies.
The Ministry/Department of Trade or Commerce or equivalent	Appraise data from Foreign Affairs unit. Signal to relevant industries and companies. Plan export credits for projects.
The Ministry or Department of Overseas Development or equivalent	Consider financing for CDM projects in LDCs (Least-Developed-Countries). Contact commercial sources, e.g. banks, lenders, for joint-financing of CDM projects. Assess risk of projects. Project insurance. Environmental or commercial objective? If choose commercial objective, then CDM host countries with genuine need of assistance will be ignored.

Ministry/Department of Energy and Environment or equivalent	Analyse the environmental data on CDM projects. - Assist project evaluation and risks for project companies, and risks to lenders and investors. - Transport sector: liaise with vehicle industry sector: e.g. engine technology firms. Infrastructure projects: rail and subway contractors.

How Non-Annex I Countries Can Assist

Ministry/Department of Energy and Environment or equivalent	Organise CDM projects on a national or regional scale – promoting awareness and collecting data on the progress of projects. On a regional basis, ASEAN or MERCOSUR countries could rank large companies and SMEs on the basis of carbon dioxide reductions every year from 2000. In Europe, non-Annex I countries in Eastern Europe and CIS could benefit from CDM projects. In Africa, substantial reductions in carbon dioxide and methane from power, oil/gas exploration and agricultural sectors can be achieved.

Appendix 9.13
A Global Private Equity Fund – The World CDM Green Fund*

*A *modus operandi* framework document based on an equity fund to abate global warming, presented to the UNFCCC, Bonn by the CDM team of Euro-Asia Pacific Corporation Limited (UK) on 22 February 1999.

Current Sources of Financing Environmental Projects

The World Bank's Global Environmental Facility (GEF) is the largest source of funds for environmental projects, amounting to about US$ 2bn. dollars. Other sources include export credit agencies and multilateral banks.

Why a Global Private Equity Fund for CDM?

- Finite capital from current sources.
- Involvement of private sector, instead of fiscal taxes to fund climate change projects.
- Involvement at individual level; personal commitments to reverse effects of climate change.

Country Targets of the Fund

The fund should target all developing countries, regardless of the size of GHG emissions.

Main Features of a Global Equity Fund for CDM

- Competent fund manager with worldwide offices.
- Financial team must work closely with technical team.
- Staff composition to reflect ethnic diversity of developing and developed countries.
- Close liaison and contact with UNFCCC and national environmental agencies.
- As the CDM initiative started in January 2000, and rules for CDM may not be decided by end 2000, high level of research by the technical team is required for the selection of projects.
- Projects must meet both environmental and financial "additionality criteria":
  - Environmental or emissions additionality: projects must be additional to those which would have taken place in the absence of CDM.
  - Financial additionality: projects with a negative NPV at a low discount rate.
  - Sustainability criteria: projects must maintain or improve the social, economic and environmental conditions of the location region in which the projects are implemented.

Appendix 10

Islands-at-Risk from Sea-Level Rise: Evacuation and Resettlement Contingency Plans

The "EVAREST" (*EVA*cuation and *RESeT*tlement) Islands Programme

Key Objectives and Issues

As a result of global warming, contingency plans to evacuate and resettle the inhabitants of low-lying islands at risk from sea-level rise must now be made to avert possibly large-scale displacements of human migrants. Such plans should include the long-term integration of migrants into the host economy, fair treatment in the labour market, and the protection of human rights in the new environment.

In 1994, the Alliance of Small Island States (OASIS) was founded in Barbados to jointly represent the concerns of island economies which face environmental disasters and wildlife extinction. Most of AOSIS members are countries from the Caribbean and South Pacific.

UNHCR and non-governmental organisations – such as Oxfam and CARE – could take the lead in initial negotiations with potential host countries, and prepare mobilisation of resources for the evacuation of islanders. The World Wildlife Fund could administer the relocation of endangered species of fauna and flora from these islands.

The underskilled and most disadvantaged percentiles of islands-at-risk are entitled to the same rights of evacuation and resettlement assistance as those with higher incomes – it is hence recommended that fair democratic ballots be held and supervised by United Nations officials for the staged movement of inhabitants to their new abodes. Legal, medical, welfare, utilities, housing, educational, employment/reoccupational issues must be addressed well in advance. Other issues will emerge – how extra fiscal revenue should be generated and how sociological adjustment can be managed at the communal interface and marketplace.

The process of bilateral and regional consultation should be embarked upon now – any delay will precipitate an unprecendented humanitarian crisis for which the international community must assume full responsibility. At end 1999 and early 2000, torrential floods in Honduras, Venezuela and India inflicted substantial damage to these economies – it is unlikely that they will consider the influx of islanders-at-risk. In the Pacific, Australia and New Zealand may not be receptive to the idea of absorbing large numbers of new migrants.

The author has prepared statistical data table listing low-lying islands at risk – describing their demographics, political linkages and economic acitivity, and possible evacuation and resettlement strategies based on these factors.

Table A10.1. Land height and socio-economic characteristic of Islands-at-Risk, and islanders' evacuation and resettlement strategies the international community could consider

Country/ land height main islands	Islands at sea level/ immed. risk	Annual rainfall in inches	Pop./ Religion	Political affiliation/ economic activity/ energy resources	Aid	HDI/ 174	Evacuation/Resettlement/ nearest friendly countries/ potential country partnership
Antigua & Barbuda 200 m.	3	40	65,000 RC, Pr.	Backs US policies. Agro, fishing.	US, UK, Fr, EU.	55	US. Possibly Fr. Guiana, Suriname, Guyana.
Bahamas 200 m.	Many	30–60	300,000 Pr, RC.	Strong US links. Agro, fishing.	US.	26	US. Possibly Fr. Guiana, Suriname, Guyana.
Barbados 200 m.	–	50–75	300,000 Pr, RC.	Backs US policies. Agro, oil, fish.	US, EU, UK.	25	US, UK. Possibly Fr. Guiana, Suriname, Guyana.
Cape Verde 1,000–2,000 m.	6	10	400,000 RC.	EU linkages. Agro. Wind, wave, biom.	EU.	123	Senegal, Guinea-Bissau, Por, Neth, Sweden, Fr.
Comoros 500–1,000 m.	–	113	700,000 Islam.	France. Agro, fishing.	Fr, EU, OPEC.	139	France. Possibly Tanzania where 1/3 are Muslims.
Dominica 500–1,000 m.	–	78	71,000 RC.	EU, US. Agriculture, fishing, geothermal.	EU, Fr, UK.	69	Fr. Guiana. Possibly Guyana, Suriname, Panama.
Dominican Republic 1,000–2,000 m.	–	56	7.8 m. RC.	US influence. Agro, metals, fish.	US, EU.	96	US. Possibly Guyana, Fr. Guiana, Suriname, Panama.
Fiji 500–1,000 m.	Many	117	800,000 Pr, Hi.	Australia. Agro, minerals, fish.	Aus, Jpn, EU	46	Australia, NZ, UK.
Grenada 200–500 m.	Many	50–200	92,000 RC.	Backs US policies. Agro, nutmeg, fish.	UK, EU, US, Jpn	67	US, UK. Possibly Fr. Guiana, Suriname, Guyana.

Haiti 500–1,000 m.	–	53	7.2 m. RC.	US influence. Agro, fishing.	US.	148	France. Possibly Fr. Guiana, Suriname, Guyana.
Kiribati <100 m.	Many	79	77,000 RC.	US, Australia, NZ. Agriculture, fish.	Japan.	>125 (est)	Australia, New Zealand.
Maldives <100 m.	Many	59	300,000 Islam.	Pakistan, India. Agro, fishing.	Japan, Ind, Pak	118	Possibly Sri Lanka, Pak, Ind, Bangladesh.
Marshall Islands <100 m.	Many	161	52,000 Pr.	US influence. Agro, fishing.	US, Aus Taiwan.	>100 (est)	US, Australia, NZ.
Mauritius 200–500 m.	–	40	1.1 m. Hi, RC.	EU. Agro, clothing, bagasse, fishing.	EU, Fr, UK.	60	France. Possibly India, Madagascar, Mozambique.
Micronesia <100 m.	Many	194	107,000 RC, Pr.	US influence. Fishing.	US.	>100 (est)	US. Possibly Aus, NZ.
Nauru 200 m.	–	75	10,000 RC, Pr.	Non-aligned. Agriculture.	Aus, NZ, UK.	>100 (est)	Hawaii, Australia, NZ.
Palau 100–200 m.	–	150	16,200 RC, Pr.	US influence. Copra, fishing.	US.	<100 (est)	US. Possibly Aus, NZ, Japan.
St Kitts & Nevis 200–500 m.	–	51	41,000 Pr.	US, EU. Agro, fishing.	US, EU, UK.	37	US, UK. Possibly French Guiana, Suriname, Guyana.
St Lucia 200–500 m.	–	60–138	145,000 RC.	Backs US policies. Agro, geothermal.	US, EU, UK.	84	US, UK. Possibly French Guiana, Suriname, Guyana.
St Vincent & Grenadines 200–500 m.	–	60–150	111,000 Pr, RC.	US influence. Agro, fishing.	US, EU, UK.	79	US, UK. Possibly French Guiana, Suriname, Guyana.
Sao Tome & Principe 200–500 m.	–	40–150	125,000 RC.	US influence. Agro, oil explor, fishing.	EU, Por, US	133	Angola, Gabon. Possibly Por, UK, France, US.

Table A10.1. *Continued*

Country/ land height main islands	Islands at sea level/ immed. risk	Annual rainfall in inches	Pop./ Religion	Political affiliation/ economic activity/ energy resources	Aid	HDI/ 174	Evacuation/Resettlement/ nearest friendly countries/ potential country partnership
Seychelles 200–500 m.	Many	95	73,000 RC.	France. Agro, fishing.	US, UK, Fr, Aus	62	France. Poss. Madagascar, Mozambique, Kenya.
Solomon Islands 500–1,000 m.	Many	80–120	400,000 Pr, RC.	Australia. Agro, bauxite, fish.	Aus.	125	Australia, New Zealand.
Tonga <200 m.	Many	80	98,000 Pr, RC.	Australia, USA. Agro, fishing.	Aus, US NZ, EU.	<100 (est)	Australia, NZ, US.
Trinidad & Tobago 200–500 m.	–	74	1.3 m. Pr, RC.	USA, UK. Asphalt, oil/gas.	US, EU.	39	Possibly French Guiana, Suriname, Guyana.
Tuvalu <100 m.	Many	120–160	9,000 Pr, RC.	Australia. Fishing.	Aus/NZ, Jpn/Tai	>125 (est)	Australia, New Zealand.
Samoa 200–500 m.	–	112	169,000 RC, Pr.	New Zealand. Agro, fishing.	Aus/NZ, Jpn, EU	102	New Zealand, US.
Vanuatu 200–500 m.	Many	84	200,000 Pr, RC.	Australia, NZ. Agro, fishing.	Aus/NZ, Fr, UK.	119	Australia, New Zealand.

HDI Human Development Index, ranking amongst 174 countries (least developed 174th) – based on life expectancy, infant mortality, GNP/capita, daily calorie intake, literacy, education.
Pol. affil. political affiliation, past/current.
Rel major religions – Hi (Hindu), Pr (Protestant), RC (Roman Catholic).

Analysis of Statistics and Recommendations

From the table, it is clear that evacuation priority should be given to the most vulnerable island-at-risks – as characterised by their lowest rankings in Human Development Index and highest susceptibility to sea level rise.

The islands most at risk are Kiribati, Maldives, Marshall Islands, Micronesia and Tuvalu – with average land heights of less than 100 metres and annual precipitation of more than 50 inches. Fiscal provisions to the health, infant mortality and literacy sectors are amongst the lowest in the region.

How the Clean Development Mechanism can be Implemented in Islands at Risk

Renewable energy technologies like wave, wind, solar power and biomass conversion can reduce the dependency on expensive fossil fuel imports – much of which is the the form of diesel and heavy fuel oils. The baselines could be based on the current average of greenhouse gas emissions per unit power or thermal output. Revenue can be generated from selling GHG reductions at market prices as a result of CDM projects implemented within the islands and from part of proceeds of global sales of GHG contracts as mandated by the Kyoto Protocol.

Meeting the Costs of Adaptation

The total cost of preparing contingency plans right through to physical evacuation of islanders at risk and subsequent communal integration are expected to exceed the nominal contributions from emissions trading. Additional capital donations from the international community are required. The issue of who pays how much could be directed to the largest industrial emitters of greenhouse gases in the twentieth century – the USA and Western Europe. These countries may be asked to assume the moral responsibility for displacing populations from low lying islands.

Who May be Liable for Adaptation Costs of Evacuees?

If islands-at-risk invoke the "polluter pay" principle, then the largest emitters of greenhouse gases on a historical basis should meet or contribute towards the full costs of adaptation, as these islands are the innocent victims of global warming. That is the simple logic which the "North" countries may have to contend with.

Global warming will displace a total of 545,000 persons from the five most vulnerable islands identified in the author's comparative analysis. It is unlikely that the world's first "environmental refugees" will secure employment in the initial years of residence or transition in host countries – hence the obligation to meet the islanders' minimum essential living costs should fall upon the industrialised economies.

Annex I countries are the most likely grouping to assume the international community's responsibility to meet the costs of adaptation. Their relative percentages of carbon dioxide emissions in 1990 – the base year for measuring CO_2 reductions under the Kyoto Protocol – could form the basis of apportionment of environmental responsibility. Table A10.2 provides the respective percentages of carbon dioxide emissions for these countries in 1990 – 36.1% for the USA, 17.4% Russian Federation, 8.5% Japan,

7.4% Germany and the UK 4.3%. The Russian Federation may not volunteer contributions on the basis of low GNP per capita.

Annex I countries may contribute capital to non-Annex I countries towards the evacuation and resettlement of islanders if they (Annex I countries) decide to accept the full costs of adaptation. For example, the Philippines could receive a budget to build a resettlement zone for South Pacific islanders-at-risk.

The Costs of Adaptation

On the assumptions Annex I countries are willing to accept the responsibility of meeting the costs of adaptation and a low probability of employment per family in the first years of integration in a host country (or temporary residence in a host country for transfer to another country). General estimates of minimum costs of food/shelter are provided in Table A10.2.

Evacuation and Resettlement Contingency Plans

A high proportion of low and high population density coral atolls and islands in the South Pacific will submerge beneath the sea-level by the year 2050. Table 10.3 lists the must probable routes of evacuation and transportation logistics, and Table 10.4 identifies the possible options for temporary or permanent resettlement of the islanders. The international community must now act to protect and preserve these islander communities from an extinction event they should never have inherited.

Table A10.2. Countries which may be liable for meeting islands-at-risks' costs of adaptation to climate change, and estimated costs of resettlement

Annex I countries	% CO_2 1990	Year 1, US$ m		Years 1–3, US$ m	
		a	b	a	b
		597	994	1791	2982
Australasia					
– Australia	2.1	12.54	20.87	37.62	62.61
– New Zealand	0.2	1.19	1.99	3.57	5.97
Europe					
– Austria	0.4	2.39	3.97	7.17	11.91
– Belgium	0.8	4.77	7.95	14.31	23.85
– Bulgaria	0.6	3.58	5.96	10.74	17.88
– Czech Rep	1.2	7.16	11.93	21.48	35.79
– Denmark	0.4	2.39	3.97	7.17	11.91
– Estonia	0.3	1.79	2.98	5.37	8.94
– Finland	0.4	2.39	3.97	7.17	11.91
– France	2.7	16.12	26.84	48.36	80.52
– Germany	7.4	44.18	73.55	132.54	220.65
– Greece	0.6	3.58	5.96	10.74	17.88
– Hungary	0.5	2.98	4.97	8.94	14.91
– Ireland	0.2	1.19	1.99	3.57	5.97
– Italy	3.1	18.51	30.81	55.53	92.43
– Latvia	0.2	1.19	1.99	3.57	5.97
– Luxembourg	0.1	0.59	0.99	1.77	2.97
– Netherlands	1.2	7.16	11.93	21.48	35.79
– Norway	0.3	1.79	2.98	5.37	8.94
– Poland	3.0	17.91	29.82	53.73	89.46
– Portugal	0.3	1.79	2.98	5.37	8.94
– Romania	1.2	7.16	11.93	21.48	35.79
– Russian Fed	17.4	103.88	172.95	311.64	518.85
– Slovak Rep	0.4	2.39	3.97	7.17	11.91
– Spain	1.9	11.34	18.88	34.02	56.64
– Sweden	0.4	2.39	3.97	7.17	11.91
– Switzerland	0.3	1.79	2.98	5.37	8.94
– UK	4.3	25.67	42.74	77.01	128.22
North America					
– Canada	3.3	19.70	32.80	59.1	98.40
– USA	36.1	215.51	358.83	646.53	1,076.5
Japan	8.5	50.74	84.49	152.22	253.47

Notes:
The estimated costs of resettlement are shown in the table. The costs of evacuating and transporting islanders-at-risk to their new or temporary homes are not included.
Basis of humanitarian needs of 545,000 islander evacuees (including children and elderly), first year.
a = estimate of US$ 3/day per individual, total US$ 597 million (year 1 column).
b = estimate of US$ 5/day per individual, total US$ 994 million (year 1 column).
If islanders were absorbed into a host Annex I or non-Annex I country, they should continue to receive financial support from proceeds of mandatory contributions from carbon contracts/flexible mechanisms of the Protocol. Islanders should be assisted by the international community in the medium and long-term over the period of their integration into the host country.

Table A10.3. Emergency action for islands-at-risk in the South Pacific – contingency evacuation action planning for top priority islands-at-risk

Island-at-maximum risk	Est. number of evacuess: 1/10 total, 1/5 total, to simulate contingency batch ops.	Possible evacuation routes: distance from main island/ other relevant factors	Transportation media, shipping, airlift: Possible journey or connecting journey options (special arrangements required for non-routine port destinations in potential host countries)
Kiribati/ Bairiki	8,000 16,000	2,500 km to New Britain, 3,000 km to Port Moresby, Papua New Guinea.	Pacific Forum Lines,via Fiji.
		3,500 km to Cairns, Australia.	Chief Container Service (special stop).
		4,000 km to Kaitaia, N. Zealand.	Pacific Forum Lines, via Fiji (Kaitaia, special stop).
		6,000 km to Kobe, Japan.	Bali Hai Service. Shipping and air lines: Shipping Corporation of Kiribati. Air Kiribati Ltd, Tarawa.
Maldives/ Malé	30,000 60,000	800 km to Colombo, Sri Lanka. Political considerations.	
		500 km to Trivandrum, SW India 2,300 km to Karachi, Pakistan. 2,700 km to Chittagong, B'desh. 3,000 km to Penang, Malaysia. 3,000 km to Medan, Indonesia.	Shipping and air lines: Maldives Transport and Contracting Co (MTCC), Malé. Maldives Ports Authority, Huvadhoo Shipping, Island Enterprises, Maldives National Ship Management. Air Maldives Ltd, Hummingbird Island Airways, Seagull Airways.
Marshall Islands/ Majuro	5,000 10,000	2,700 km to New Britain, 3,200 km to Port Moresby, Papua New Guinea.	Pacific Forum Lines.
		3,400 km to Honolulu, USA.	Nauru Pacific Line.
		3,800 km to Cairns, Australia.	Nauru Pacific Line (special stop).
		4,600 km to Kaitaia, N. Zealand.	Pacific Forum Lines, via Fiji (Kaitaia, special stop). Shipping and air lines: Pacific Micronesian Lines, Philippine Micronesia and Orient Navigation (PM&O), Tiger Line, NYK and Matson Line. Air Marshall Islands, Majuro.
Micronesia/ Palikir	11,000 22,000	3,400 km to Davao, Mindanao, Philippines.	China Navigation Company

Table A10.3. *Continued*

Island-at-maximum risk	Est. number of evacuess: 1/10 total, 1/5 total, to simulate contingency batch ops.	Possible evacuation routes: distance from main island/ other relevant factors	Transportation media, shipping, airlift: Possible journey or connecting journey options (special arrangements required for non-routine port destinations in potential host countries)
Micronesia/ Palikir	11,000 22,000	1,700 km to New Britain, 2,300 km to Port Moresby, Papua New Guinea.	Shipping to be arranged with commercial operators in the region.
		3,100 km to Cairns, Australia	Micronesia Transport Line (special stop).
			Nauru Pacific Line (special stop).
		4,000 km to Taiwan.	Kyowa Line.
		4,500 km to Hong Kong, China.	Kyowa Line.
		4,700 km to Honolulu, USA.	Nauru Pacific Line.
		5,000 km to Kaitaia, N. Zealand	Pacific Forum Lines, via Fiji (Kaitaia, special stop).
			Shipping and air lines: Micronesian Maritime Authority. Pacific Missionary Aviation, Continental Micronesia, Air Nauru and Continental Airlines (US).
Tuvalu/ Funafuti	900 1,800	3,000 km to Kaitaia, N. Zealand.	Pacific Forum Lines, via Fiji (Kaitaia, special stop).
		3,000 km to New Britain, 3,300 km to Port Moresby, Papua New Guinea.	Via Solomon Islands: Sofrana-Unilines and NGAL/PNGL/CONPAC shipping group. Pacific Forum Lines.
		3,500 km to Cairns, Australia.	Shipping to be arranged with commercial operators in the region.
			Shipping and air lines: Deep sea lagoon at Nukufetau.
			Air Marshall Islands – Funafuti, Nadi (Fiji) and Tarawa (Kiribati).

Notes:
Capital requirements per day, US$, for food, shelter, water supply, sanitation, medical aid to the elderly and children need to be estimated for the environmentally displaced in potential host countries.
Micronesia Transport Line's port stops: Yap, Chuuk and Pohnpei (Micronesia); Sydney.
Pacific Forum Line's port stops: Auckland, Fiji, Tuvalu, Kiribati, Marshall Is, Papua New Guinea.

Table A10.4. Coordinating departments and organisations in potential host countries for the evacuation and resettlement of islands-at-risk in the South Pacific and Indian Ocean

Potential host country	Ministries, departments
Australia	Department of Health and Family Services
	Department of Environment, Sport and Territories
	Department of Immigration and Multicultural Affairs
	Department of Social Security
Bangladesh	Ministry of Health and Family Welfare
	Ministry of Finance and Planning
	Ministry of Land
	Ministry of Local Government and Rural Development
China (via Hong Kong)	State Development Planning Commission
Chinese Taipei (Taiwan)	Ministry of Economic Affairs
European Union	Directorate General I (External Relations)
India	Ministry of External Affairs
Indonesia	Ministry of Foreign Affairs
	Ministry for Population and the Environment
	Ministry of Public Housing
Japan	Economic Planning Agency
	Ministry of Foreign Affairs
Korea	Ministry of Foreign Affairs
	Ministry of Health and Welfare
Malaysia	Ministry of Foreign Affairs
	Ministry of Health
	Ministry of Housing and Local Government
	Ministry of Land and Cooperative Development
New Zealand	Department of Social Welfare
	Ministry of Foreign Affairs and Trade
	Ministry of Health
	Ministry of Pacific Island Affairs
Pakistan	Ministry of Foreign Affairs
	Ministry of Health, Specail Education and Social Welfare
	Ministry of Housing and Works
	Ministry of Planning and Development
	Ministry of Water and Power
Papua New Guinea	Department of Health
	Department of Land and Physical Planning
	Department of Provincial Affairs and Village Development
	Department of Transport, Civil Aviation and Works
	National Housing Association
Philippines	Department of Health
	Department of Foreign Affairs
	Department of Social Welfare and Development
Sri Lanka	Ministry of Foreign Affairs
	Ministry of Health
	Ministry of Housing and Urban Development
	Ministry of Planning
	Ministry of Social Services
Thailand	Ministry of Foreign Affairs
	Ministry of Public Health
United Arab Emirates	Ministry of Foreign Affairs
USA (State of Hawaii)	Department of State, Washington D.C.
	Office of Hawaiian Affairs, Honolulu

Table A10.5. Information table on five islands at maximum risk from sea-level rise

Information Table on Five Islands at Maximum Risk from Sea-Level Rise							
Country/ land ht main islands	Islands at sea level/immed. risk	Annual rainf. in inches	Pop./ Rel.	Pol. affiliation/ economic activity/ energy resources	Aid	HDI/ 174	Evacuation/Resettlement/ near. friendly countries/pot. country partnership
Kiribati <100 m.	Many	79	77,000 RC.	US, Australia, NZ. Agriculture, fish.	Japan.	>125 (est)	Australia, New Zealand.
Maldives <100 m.	Many	59	300,000 Islam.	Pakistan, India. Agro, fishing.	Japan, Ind, Pak	118	Possibly Sri Lanka, Pak, Ind, Bangladesh.
Marshall Islands <100 m.	Many	161	52,000 Pr.	US influence. Agro, fishing.	US, Aus Taiwan.	>100 (est)	US, Australia, NZ.
Micronesia <100 m.	Many	194	107,000 RC, Pr.	US influence. Fishing.	US.	>100 (est)	US. Possibly Aus, NZ.
Tuvalu <100 m.	Many	120–160	9,000 Pr, RC.	Australia. Fishing.	Aus/NZ, Jpn/Tai	>125 (est)	Australia, New Zealand.
Total			545,000 islanders				

HDI Human Development Index, ranking amongst 174 countries (least developed 174th) – based on life expectancy, infant mortality, GNP/capita, daily calorie intake, literacy, education.
Pol. affil. political affiliation, past/current.
Rel major religions – Hi (Hindu), Pr (Protestant), RC (Roman Catholic).

Annexes

Annex 1 Annex A of the Kyoto Protocol

Greenhouse gases

Carbon dioxide (CO_2)
Methane (CH_4)
Nitrous oxide (N_2O)
Hydrofluorocarbons (HFCs)
Perfluorocarbons (PFCs)
Sulphur hexafluoride (HF_6)

Sectors/source categories

Energy
- Fuel combustion
  - Energy industries, Manufacturing industries and construction, Transport
  - Other sectors
- Other
  - Fugitive emissions from fuels, Solid fuels, Oil and natural gas
  - Other

Industrial processes
- Mineral products, Chemical industry, Metal production, Other production
- Production and consumption of halocarbon and sulphur hexafluoride
- Other

Solvent and other product use

Agriculture
- Enteric fermentation
- Manure management, Rice cultivation, Agricultural soils
- Prescribed burning of savannas, Field burning of agricultural residues
- Other

Waste
- Solid waste disposal on land, Wastewater handling, Waste incineration
- Other

Annex 2 Annex B of the Kyoto Protocol

	Quantitative emission limitation and reduction commitment (percentage of base year or period)
Australia	108
Austria	92
Belgium	92
Bulgaria	92
Canada	95
Croatia	92
Czech Republic	92
Denmark	92
European Community	92
Estonia	92
Finland	92
France	92
Germany	92
Greece	92
Hungary	94
Iceland	110
Ireland	92
Italy	92
Japan	94
Latvia	92
Liechtenstein	92
Lithuania	92
Luxembourg	92
Monaco	92
Netherlands	92
New Zealand	100
Norway	101
Poland	94
Portugal	92
Romania	92
Russian Federation	100
Slovak Republic	92
Slovenia	92
Spain	92
Sweden	92
Switzerland	92
Ukraine	100
United Kingdom	92
USA	93

Annex 3 Annex I of the UNFCCC

Country	*CO_2 emissions in 1990 ×1000 metric tonnes**	*Percentage**
Australia	288,965	2.1
Austria	59,200	0.4
Belarus	n/a	n/a
Belgium	113,405	0.8
Bulgaria	82,990	0.6
Canada	457,441	3.3
Czech Republic	169,514	1.2
Denmark	52,100	0.4
Estonia	37,797	0.3
Finland	53,900	0.4
France	366,536	2.7
Germany	1,012,443	7.4
Greece	82,100	0.6
Hungary	71,673	0.5
Iceland	2,172	<0.1
Ireland	30,719	0.2
Italy	428,941	3.1
Japan	1,173,360	8.5
Latvia	22,976	0.2
Lithuania	208	<0.1
Luxembourg	11,343	0.1
Monaco	71	<0.1
Netherlands	167,600	1.2
New Zealand	25,530	0.2
Norway	35,533	0.3
Poland	414,930	3.0
Portugal	42,148	0.3
Romania	171,103	1.2
Russian Federation	2,388,720	17.4
Slovak Republic	58,278	0.4
Spain	260,654	1.9
Sweden	61,256	0.4
Switzerland	43,600	0.3
Ukraine	n/a	n/a
United Kingdom	584,078	4.3
USA	4,957,022	36.1
Total	13,728,306 (13,728 million metric tonnes)	100.0

Notes: * emissions volumes of countries whose votes if equal or excess of 55% of total emissions will enable the Kyoto Protocol to be ratified as per Article 25 of the Protocol.
n/a – emissions from these countries are not subject to the "ratification volume vote".

Annex 4 List of Countries Signing and Ratifying the Kyoto Protocol, at end September 2000

Country	*Date signed*	*Date ratified or acceded*
Afghanistan (LDC)		
Albania		
Algeria		
Andorra		
Angola (LDC)		
Antigua and Barbuda	16 March 1998	3 November 1998
Argentina	16 March 1998	
Armenia		
Australia	29 April 1998	
Austria	29 April 1998	
Azerbaijan		28 September 2000
Bahamas		9 April 1999
Bahrain		
Bangladesh (LDC)		
Barbados		7 August 2000
Belarus	Yes, late 1999	
Belgium	29 April 1998	
Belize		
Benin (LDC)		
Bhutan (LDC)		
Bolivia	9 July 1998	30 November 1999
Bosnia & Herzegovina		
Botswana		
Brazil	29 April 1998	
Brunei		
Bulgaria	18 September 1998	
Burkina Faso (LDC)		
Burundi (LDC)		
Cambodia (LDC)		
Cameroon		
Canada	29 April 1998	
Cape Verde (LDC)		
Central African Rep (LDC)		
Chad (LDC/UNCTAD)		
Chile	17 June 1998	
China	29 May 1998	
Colombia		
Comoros (LDC)		
Congo, Dem. Rep. (LDC)		
Congo, Rep.		
Cook Islands	16 September 1998	
Costa Rica	27 April 1998	

Cote d'Ivoire		
Croatia	11 March 1999	
Cuba	15 March 1999	
Cyprus		16 July 1999
Czech Republic	23 November 1998	
Denmark	29 April 1998	
Djibouti (LDC)		
Dominican Republic		
Ecuador	15 January 1999	13 January 2000
Egypt	15 March 1999	
El Salvador	8 June 1998	30 November 1998
Equatorial Guinea (LDC)		16 August 2000
Eritrea (LDC/UNCTAD)		
Estonia	3 December 1998	
European Community	29 April 1998	
Ethiopia (LDC)		
Fiji	17 September 1998	17 December 1998
Finland	29 April 1998	
France	29 April 1998	
Gabon		
Gambia (LDC)		
Georgia		16 June 1999
Germany	29 April 1998	
Ghana		
Greece	29 April 1998	
Grenada		
Guatemala	10 July 1998	5 October 1999
Guinea (LDC)		7 September 2000
Guinea-Bissau (LDC)		
Guyana		
Haiti (LDC)		
Honduras	25 February 1999	17 July 2000
Hungary		
Iceland		
India		
Indonesia	13 July 1998	
Iran		
Iraq		
Ireland	29 April 1998	
Italy	29 April 1998	
Israel	16 December 1998	
Jamaica		28 June 1999
Japan	28 April 1998	
Jordan		
Kazakhstan	12 March 1999	
Kenya		
Kiribati (LDC, IRISK)		7 September 2000

Korea, D.P.R (North)		
Korea, Rep. (South)	25 September 1998	
Kuwait		
Kyrgyz Republic		
Laos, P.D.R. (LDC)		
Latvia	14 December 1998	
Lebanon		
Lesotho (LDC)		6 September 2000
Liberia (LDC)		
Libya		
Liechtenstein	29 June 1998	
Lithuania	21 September 1998	
Luxembourg	29 April 1998	
Macedonia		
Madagascar (LDC)		
Malawi (LDC)		
Malaysia	12 March 1999	
Maldives (LDC, IRISK)	16 March 1998	30 December 1998
Mali (LDC)	27 January 1999	
Malta	17 April 1998	
Marshall Islands (IRISK)	17 March 1998	
Mauritania (LDC)		
Mauritius		
Mexico	9 June 1998	7 September 2000
Micronesia (IRISK)	17 March 1998	21 June 1999
Moldova		
Monaco	29 April 1998	
Mongolia		15 December 1999
Morocco		
Mozambique (LDC)		
Myanmar (LDC)		
Namibia		
Nauru		
Nepal (LDC)		
Netherlands	29 April 1998	
New Zealand	22 May 1998	
Nicaragua	7 July 1998	18 November 1999
Niger (LDC)	23 October 1998	
Nigeria		
Niue	11 December 1998	6 May 1999
Norway	29 April 1998	
Oman		
Pakistan		
Palau		10 December 1999
Palestine (W.Bank/Gaza)		
Panama	8 June 1998	5 March 1999
Papua New Guinea	2 March 1999	

Paraguay	25 August 1998	27 August 1999
Peru	13 November 1998	
Philippines	15 April 1998	
Poland	15 July 1998	
Portugal	29 April 1998	
Qatar		
Romania	5 January 1999	
Russian Federation	11 March 1999	
Rwanda (LDC)		
Saint Kitts and Nevis		
Saint Lucia	16 March 1998	
Saint Vincent & Grenadines	19 March 1998	
Samoa (LDC)	16 March 1998	
San Marino		
Sao Tome & Principe		
Saudi Arabia		
Senegal		
Seychelles	20 March 1998	
Sierra Leone (LDC)		
Singapore		
Slovak Republic	26 February 1999	
Slovenia	21 October 1998	
Solomon Islands (LDC)	29 September 1998	
Somalia (LDC)		
South Africa		
Spain	21 April 1998	
Sri Lanka		
Sudan (LDC)		
Suriname		
Swaziland		
Sweden	29 April 1998	
Switzerland	16 March 1998	
Syrian Arab Rep.		
Tajikstan		
Tanzania, U. Rep (LDC)		
Thailand	2 February 1999	
Togo (LDC)		
Trinidad and Tobago	7 January 1999	28 January 1999
Tunisia		
Turkey		
Turkmenistan	25 September 1998	11 January 1999
Tuvalu (LDC, IRISK)	16 November 1998	16 November 1998
Uganda (LDC)		
Ukraine	15 March 1999	
United Arab Emirates		
United Kingdom	29 April 1998	
United States of America	12 November 1998	

Uruguay	29 July 1998	
Uzbekistan	20 November 1998	
Vanuatu (LDC)		
Venezuela		
Vietnam	3 December 1998	
Yemen (LDC)		
Yugoslavia, F.R.		
Zambia (LDC)	5 August 1998	
Zimbabwe		
Total	*84*	*30*

Notes: Countries with LDC designation are Least Developed Countries, as categorised by UNCTAD; and countries with IRISK classification are the five most vulnerable islands-at-risk from sea level rise as identified by the author – these countries should deserve attention with regard to equitable financial flows and share of financial proceeds of Kyoto Protocol projects, and increased aid in terms of humanitarian as well as material resources.

Units, Definitions and Conversion Factors; Short Forms Used Throughout the Book

BTU	British Thermal Unit, heat energy, equivalent to 252 calories or 1,055 J
J	Joule, unit of energy, equivalent to 0.2390 calorie or 0.000948 BTU
G	Giga, 10^9
k	kilo, 10^3
kW_e	kilowatt-hour electric, 1×10^6 Watts, energy in the form of electricity
kW_t	kilowatt-hour thermal, 1×10^6 Watts, energy in hot water or steam
M	Mega, 10^6
Therm	Unit of heat energy, equivalent to 1×10^6 BTUs or 105.506 MJ
tonne	1 metric tonne, 1,000 kilogrammes
W	Watt, unit of energy release, equivalent to 4.2 Joules per sec
Annex I	Annex I (to the UNFCCC) is a listing of countries which are industrialised or in the process of industrialisation.
Non-Annex I countries	Countries which do not appear on the Annex I listing – they include LDCs, developing countries and some countries which have moder - ate to significant oil and gas resources.
Annex B	Annex B (to the Kyoto Protocol) – a listing of countries permitted to trade emissions, and contains a schedule of quantitative reduction commitments for the listed countries.
Convention	United Nations Framework Convention on Climate Change.
East-West	Refer to relationships between OECD/industrialised countries and the former "East" Bloc countries in the Soviet era, e.g. Germany > Romania. The term is being progressively replaced by "North-South" as ex-Soviet countries are undergoing transition to near market or market economies.
North-North	Refer to relationships between OECD/industrialised or industrialising countries.
North-South	Refer to OECD/industrialised – developing/LDC country relationships; e.g. USA > China, Netherlands > Indonesia.
Parties	Parties to the UNFCCC – governments participating in the Conference of the Parties, which are held annually.

Fuel Conversion to Electricity

Coal: 1 million tonnes of coal produces about 2,400–2,550 GWh_e in a modern power plant.

Natural gas: 1 million cubic metres of natural gas produces about 4.4 GWh_e in a combined cycle plant or about 3.6 GWh_e in an open cycle plant.

Oil: 1 million tonnes of oil produces about 4,000 GWh_e in a modern power plant.

Notes: Average electrical output for each fuel is quoted; exact output depends on the calorific value of the fuel type – which varies in composition worldwide – and the thermal efficiency of the power plant.

Units of Greenhouse Gas Emissions

Greeenhouse gas (GHG)	*Global Warming Potential (GWP)*	*To obtain carbon dioxide equivalent (mass)*	*To obtain carbon equivalent (mass)*
Carbon dioxide	1	Multiply by 1	Divide by 3.67
Methane	21	Multiply by 21	Divide by 3.67
Hydrofluorocarbons	140–11,700	*	Divide by 3.67
Nitrous oxide	310	Multiply by 310	Divide by 3.67
Perfluorocarbons	6,800 (av)	*	Divide by 3.67
Sulphur hexafluoride	3,200	*	Divide by 3.67

* Multiply by GWP of specific gas – refer to IPCC's Second Assessment Report. For HFCs, GWP of most common species HFC 134a is 1,300.

Example of Conversion of GHG Units Using Data on the United Kingdom

UK greenhouse gases inventory

1990 reference year for CO_2, CH_4 and N_2O; 1995 for other gases

	Actual emissions	*Carbon dioxide equivalent units (GWP basis)*	*Carbon* equivalent units*
Carbon dioxide	616 mt	616 mt	168 mt
Methane	4.4 mt	91.75 mt	25 mt
Hydrofluorocarbons	0.012 mt	15.41 mt	4.2 mt
Nitrous oxide	0.213 mt	66.06 mt	18 mt
Perfluorocarbons	0.000108 mt	0.734 mt	0.2 mt
Sulphur hexafluoride	0.000229 mt	0.734 mt	0.2 mt

Source: Economic Instruments and the Business Use of Energy: A Report by Lord Marshall, November 1998, (page 6), Her Majesty's Treasury Department, United Kingdom.

Notes: * Many sources on climate change reports and publications use "Mt C" or "Mt C equiv" units to express emissions in mass terms – these should bear the same value unless indicated otherwise.

Figure for HFCs and PFCs are author's estimates.

In emissions trading, a seller of one tonne of methane reduction could, in theory, command a price 21 times that of one tonne reduction of carbon dioxide. However, in practice, the price of each greenhouse gas may be based on the cost of control and not on GWP dilution advantages.

Energy Equivalents

1 kWh_e = 3,412 BTU
1,000 kWh_e = 3.6 GJ
1 tonne of oil equivalent (t.o.e) = 40 million BTUs

Range of net calorific values of premium carbon-based fuels

Coal:	bituminous/dry steam coal: 27–33 MJ/kg
Fuel oil:	light, 41–43 MJ/kg; heavy, 40 MJ/kg
Natural gas:	46–48 MJ/kg or 33–35 MJ/m^3

Average prices of premium fuels, July 1999

Coal:	14 cents per Therm (marker price)
Crude oil:	33 cents per Therm
Natural gas:	15 cents per Therm

Note: Producer prices are quoted – transportation and transaction costs and margins are not included in the market prices.

Figures Index

Subject Index

Country, Place Index

Note
Page 90 referrals relate to specific baseline years of greenhouse gas emissions for East European countries.

Persons Index

Note

Persons Index is principally based on the original group of individuals meeting in London to discuss the projects of the Clean Development Mechanism of the Kyoto Protocol (late 1998, 1999).

MIX
Papier aus verantwortungsvollen Quellen
Paper from responsible sources
FSC® C105338

If you have any concerns about our products,
you can contact us on
ProductSafety@springernature.com

In case Publisher is established outside the EU,
the EU authorized representative is:
Springer Nature Customer Service Center GmbH
Europaplatz 3, 69115 Heidelberg, Germany

Printed by Libri Plureos GmbH
in Hamburg, Germany